High-Pressure Shock Compression of Condensed Matter

Springer
New York
Berlin
Heidelberg
Barcelona
Budapest
Hong Kong
London
Milan
Paris
Santa Clara
Singapore
Tokyo

High-Pressure Shock Compression of Condensed Matter

J. Asay and *M. Shahinpoor* (Eds.): High-Pressure Shock Compression of Solids
A.A. Batsanov: Effects of Explosion on Materials: Modification and Synthesis Under High-Pressure Shock Compression
R. Cherét: Detonation of Condensed Explosives
L. Davison, D. Grady, and *M. Shahinpoor* (Eds.): High-Pressure Shock Compression of Solids II
L. Davison, Y. Horie, and *M. Shahinpoor* (Eds.): High-Pressure Shock Compression of Solids IV
L. Davison and *M. Shahinpoor* (Eds.): High-Pressure Shock Compression of Solids III
R. Graham: Solids Under High-Pressure Shock Compression
M. Sućeska: Test Methods for Explosives
J.A. Zukas and *W.P. Walters* (Eds.): Explosive Effects and Applications
J.N. Johnson and *R. Cherét* (Eds.): Classic Papers in Shock Compression Science

James N. Johnson Roger Chéret
Editors

Classic Papers in Shock Compression Science

With 35 Illustrations

Springer

James N. Johnson
Theoretical Division
Los Alamos National Laboratory
Los Alamos, NM 87545
USA

Roger Chéret
Commissariat a l'Energie Atomique
Paris Cedex 15, F-75752
France

Editor-in-Chief:
Robert A. Graham
Director of Research
The Tomé Group
383 Entrada Road
Los Lunas, NM 87031
USA

Library of Congress Cataloging-in-Publication Data
Johnson, James N.
Classic papers in shock compression science / James N. Johnson, Roger Chéret.
p. cm. — (High-pressure shock compression of condensed matter)
Includes index.
ISBN 0-387-98410-0 (hardcover : alk. paper)
1. Condensed matter. 2. Shock waves. I. Chéret, Roger. II. Title. III. Series.
QC173.4.C65J64 1998
531′.33—dc21 97-43747

Printed on acid-free paper.

Production managed by Karina Mikhli; manufacturing supervised by Jeffrey Taub.
Photocomposed copy prepared using the authors' Microsoft Word files.
Printed and bound by Braun-Brumfield, Inc., Ann Arbor, MI.
Printed in the United States of America.

9 8 7 6 5 4 3 2 1

ISBN 0-387-98410-0 Springer-Verlag New York Berlin Heidelberg SPIN 10524585

Preface

The field of shock compression science has a long and rich history involving contributions of mathematicians, physicists and engineers over approximately two hundred years. The middle of the nineteenth century was an especially active period with the contributions of Riemann, Rankine and Hugoniot, among others. The middle of the twentieth century saw another increase in activity related to shock compression of condensed matter as a result of military applications. It was also recognized that shock compression provided a means of subjecting solids and liquids to extreme states of temperature and pressure difficult to achieve by static means. It has thus become an academic study in its own right. The principal modern contributions to this science were summarized in the landmark paper by Rice, McQueen and Walsh [*Solid State Physics*, Vol. 6, pp. 1-63, 1958].

As this field has continued to mature, interest has increased in tracing the early papers that have served as the foundations of the field. Chéret [*Shock Compression of Condensed Matter - 1989*, Elsevier Sci. Pub. B.V., pp 11-19, 1990] has contributed to this literature with his review of the life of Hugoniot on the one-hundredth anniversary of the publication of Hugoniot's classic paper on the propagation of discontinuous waves in gases. This contribution prompted additional historical investigation involving the precursors to Hugoniot. Of course this involves the work, about ten years earlier, of Rankine, the Scottish engineer, along with the contributions of Riemann and Poisson; the latter two works were published in German and French, respectively. When one also considers the valuable contributions of Bethe on the theory of shock waves for an arbitrary equation of state, published as an obscure report for the Office of Scientific Research and Development during the Second World War, it becomes reasonable to think about collecting all these papers, in English, into one volume. This publication is the result of those ideas. We have consciously omitted classical papers on detonation science; this would be an obvious related subject for an independent collection. We have also omitted modern works easily found in most scientific libraries.

A question asked of one of us as we sought a publisher for this collection, was what would this volume provide that was not contained in the reference book *Supersonic Flow and Shock Waves* by Courant and Friedrichs [Interscience Publishers, Inc., New York, 1948]? At the time the question was asked, we had not yet collected all the material and were not sure of the answer. At the present time the answer would seem to be that the collection published here does not yield any important theoretical results that were overlooked in the decades and centuries since their original publication. What does emerge, as it must in all scientific fields, is the rich and complex process by which incremental discoveries are made and fundamental understanding is advanced. If one wonders exactly what Hugoniot did, and how it illuminated the science of his time, it is now readily available in English. If you wonder what else Bethe included in his OSRD report, but were unable to locate a copy, now you can refer to it in a volume which also contains Weyl's classical paper that treats the same subject material.

Another question which arose in the course of preparing these articles is the following: How much of Poisson's and Hugoniot's work do we include in order to provide a proper summary of their contribution to shock compression science? The answer is that we must include it all. It is obvious that some measurable fraction of the material collected here is marginal to our understanding of the shock process. However, if we were to selectively present those portions considered to be relevant only to shock compression, we would open ourselves to possible later criticism that we had omitted something important. Included here is the early prominent literature laying the foundation for shock compression science in its unabridged form, along with an often quoted, difficult-to-find government report from 1942.

Concerning completeness, there is a qualification. In discussing the genesis of the report "On the Theory of Shock Waves for an Arbitrary Equation of State," Prof. Bethe pointed out that he thought a more important contribution to the field was the work done with Edward Teller in 1940 on "Deviations from Thermal Equilibrium in Shock Waves (in Gases)." This report, NP-4898, U. S. Army Ballistic Research Laboratory, 1940, was obtained and retyped. However, space limitations have prevented it being included in this collection. Because of Prof. Bethe's comments and the fact that we have a good copy of the report in modern format, one of the editors (JNJ) will gladly provide photocopies of this report upon request. This report should also be available from traditional U. S. Government sources according to the reference above.

One of the surprises in putting together this collection was the realization of how close G. G. Stokes came to the correct treatment of the shock transition in his paper published in 1848. In Stokes' collection of *Mathematical and Physical Papers* (1883), he describes how Sir William Thompson and Lord Rayleigh later (after 1848) dissuaded him from considering the propagation of discontinuities as a means for dealing with nonlinear wave propagation. The problem came in the confusion (at the time) regarding the necessary generalization of the equation of state and irreversible nature of the shock transition. The object lesson here is that scientific discovery and progress are slow, even for the recognized greats in the field.

The editors of this volume wish to thank numerous persons who have contributed to its publication. We thank Dr. Galen K. Straub and Dr. Richard C. Slansky of Los Alamos National Laboratory for their support of this project. We especially thank Jo Ann Brown for her typing of many of the articles and translations. Prof. George E. Duvall pointed out that their library in the Shock Dynamics Laboratory, Washington State University, had an English translation of Hugoniot's papers. Prof. Yogendra M. Gupta kindly had this translation copied and sent to us. However, one of us (RC) had already made considerable progress in the French-to-English translation of this work and thus continued to prepare an independent translation. We do, however, thank Profs. Duvall and Gupta for their kindness.

The editors have made no attempt to reach some kind of uniform use of symbols; in other words, each paper appears with its original notations. Surprisingly enough this fidelity to basic writing involves no serious difficulty. On the other hand, one of us (R.C.) did achieve a quite small number of corrections of original text, whenever an obvious typographical error was detected in the French papers.

A useful supplement to this collection is the review article by Band and Duvall [Am. J. Phys. **29**, pp. 780-785, 1961].

James N. Johnson
Los Alamos, New Mexico

Roger Chéret
Paris

September 1997

Contents

List of Photos and Biographies

Photos and Portraits

Biographies

Poisson
courtesy of Ecole Polytechnique Archives

SIMÉON-DENIS POISSON

Poisson, the French scientist, was born in Pithiviers on June 21, 1781. He came from modest means and learned to read and write from his father. After some unsuccessful training in the art of surgery – where he displayed clumsiness and lack of interest – he attended *École Centrale de Fontainebleau* newly established by the republican *Convention* where he quickly demonstrated his intellectual capacity, particularly in mathematical analysis. In 1798, he was admitted to *École Polytechnique* in the first rank. Immediately upon his graduation in 1800, he joined the staff as a *répétiteur*. Poisson received his appointment largely through the help of Laplace, who maintained this support throughout his own lifetime; *per contra*, so to say, Poisson remained the most faithful follower of Laplace's mechanico-molecular view of physics.

Upon Laplace's death in 1827 and Cauchy's exile 1830, he took upon himself the role of France's leading mathematician. The latter years of Poisson's career consisted of many administrative duties in the *Académie des Sciences*, and in the *Conseil Royal de l'Université*, as well as publishing numerous research articles and books. In the attempt to complete the many projects he had set for himself, his health failed, and he died in 1840.

Poisson's unwavering support to science mostly explains his successful career spanning the tumultuous period in France's history including *Consulat, Empire, Restauration* and *Monarchie de Juillet.* However, he did not escape, even in his time, sharp criticism for lack of originality, and for unduly high weight given to mathematical analysis in physics. In spite of these reservations – which are not groundless but which the history of science is inclined to moderate – today's science makes his name almost a household word: consider Poisson ratio, Poisson brackets, Poisson distribution and Poisson's equation.

His contribution to the field of what is now called *shock compression science* occurred quite early in his career: the work was read to the Institute on August 17, 1807, and published in *Journal de l'École Polytechnique* in 1808. It mainly consists in the attempt to find solutions to the nonlinear wave equation in which convective terms give rise to solutions of the form

$$\frac{d\varphi}{dx} = f\left[x - \left(a + \frac{d\varphi}{dx}\right)t\right]$$

for simple waves propagating in the positive x direction. Here φ is the velocity potential such that $u = d\varphi / dx$. The sound speed a is assumed to be constant as determined by isothermal conditions and Boyle's law.

A Paper on the Theory of Sound

Siméon-Denis Poisson, *Translated from the Journal de l'École Polytechnique, Vol. 14, No. 7, pp. 319-392 (1808)*

It has been known for some time that sound is produced by a sound-generating body causing the air to vibrate; *Lagrange* was the first to subject this motion to mathematical analysis and deduce the principal elements comprising the theory of sound. Geometricians are so familiar with his elegant research on this subject that there is little need to repeat his findings here. However, at the time of their publication, very little was known about the use of partial differential equations on which the solution for these types of problems depend. There was disagreement on the use of *discontinuous functions* which are nevertheless fundamental for representing the status of the air at the origin of the motion: thankfully, these difficulties have been removed with the progress made in the analysis, whilst those which persist relate to the nature of the problem. Much work still needs to be undertaken before all these can be overcome, thus the principal aim of this paper is to prove several general theorems which would appear to be of interest to both physicists and geometricians.

These theorems are independent of both the specific motion of the air molecules and the cause producing the sound, relating instead to sound propagation and reflection which, up until now, have not been considered from a more general standpoint. Retaining the three-dimensional aspect of air, *Lagrange* and *Euler* assumed that the velocity of the air molecules is dependent only on the distance to the origin of the motion, in other words that the intensity of the sound is the same at all points of the sound wave, whereby *sound wave* is understood to mean that portion of the air moving at any given instant. In this particular case, the known equation describing the theory of sound may be finitely integrated, and the problem may be solved completely satisfactorily. However, in the general case where sound intensity varies in a random manner between any two points along a sound wave, the sound equation can no longer be finitely integrated. Thus, using a simple method and without resorting to the use of series, it may be shown that, for this general case, the velocity of sound remains the same along the various sound radius vectors, implying that the sound wave always has a spherical form the centre of which is that of the original vibrating source. The law which describes the change in sound intensity along a single sound wave is dependent on the cause producing the sound. If it is produced by compressing or expanding a portion of the air mass, then by suddenly re-establishing communication between that portion of the air mass and the total air mass, it may be easily proven that the intensity of the sound will be the same at all points of the sound wave, and at least when this wave has travelled a large distance away from the origin of the motion. Indeed, by breaking down the portion of air compressed or expanded into an infinite number of infinitesimally small parts, each of the individual particles, if either expanded or compressed alone, would produce a wave whose centre would be the particle, and which would spread in a uniform manner in all directions. However, according to *Daniel Bernoulli* on the coexistence of small oscillations, all these undulations could occur at the same time without any form of inter-degradation. In addition, it may be seen that when the radius vectors of all these partial waves become very large with respect to the dimensions of the portion of compressed or expanded air, all the waves merge into one. It can thus be concluded that, irrespective of the shape of the portion of air compressed or expanded, and similarly irrespective of the law of density within the given portion of air, the intensity of the sound would soon be the same at all points on the sound wave. However, when the portion of air being compressed or expanded receives at the same time velocities which may be different in size and direction, for all molecules making it up, it is no longer possible to state how this type of vibration will spread through the fluid mass. Thus for example, the portion of air placed at the end of an instrument such as a megaphone, is simultaneously compressed and set in motion by vibrations in the internal air. Experience proves that the intensity of sound is greatest at the front of the megaphone along the line of the axis; however, the relationship which exists between this intensity and that occurring in another direction equidistant from the origin of the motion is a problem which has still to be resolved. This is not the case for the intensity of sound along a single radius vector, in other words along the same straight line drawn through the centre of the original vibrating source. Irrespective of the vibration, the velocity of the air molecules located along the same radius vector is inversely proportional to the distances to the centre of the motion when the said distances are very large with respect to the dimensions of the portion of the air originally vibrated. If sound is therefore compared to the impact of a fluid against a fixed obstacle, in this case the ear, and if it is assumed that the impact is proportional to the square of the velocity of the fluid, then the intensity of the sound along the same radius vector must decrease proportionally to the square of the distance to the body producing the sound, especially when the distance becomes large. The angle made by the direction of the velocity of an air molecule with the sound radius vector passing through this molecule, decreases as the radius increases, such that the angle may soon be regarded as zero. However, it is natural to believe that it is the direction of the velocities of the air molecules which allow the direction of the sound to be determined, as it is their size which determines the intensity. This therefore explains why the direction of sound can provide an indication of the location of the sound source.

These results for the direction and fall-off in velocity of the air molecules were obtained by developing the integral of the sound equation using a methodical series approach based on the decreasing powers of the distance to the centre of the motion. Once this series was established, it was then transformed into a definite integral, revealing a very extended class of particular, finite integrals which satisfy the sound equation. Each of these integrals satisfies a particular hypothesis on the nature of the original vibrating source; they have the advantage of demonstrating how such vibration, which is generated in a very small portion of air, can spread throughout the whole mass in such a way that the velocity of the air molecules is not the same in all directions. It is also possible to confirm that, for each of these and despite this difference in velocity, the velocity of sound is always the same along all the sound radius vectors, as has been proven in general terms.

Where the mass of air is delimited by a surface in a fixed position, experience has shown that the sound arriving at this surface is reflected by it. *Lagrange* considered the case of sound reflection along an infinitesimally narrow cylindrical pipe; however, this still remains to be determined satisfactorily whilst retaining air's three-dimensional aspect, and assuming that the sound radius vectors traverse the reflecting surface at all possible angles. Where such a surface is an indefinite plane, it can be shown that sound is reflected in the same way as a plane mirror reflects light, in other words that the reflected sound is the same for the direction and the intensity, as if the original vibrating source was positioned behind the flat surface at a distance from it equal to that of the true original source in front of the surface, and in such a way that the straight line joining the two centres is perpendicular to the said plane. From this it can be seen that each air molecule is set in motion at two different times; each time it describes a small straight line, the first time along the direct radius vector, and the second time along the reflected radius vector. The two motions are separated by a time interval which depends on the distance of the molecule from the fixed plane. When the molecule is very close to the plane, the two motions partially coincide, with one commencing where the other ends. Consequently this molecule describes a small curve the type of which depends on that of the original vibrating source. It may be assumed that this vibration is such that the curves mentioned have several points of inflexion; this is the case where the air molecules move backwards and forwards several times along the direct and the reflected radius vectors. However, whatever the types of curves, the molecules not touching the flat surface may approach it indefinitely while never reaching it, and the lines they describe will always be terminated by two straight lines, one drawn along the direct radius vector, the other along the reflected radius vector. On the other hand, the molecules touching the flat surface do not exit the plane, instead moving only within it.

In the case of the light emission system, one only needs to know how the light is reflected by a flat surface in order to determine how it is reflected by any given surface; as it is possible to consider each ray of light in isolation, and substitute the reflecting surface with its tangent plane at the point where this ray falls, it is then but a simple geometric exercise to determine the intensity of the light reflected at a given point, or the number of reflected beams which cross at the said point, which in effect adds up to the same thing. However, to ascertain how the waves produced in an elastic fluid or an incompressible fluid are reflected by a surface delimiting the said fluid, one needs to consider the reflection of the complete wave, and determine the shape and velocity of the reflected wave. This is the root of the difficulty of the problem, a problem which can only be solved for each reflecting surface on a case by case basis.

Assuming that the sound leaves one of the foci of an ellipsoid of revolution, and that it is then reflected by the said surface, it may be demonstrated that the reflected sound forms a sound wave the centre of which is at the other focus of the ellipsoid, and which continuously approaches this other focus while maintaining its spherical shape.

The radius vectors of the direct sound wave and the reflected sound wave, which end at the same point on the reflecting surface, thus form angles which are equal to the normal at this point. Hence it is possible to conclude that the angle of incidence is equal to the angle of reflection, as in the reflection of light, although it is not the result of the same cause. In addition, the intensity of reflected sound is deduced from that of direct sound, and the calculation shows that the ratio between these intensities for two sound radius vectors which arrive at the same point on a surface is the same as that which would have occurred under the same conditions between the direct light and the reflected light. Along a single sound radius vector, the intensity of the reflected sound increases as it approaches the second focus of the ellipsoid such that, at points close to this focus, this intensity is much greater than that of direct sound. This result is confirmed by experimentation, as it is widely known that if one talks quietly at the focus of an elliptical vault, one's voice can be heard distinctly at the other focus, while it cannot be heard at any other point. The velocity of reflected sound is the same as for direct sound; hence sound following a zig-zag path will arrive at a given point at the same time as direct sound following a straight path of equal length to the zig-zag path.

The case of an ellipsoid of revolution having a given distance between the foci equally applies to a paraboloid, assuming that the given distance is infinite. The sound produced at the focus of a paraboloid of revolution will therefore be reflected by its surface, in parallel to its axis and *vice versa*.

Consideration has also been given to the sound produced at one of the foci of a hyperboloid of revolution, and reflected by either its concave or convex surface; such reflection exhibits properties analogous to those of reflection off an ellipsoid. Finally, if one dimension of the air is ignored, the sound produced at the focus of a conical section is found to be reflected towards the other focus.

A significant difference exists between the velocity of sound calculated using theory, and that found by experimentation. All physicists who have measured this velocity directly have found that it is greater than the calculated velocity. Members of the French Academy of Sciences have found that sound travels at a rate of 337 metres per sexagesimal second; however, if the most recent experimentally-found data on the elasticity and density of air are employed, it is found that the velocity of sound as calculated using the theory equals only 282 metres per second, which is *one-sixth* less than the experimental value.

Although it is superfluous to reproduce here all the hypotheses used to attempt to make the calculated and observed results agree in this regard, one nevertheless is of interest to physicists.

Lagrange has already noted (*Mémoires de Turin*, Volume 2) that it is possible to deduce the observed velocity by analysis, assuming that when compressing air, its elasticity increases more rapidly than its density. However, he himself has observed that this assumption is inadmissible, since *Mariotte* and all others who have repeated his experiments have found that air density increases in proportion to the weights used to compress it, whilst the temperature remains unchanged. Nevertheless, *Mariotte's* law can be reconciled with the increase in elasticity which the velocity of sound requires: this is exactly what *Laplace* has done by taking into consideration the production of heat accompanying the compression of air. Indeed, it is now well understood that, when a given volume of gas is compressed, a volume of heat is liberated which is sufficiently large to ignite a flammable body, but which is dissipated quickly as radiation or otherwise. However, the elasticity of air increases or decreases with temperature at a given fixed density. The law which dictates that density is proportional to pressure therefore assumes that the compressed air has been allowed sufficient time to dissipate any increase in heat produced from compression; and in the first instance, elasticity must increase to a greater degree than density. There are insufficient direct experimental results available to determine the quantity of sensible heat obtained from the compression of air. In addition, in the production of sound, each layer of air is compressed between two other layers of air, and it may be that, for the same method of compression, the same amount of sensible heat is not produced as when a volume of air is compressed in a closed vessel: it is therefore impossible to determine *a priori* the increase in elasticity due to the production of heat accompanying the propagation of sound. Be that as it may, this increase is no less indisputable, and it undoubtedly influences the velocity of sound. However, the question remains whether the difference which exists between the calculation and the observations is explained by this single cause. It would appear valid to conclude so, where no objection is raised against the analysis used to deduce the velocity of sound, and if, at the same time, all physical factors which could affect the result have been taken into account. Thus by comparing the calculated velocity with the observed velocity, it is possible to determine the quantity of sensible heat produced in the generation of sound and also used to increase the elasticity of the air. This paper gives the results of a comparison of these two velocities.

In the calculation used to deduce the velocity of sound, the velocity of the air molecules is considered as being very low, and the square and higher powers of this velocity are ignored. *Euler* considered that the difference between the calculation and the observation stemmed from this consideration, and that the velocity of the air molecules producing the sound is not so small as to enable its square to be ignored without introducing a significant error into the equation, especially when the intensity of the sound is very high, as in the cases where its velocity was determined by observation. To remove this difficulty, the propagation of sound was considered along a line of air, without eliminating any terms in the motion equation. The equation thus is no longer linear, nor can it be finitely integrated; instead it has in this form, a particular integral which contains an arbitrary function, and which is sufficient to determine the motion of air when it is assumed that the velocity of the molecules, without being very slow, is simply less than that of sound. Using this integral, it can be rigorously demonstrated that the velocity of sound is independent of both that of air molecules and of the cause producing the sound. Irrespective of whether the sound is loud or soft, it is always transmitted at the same velocity, an observation which has been corroborated by experimentation.

All of the above assumes that the temperature and density are constant throughout the whole air mass. However, as the velocity of sound equals the square root of the ratio between the elasticity and the density of air (and ignoring any correction for the production of heat), it follows that as long as this ratio does not change, the velocity of sound will not vary. From this, and assuming all layers within the atmosphere are at the same temperature, it may be concluded that the velocity of sound will be the same, just as if the air density did not vary when moving from one atmospheric layer to the next, as elasticity varies in proportion to density irrespective of the law of gravity. This will not be the case, however, if the temperature varies at the same time as the density; instead, the sound transmission will be non-uniform and its velocity will not be the same for all radius vectors, which means that the sound wave will no longer have a spherical shape, as would be the case if the temperature was constant. As an example, let us assume that temperature decreases in proportion to the vertical height as one moves away from the earth's surface, as is in fact the case in nature as can be seen if we compare refraction theory with experimental results (*cf.* '*la Mécanique céleste*', Book X). Under these circumstances, if one considers a sound wave which is emitted from a point high in the atmosphere down to the earth's surface, the temperature, and therefore the air elasticity to density ratio, will increase along the radius vector in proportion to its length multiplied by the cosine of the angle it makes with the vertical. From this it may be concluded that the motion of sound along each radius vector, will have the same characteristics as a heavy body sliding along this radius vector as it would along a flat, tilted surface, leaving the origin of the radius vector at a given velocity. Thus the closer the sound radius vector is to the vertical, the faster the velocity of sound, while the velocity along any given radius vector will accelerate proportionally with the time lapsed since the origin of the motion.

The partial differential equation which represents the theory of sound changes form when the weight of the air and the variation in temperature are taken into account. To directly determine the velocity at which sound propagates, the integral of this equation must be used, expressed using a definite integral, thus leading to the expression of the velocity as per the previously stated conditions. This same integral also yields a method for determining the intensity of the sound produced at different heights in the atmosphere. When the temperature is assumed constant, a noteworthy result is obtained, namely that the intensity of the sound depends only on the distance it has travelled and the density of the atmospheric layer from which it originated. This means that, in all cases, the intensity remains the same, just as if the atmosphere was homogeneous and had the same density as that of the said layer. It follows therefore that individuals flying high in a balloon will hear a noise generated at the earth's surface as clearly as if they had remained on the ground, while a noise generated in an upper layer of the atmosphere will be as poorly heard at the earth's surface as in the same atmospheric layer, all distances being equal. It can be easily proven that the variation in temperature between different layers of the atmosphere will not significantly alter this result, which would appear to conform with experimentally-obtained results.

In general, it is easy to obtain definite integrals which satisfy a given partial differential equation. However, these integrals are more often than not a simple analytical exercise, giving no useful information on the nature of the function to be determined and leading us no further towards solving the problem at hand. Those demonstrated by *Laplace* for the integration of second-order partial differential equations having three variables (*Mémoires de l'Académie*, 1779), are of a different type; on the contrary, it is integrals of this form which give the velocity and intensity of sound, where the density and temperature of air are considered variable. *Laplace* has already used these to determine the velocity of sound where one dimension of the air is disregarded, and where it is assumed that the intensity depends only on the distance to the centre of the original vibrating source. Many other problems may be resolved using definite integrals, such as the transmission of motion through a heavy chain, the propagation of sound along a tube of variable width, and vibrations in cords of unequal thicknesses. As an example, the solution to the heavy chain problem has been provided at the end of this paper, making the assumptions that the chain is uniform and of constant thickness along its whole length.

Differential Equations Describing the Motion of Air

1. **TO** examine the question from the most general viewpoint, let us consider an elastic fluid wherein all the molecules are subject to the action of various forces. Let m represent any given molecule, x, y, z the three rectangular coordinates for the given molecule at t, the time elapsed since the motion was first initiated, u, v, v' the three velocities of m at the same instant parallel to axes x, y, z, respectively, and finally, still at time t, let ρ and p represent the density of m and the pressure the molecule withstands by virtue of its elasticity, this pressure being expressed per unit surface area. If the temperature were the same throughout the fluid mass, p would be proportional to ρ as given by *Mariotte's* law. In general, p is dependent on the temperature and density of m, although it is assumed in this case that the fluid is such that molecules of equal density also have the same temperature, thus allowing p to be considered as a function of ρ alone.

Next, let us break down the various forces applied to the molecule m into three other forces designated as X, Y and Z which act in the directions given by coordinates x, y, z respectively, and with a tendency to increase these coordinates. The formula $Xdx + Ydy + Zdz$ is an exact differential as long as the forces applied to point m are directed towards the fixed or moving centres, and each force is a function only of the distance from the point m to the centre towards which it is driven, which is the case for all forces in nature. Thus:

$$Xdx + Ydy + Zdz = dq,$$

where q is a function of x, y, z. It is also assumed that the formula $udx + vdy + v'dz$ is an exact differential, thus giving:

$$udx + vdy + v'dz = d\varphi,$$

where φ is a function of x, y, z and t, and where the differential $d\varphi$ is used only with respect to x, y, z. This assumption is valid for all occasions where the fluid oscillates only very slightly around the steady state and where only the first powers of the velocities u, v, v' are used in the calculation. It is in particular this type of motion which will be examined later in this paper. In addition, it is shown that the formula $udx + vdy + v'dz$ is an exact differential for all values of t where this is the case for a single specific value of this variable. Thus, to validate the assumption made, it is sufficient to imagine that the original vibrating source of the fluid was such that the formula in question was an exact differential, which includes the case where the initial velocities of the molecules are all zero, and numerous other cases.

This declared, the fluid motion equations can be reduced to two *(cf. la Mécanique analytique)*, namely:

$$q - \int \frac{dp}{\rho} = \frac{d\varphi}{dt} + \frac{1}{2} \cdot \left(\frac{d\varphi^2}{dx^2} + \frac{d\varphi^2}{dy^2} + \frac{d\varphi^2}{dz^2} \right) \qquad \text{(a)}$$

$$\frac{d\rho}{dt} + \frac{d \cdot \rho \frac{d\varphi}{dx}}{dx} + \frac{d \cdot \rho \frac{d\varphi}{dy}}{dy} + \frac{d \cdot \rho \frac{d\varphi}{dz}}{dz} = 0 \qquad \text{(b)}$$

As p is a given function of ρ, these two equations include only the two unknown quantities ρ and φ. By integrating these, values for φ and ρ are obtained as functions of x, y, z and t, hence giving u, v, v' using:

$$u = \frac{d\varphi}{dx}, \quad v = \frac{d\varphi}{dy}, \quad v' = \frac{d\varphi}{dz};$$

and thus the problem may be solved completely as, at any instant and at any point whatsoever in the fluid mass, one knows the density, the pressure, the velocity and the direction of the fluid. General solutions for these equations are impossible although they are readily simplified when considering the particular type of motion in air which produces sound. While it is still not possible to carry out a finite integration, it is nevertheless possible to determine the most important properties of this motion, namely those which relate to its propagation and reflection.

The Propagation of Sound in a Mass of Air of Constant Density and Temperature

2. **THIS** article ignores the weight of air and assumes that the temperature and density are constant throughout the whole of the fluid mass. Thus, let us consider an equal mass of air at steady state and then assume that a very small portion is either compressed or expanded, whilst at the same time transferring, by whatever method, velocity to the molecules which make up the said portion of air: this partial vibration will then spread gradually as waves throughout the whole mass. The task is therefore to determine the law governing such *propagation*.

For clarity, let us assume that the portion of originally vibrated air is a sphere of radius α, at the centre of which is placed the origin of the coordinates x, y, z. The contraction and velocity of the molecules vary at random within this sphere such that if the portion of air vibrated is not a sphere, it can still be assumed to have such a shape, provided it is assumed at the same time that the velocities and contraction are zero in some parts of this sphere. The term *sound wave* is used to describe that part of the air in motion at any instant, while *sound radius vector* is used to designate the lines radiating from the centre of the original vibration source to the various points of the sound wave. Finally, it is assumed that the contraction and velocities of the air molecules producing the sensation of sound are very small, while the products and the squares of these quantities are removed from the calculation in keeping with the practices of all those working on the theory of sound. Even where the magnitudes of the initial contraction and velocities are not very small, they very soon become so, since the original vibrating source is situated in a portion of air having very small dimensions and also because the sound has to spread in all directions within a non-delimited mass of air.

3. **LET** D be the density and h the height of the barometer for the whole of the air mass before the motion commences, then $\rho = D\,(1+\gamma)$, such that γ represents the contraction (whether negative or positive) of the air molecules during the motion. By designating gravity as g and using the density of mercury as the basic unit, gh represents the pressure corresponding to the density D, where the pressure provides a measure of the elasticity of the air. According to *Mariotte's* law, during motion the pressure becomes $p = gh\,(1+\gamma)$ at a constant temperature. However, as the motion is accompanied by either compression or expansion of the air, the temperature must rise or fall accordingly, resulting in a change in elasticity. In addition, as the contraction γ is very small, it may be assumed that the change in temperature is proportional to this contraction. Hence, a term needs to be added to $gh\,(1+\gamma)$ such as $gh\cdot k\gamma$ to obtain the elasticity of air during the motion, k being a constant coefficient the value of which is determined later in this paper (*cf.* the section entitled '*Examination of the various physical circumstances which can influence the velocity of sound*').

Thus this gives $p = gh\,(1+\gamma+k\gamma)$ and $\int \frac{dp}{\rho} = a^2\cdot\log\cdot\left(1+\gamma\right)$, which abbreviates to $\frac{gh}{D}\cdot(1+k) = a^2$. Equations (a) and (b), section 1, thus become:

$$a^2\gamma + \frac{d\varphi}{dt} = 0\,,$$

$$\frac{d\gamma}{dt} + \frac{d^2\varphi}{dx^2} + \frac{d^2\varphi}{dy^2} + \frac{d^2\varphi}{dz^2} = 0\,;$$

ensuring that, as no force acts on the air molecules, $q = 0$, and in the absence of second-order terms involving the contraction γ and the velocities $\frac{d\varphi}{dx}$, $\frac{d\varphi}{dy}$, $\frac{d\varphi}{dz}$. Eliminating γ from these two equations produces:

$$\frac{d^2\varphi}{dt^2} = a^2\left(\frac{d^2\varphi}{dx^2} + \frac{d^2\varphi}{dy^2} + \frac{d^2\varphi}{dz^2}\right) \quad \text{..........................} (1).$$

This is the well known equation which represents the theory of sound. Once the function φ has been determined, these partial differentials with respect to x, y, z, will yield the velocity of the air molecule for x, y, z, parallel to these three co-ordinates respectively, whereas the difference $\frac{d\varphi}{dt}$ will give the contraction for the same molecule by virtue of the equation $a^2\gamma + \frac{d\varphi}{dt} = 0$.

Equation (1), despite ignoring one dimension of air, thus reducing it to:

$$\frac{d^2\varphi}{dt^2} = a^2\left(\frac{d^2\varphi}{dx^2} + \frac{d^2\varphi}{dy^2}\right),$$

cannot be finitely integrated, in other words its general solution can only be expressed as an infinite series. *Legendre* has directly demonstrated this (*cf. Mémoires de l'Academie*, 1787), and it is generally recognised that the last equation cannot be finitely integrated in the special case where φ is only a function in t and $\sqrt{(x^2 + y^2)}$ only, from which it may be safely concluded that this is not the case when φ is a function of t, x, y.

It is easy to obtain particular finite integrals which satisfy equation (1). These integrals are consistent with particular hypotheses on the nature of the original vibrating source, although up until now, only two have been found which allow the source to be enclosed in a very small portion of the air mass, and which therefore could be used in sound theory, where one needs to know how such a vibration spreads throughout the whole air mass. These two integrals relate to the case where two dimensions of the air are ignored, and to the case where the quantity φ is a function in t and $\sqrt{(x^2 + y^2 + z^2)}$ only; there also exist very many others which will also be considered later.

4. **TRANSFORMING** the coordinates x, y, z, into polar coordinates having the same origin, let r equal the radius vector, u the cosine of the angle which the said radius vector makes with the axis z, and ω the angle of projection for the same radius vector along the plane of x and y, made with the x-axis. Then:

$$z = ru, \quad x = r \cdot \sqrt{(1 - u^2)} \cdot cos.\ \omega, \quad y = r \cdot \sqrt{(1 - u^2)} \cdot sin.\ \omega;$$

and equation (1) then becomes:

$$\frac{d^2 \cdot r\varphi}{dt^2} = a^2 \cdot \left[\frac{d^2 \cdot r\varphi}{dr^2} + \frac{d \cdot \left(1 - u^2\right) \frac{d \cdot r\varphi}{du}}{r^2 \cdot du} + \frac{1}{r^2\left(1 - u^2\right)} \cdot \frac{d^2 \cdot r\varphi}{d\omega^2} \right] \text{.........................(2).}$$

In addition, one obtains:

$$\frac{d\varphi}{dr} = \frac{d\varphi}{dz} \cdot u + \frac{d\varphi}{dx} \cdot \sqrt{\left(1 - u^2\right)} \cdot \cos.\omega + \frac{d\varphi}{d\gamma} \cdot \sqrt{\left(1 - u^2\right)} \cdot \sin.\omega \;\; ;$$

from which it may be deduced that $\frac{d\varphi}{dr}$ gives the velocity along radius vector r, as the three terms for the value of $\frac{d\varphi}{dr}$ are in fact the three velocities $\frac{d\varphi}{dx}$, $\frac{d\varphi}{dy}$, $\frac{d\varphi}{dz}$, factorized as per the said radius vector.

It is also found that:

$$\frac{d\varphi^2}{dx^2} + \frac{d\varphi^2}{dy^2} + \frac{d\varphi^2}{dz^2} = \frac{d\varphi^2}{dr^2} + \frac{d\varphi^2}{du^2} \cdot \frac{1 - u^2}{r^2} + \frac{d\varphi^2}{d\omega^2} \cdot \frac{1}{r^2\left(1 - u^2\right)} \;\; ;$$

and it is this equation which provides the value of the velocity in its own direction, such that if m is the angle which this direction forms with the radius r, then one obtains:

$$\cos.m = \frac{\frac{d\varphi}{dr}}{\sqrt{\left(\frac{d\varphi^2}{dr^2} + \frac{d\varphi^2}{du^2} \cdot \frac{1 - u^2}{r^2} + \frac{d\varphi^2}{d\omega^2} \cdot \frac{1}{r^2\left(1 - u^2\right)}\right)}} .$$

5. **MULTIPLYING** all the terms in equation (2) by $du\ d\omega$, and then integrating with respect to u and ω, between $u = 1$ and $u = -1$, and between $\omega = 0$ and $\omega = 2\pi$ (where π represents the ratio of the circumference to the diameter), gives:

$$\frac{d^2 \cdot rR}{dt^2} = a^2 \frac{d^2 \cdot rR}{dr^2},$$

where, for clarity, R represents the integral $\int \cdot \varphi\ du\ dw$. Indeed, the term $\dfrac{d \cdot \left(1 - u^2\right)\dfrac{d \cdot r\varphi}{du}}{du}$, multiplied by du and then integrated with respect to u, gives $\left(1 - u^2\right)\dfrac{d \cdot r\varphi}{du}$, which equals zero at the two limits $u = 1$ and $u = -1$, and the term $\dfrac{d^2 \cdot r\varphi}{d\omega^2}$ multiplied by $d\omega$ and then integrated with respect to ω, gives $\dfrac{d \cdot r\varphi}{d\omega}$, which has the same value at both limits $\omega = 0$ and $\omega = 2\pi$.

The integral of the last equation is given as:

$$R = \frac{f(r - at)}{r} + \frac{F(r + at)}{r},$$

where f and F are two arbitrary functions. From this can be extracted:

$$r^2 \cdot \frac{dR}{dr} = r \cdot f'(r - at) - f(r - at) + r \cdot F'(r + at) - F(r + at),$$

$$r^2 \cdot \frac{dR}{dr} = ar \cdot F'(r + at) - ar \cdot f'(r - at),$$

where $f'(r - at)$ and $F'(r + at)$ represent the differentials of $f(r - at)$ and $F(r + at)$ taken with respect to r and divided by dr. However, if one considers a spherical surface of radius r, with a centre at the origin of the co-ordinates, and if one factorizes the velocity of each of the air molecules located at its surface along radius r which ends at this molecule, it is evident that $r^2 \cdot \dfrac{dR}{dr}$ will be the sum of all these factorized velocities, and that $-\dfrac{r^2}{a^2} \cdot \dfrac{dR}{dt}$ will express the sum of the contractions for the same molecules. However, at the origin of the motion, the air has only been vibrated in a portion which extends from $r = 0$ to $r = \alpha$, around the origin of the co-ordinates placed at the centre of the original vibrating source; the two functions $\frac{dR}{dr}$ and $\frac{dR}{dt}$ must therefore equal zero when $t = 0$ for all values of r exceeding α, a condition which is met where the functions fr and Fr are zero for all these values of r. Thus the function $f(r - at)$ will equal zero as long as $r > at + \alpha$ and the function $F(r + at)$ will equal zero at all times $r > \alpha - at$. In addition, $f(r - at)$ will also be zero as long as $r < at$. This may be demonstrated by considering a sphere with a radius of $\alpha' < \alpha$, with a centre at the origin of the co-ordinates, and assuming that the air molecules making up the said sphere have not been subjected to either contraction or velocities at the origin of the motion, then $\frac{dR}{dr}$ and $\frac{dR}{dt}$ must equal zero when $t = 0$ for all values of $r < \alpha'$; fr will then be zero for values of $r < \alpha'$ as well as for values of $r > \alpha$. Thus $f(r - at)$ will be zero when $r < at + \alpha'$, a case which holds however small the radius α' and which still applies when $\alpha' = 0$, and from which it follows that $f(r - at)$ shall equal zero at all times $r < at$. Hence, the above values for $\frac{dR}{dr}$ and $\frac{dR}{dt}$ will equal zero as long as $r > at + \alpha$; whereas at the instant $r = at + \alpha$, the quantities $\frac{dR}{dr}$ and $\frac{dR}{dt}$ will take values other than zero, and will return immediately to zero as soon as $r = at$. The air molecules, located at distance r from the origin of the co-ordinates, will therefore not begin to vibrate until $r = at + \alpha$, and they will only continue to vibrate as long as $r = at$, in order words for a period of time equal to $\frac{\alpha}{a}$. The molecules will all vibrate more or less at the same time, as the vibration duration is very short, α always being very much smaller than a. Thus, irrespective of the method used to produce the sound, *its velocity will be the same along all sound radius vectors, the sound wave will always maintain its spherical form, and the velocity of the said wave will be constant and equal to a.*

It should be noted that the conditions $\frac{dR}{dr} = 0$ and $\frac{dR}{dt} = 0$ are met when $t = 0$, without the need to use $fr = 0$ and $Fr = 0$. In fact these equations give:

$$rf'r - fr + rF'r - Fr = 0,$$

$$F'r - f'r = 0,$$

from which is obtained:

$$fr = b + cr, \quad Fr = -b + cr,$$

where b and c are arbitrary constants. This then gives:

$$f(r-at)=b+cr-at, \qquad f'(r-at)=c \quad ,$$

$$F(r+at)=-b+cr+at, \qquad F'(r-at)=c \quad ;$$

although when these values are used, it is found that those of $\frac{dR}{dr}$ and $\frac{dR}{dt}$ equal zero for all values of t, as if $fr=0$ and $Fr=0$ had been used. This observation is useful in order to avoid any criticism being levelled against the thoroughness of the above demonstration. It may also be feared that the functions $\frac{dR}{dr}$ and $\frac{dR}{dt}$ equal zero without the velocities and contractions of the air molecules located at a distance r from the origin of the coordinates also equalling zero. In such a situation, these molecules would begin vibrating before $\frac{dR}{dr}$ and $\frac{dR}{dt}$ take any values, or after the said functions have returned to zero, thus invalidating the above demonstration. However, it is evident that this can only occur if a particular assumption on the nature of the original vibrating portion of source is made, and by placing the origin of the coordinates in a suitable position, such that by moving the said origin to another point in the vibrating portion of air, it follows that it will not be possible for the functions $\frac{dR}{dr}$ and $\frac{dR}{dt}$ to equal zero without the velocities and contractions also equalling zero.

6. **NOW,** in order to determine the size and direction of the velocity of the air molecules at each point of the sound wave, a series integration of equation (2) is required. To do this, let us put $x=r-at$ and:

$$r\varphi = f(x,\, u,\, \omega) + \frac{1}{r}\cdot\int\cdot f_1(x,u,\omega)dx + \frac{1}{r^2}\cdot\int^2\cdot f_2(x,u,\omega)dx^2 + etc. \;........(f);$$

where f, f_1, f_2 *etc.* are functions of x, u, ω, which require solving. Substituting $r\varphi$ from equation (2) with this series and comparing the terms multiplied by similar powers of r, gives the general term:

$$r(n+1)\cdot f_{n+1} = n\cdot n+1\cdot f_n + \frac{d\cdot\left(1-u^2\right)\frac{df_n}{du}}{du} + \frac{d^2\cdot f_n}{d\omega^2}\cdot\frac{1}{1-u^2} \quad ,$$

where n is an unspecified subscript. The values f_1, f_2, f_3 *etc.*, may then be successively extracted from this equation using f which remains an arbitrary function.

By making $x' = r + at$, equation (2) is again satisfied using the series:

$$r\varphi = F(x', u, \omega) + \frac{1}{r} \cdot \int \cdot F_1(x', u, \omega)\, dx' + \frac{1}{r^2} \cdot \int^2 F_2(x', u, \omega)\, dx'^2 + etc. \ldots\ldots\ldots (F);$$

where F is an arbitrary function of x', u, ω, and F_1, F_2, F_3, *etc.*, are other functions of x', u, ω, deduced from the first using the general equation:

$$r(n+1) \cdot F_{n+1} = n \cdot n+1 \cdot F_n + \frac{d \cdot \left(1-u^2\right) \frac{dF_n}{du}}{du} + \frac{d^2 \cdot F_n}{d\omega^2} \cdot \frac{1}{1-u^2},$$

where n is an unspecified subscript. The sum of these two series then gives the complete integral of equation (2).

7. **THE** series integral of equation (2) as presented in this form, can be used to determine the velocity of sound, as found using another method. In fact, the portion of vibrating air at the origin of the motion is assumed to extend from $r = 0$ to $r = \alpha$, in which case the four partial differentials $\frac{d\varphi}{dt}$, $\frac{d\varphi}{d\omega}$, $\frac{d\varphi}{du}$ and $\frac{d\varphi}{dr}$ equal zero when $t = 0$, for all values of r greater than α. This condition is met by assuming:

I. That the functions $f(r, u, \omega)$ and $F(r, u, \omega)$ are zero for all values of r greater than α, irrespective of whatever other values u and ω take, meaning that the partial differentials for these functions with respect to u and to ω, and hence all functions f_1, f_2, f_3, *etc.*, and F_1, F_2, F_3 *etc.*, will also equal zero for the same values of r;

II. That the integrals $\int \cdot f_1(r,u,\omega)\, dr$ and $\int \cdot F_1(r,u,\omega)\, dr$ are used in the case where the value for r is greater than α up to the variable value for r, such that these integrals will equal zero once this second value for r is also greater than α;

III. That the double integrals $\int^2 \cdot f_2(r,u,\omega)\, dr^2$ and $\int^2 \cdot F_2(r,u,\omega)\, dr^2$ are taken in the first instance using a value for r chosen arbitrarily but provided it is greater than α, up to the variable value for r, and a second time from a value for $r > \alpha$ up until the variable value for r, such that these double integrals will equal zero once the last value for r is also greater than α; and so on.

All terms in the (f) series will therefore equal zero so long as $x > \alpha$ or $r > at + \alpha$, and similarly all terms in the (F) series will equal zero so long as $r + at < \alpha$; thus the function φ and its partial differentials shall equal zero so long as r is greater than $at + \alpha$. Therefore at the instant when $r = at + \alpha$, the air molecules at distance r from the origin of the coordinates will begin to move, and the velocity of sound will be the same along all sound radius vectors, while along each radius vector it will be constant and equal to a as already seen in Section 5.

It may also be shown that the vibration of molecules located at the extreme end of radius r will cease when $r = at$. For this, one must imagine (as above in Section 5) a sphere which has its centre at the origin of the co-ordinates and whose radius equals $\alpha' < \alpha$, and assume that all air molecules within the boundary of such a sphere have not been subjected to either contraction or any velocities at the origin of the motion. Under this hypothesis, the functions f, f_1, f_2 *etc.*, and F, F_1, F_2 *etc.*, will equal zero when $t = 0$ for all values of $r < \alpha'$ as well as for all values of $r > \alpha$; and as this must indeed be the case, however small the radius α', it follows that when α' is zero or infinitely small, the functions f, f_1, f_2 *etc.* will also equal zero, after a given time period, if the value of x or $r - at$ is negative. From this, one may conclude that all terms from the (f) series will equal zero if at the same time $r + at < \alpha$, and similarly all terms from the (F) series will equal zero if $r > \alpha$: the function φ and its partial differentials then return to zero as soon as $r < at$.

8. **ASSUMING** the case that the air molecules are outside the limits of the original vibrating source, and that therefore $r > \alpha$, it is evident that all terms of the (F) series will equal zero, irrespective of the time t, such that, for these molecules, the value of $r\varphi$ will reduce to the (f) series. In addition, if r is much larger than α, as will be the case for molecules which are very remote from the location the sound was produced, and if the quantities of the order $\frac{\alpha}{r}$ are ignored, the (f) series will then reduce to its first term. In fact, as the function $f_1(x, u, \omega)$ only takes values from $x = 0$ up to $x = \alpha$, it follows that the term $\frac{1}{r} \cdot \int \cdot f_1(x, u, \omega)\, dx$ has a value of the order of $\frac{\alpha}{r}$; similarly, as the function $f_2(x, u, \omega)$ equals zero for all values of x which fall outside the limits $x = 0$ and $x = \alpha$, the term $\frac{1}{r^2} \cdot \int^2 \cdot f_2(x, u, \omega)\, dx^2$ will have a value of the order of $\frac{\alpha^2}{r^2}$, and so on for the other terms of the (f) series.

This therefore gives very simply:

$$r\varphi = f(x, u, \omega);$$

which, when divided by r and then differentiated with respect to r, gives:

$$\frac{d\varphi}{dr} = \frac{1}{r} \cdot \frac{d \cdot f(x,\omega,u)}{dr} - \frac{1}{r^2} \cdot f(x, u, \omega),$$

or more simply:

$$\frac{d\varphi}{dr} = \frac{1}{r} \cdot \frac{d \cdot f(x,u,\omega)}{dr},$$

by ignoring the term divided by r^2. One also finds that:

$$\frac{d\varphi}{du} = \frac{1}{r} \cdot \frac{d \cdot f(x,u,\omega)}{du},$$

$$\frac{d\varphi}{d\omega} = \frac{1}{r} \cdot \frac{d \cdot f(x,u,\omega)}{d\omega};$$

and hence, from Section 4:

$$\cos.m = \frac{\dfrac{df}{dr}}{\sqrt{\left[\left(\dfrac{df}{dr}\right)^2 + \left(\dfrac{df}{du}\right)^2 \cdot \dfrac{1-u^2}{r^2} + \left(\dfrac{df}{d\omega}\right)^2 \cdot \dfrac{1}{r^2\left(1-u^2\right)}\right]}},$$

where it can be seen that as r increases, the cosine of the angle m will approach one, in other words the angle formed by the direction of the velocity of each air molecule, with the sound radius vector which terminates at this molecule, tapers off indefinitely.

If one therefore considers the propagation of sound over a large distance from the place where it is generated, one could suppose that the air molecules, over the whole period they are vibrating, do not leave the sound radius vector on which they are situated. Hence, their velocity will equal $\frac{d\varphi}{dr}$, and because $\frac{d\varphi}{dr} = \frac{1}{r} \cdot \frac{df}{dr}$, it can be seen that this velocity, for all molecules located along a same radius vector, is inversely proportional to the distance to the centre of the original vibrating source. Hence, if it is accepted that the intensity of sound is proportional to the square of this velocity, it follows that this intensity will fall off as the square of the distances increases, when these distances become much larger than the size of the portion of originally vibrated air.

Moreover, this intensity may vary at random, moving from one point to another along the same sound wave, as $\frac{df}{dr}$ is a function of u and ω, the form of which depends on the nature of the original vibration source or the cause producing the sound.

9. **THE** (f) series may be expressed using a definite integral relative to a new variable h. Indeed, it is easy to confirm that:

$$\int^n \cdot f_n(x,u,\omega)\, dx^n = \int \cdot \frac{f_n(x,u,\omega) \cdot (x - h)^{n-1} \cdot dh}{1 \cdot 2 \cdot 3 \ldots\ldots n - 1};$$

is the general case, provided that we make $h = x$ after integration. Therefore by using $\int \cdot f(x,u,\omega)dx$ instead of $f(x,u,\omega)$, a wholly acceptable operation, and by reducing it to $\frac{x - h}{r} = \theta$, the (f) series becomes:

$$\int \cdot \left[f(h,u,\omega) + \theta f_1(h,u,\omega) + \frac{\theta^2}{2} \cdot f_2(h,u,\omega) + etc. \right] dh,$$

or:

$$y = f + \theta \cdot f_1 + \frac{\theta^2}{2} \cdot f_2 + \frac{\theta^3}{2 \cdot 3} \cdot f_3 + \frac{\theta^4}{2 \cdot 3 \cdot 4} \cdot f_4 + etc.$$

To simplify matters, let the character δ, placed in front of any function ψ of u and ω, indicate the following expression:

$$\delta\psi = \frac{d \cdot \left(1 - u^2\right) \frac{d\psi}{du}}{du} + \frac{d^2\psi}{d\omega^2} \cdot \frac{1}{1 - u^2};$$

which gives:

$$\delta y = \delta f + \theta \cdot \delta f_1 + \frac{\theta^2}{2} \cdot \delta f_2 + \frac{\theta^3}{2 \cdot 3} \cdot \delta f_3 + etc.;$$

and hence:

$$\delta f_n = 2\,(n + 1) \cdot f_{n+1} - n \cdot n + 1 \cdot f_n,$$

because of the general relationship provided in section 6. The general term for this latter series is:

$$\frac{\theta^n}{2 \cdot 3 \ldots n} \cdot \left[2\,(n + 1) f_{n+1} - n(n+1) \cdot f_n\right];$$

from which one may derive, by comparing this series with that yielding the value for y, the following equation:

$$\delta y = \frac{d \cdot \left(2\theta - \theta^2\right) \frac{dy}{d\theta}}{d\theta},$$

or, by substituting δy with its long form:

$$\frac{d \cdot \left(1 - u^2\right) \frac{dy}{du}}{du} + \frac{d^2 y}{d\omega^2} \cdot \frac{1}{1 - u^2} = \frac{d \cdot \left(2\theta - \theta^2\right) \frac{dy}{d\theta}}{d\theta} \quad \text{................(3).}$$

This equation provides a value for y which contains two arbitrary functions. However, one of these functions may be solved by taking the values of y and $\frac{dy}{d\theta}$ relative to $\theta = 0$, and by noting that since the value for y is used for the function $f(h, u, \omega)$, that of $\frac{dy}{d\theta}$ must be $f_1(h, u, \omega)$ which is equal to $\frac{1}{2} \cdot \delta f(h,u,\omega)$. Having reduced the two arbitrary functions to one in this way, $\int ydh$ then yields the sum of the (f) series by replacing h with x following integration. The (F) series can then be transformed in the same way.

This result is analogous to that found by *Laplace* (*cf. Mémoires de l'Académie*, 1779) with respect to the second-order partial differential equations involving three variables: the integrals of these equations are expressed by definite integrals relative to a new variable, and the function which then takes the integral sign, depends on a two-variable differential equation. Similarly here, the integral of the equation presented with its five variables φ, t, r, u and ω, is expressed using definite integrals, and the function which takes the integral sign depends on a four-variable equation. If one assumes that the function φ is independent of ω, then equation 2 would only contain four variables and equation 3, on which the value for y depends, would only contain three variables. By making $\theta = 1 + v$, this equation then takes the form:

$$\frac{d \cdot \left(1 - u^2\right) \frac{dy}{du}}{du} = \frac{d \cdot \left(1 - v^2\right) \frac{dy}{dv}}{dv};$$

which is very much simplified, but which nevertheless cannot be finitely integrated.

10. **IF**, in short, $\int \cdot y\, du\, d\omega = Y$ is the integral, taken from $u = 1$ up to $u = -1$, and from $\omega = 0$ up to $\omega = 2\pi$, with π being the ratio of the circumference to the diameter, equation (3) becomes:

$$\frac{d \cdot \left(2\theta - \theta^2\right) \frac{dY}{d\theta}}{d\theta} = 0;$$

from which:

$$\left(2\theta - \theta^2\right) \frac{dY}{d\theta} = const.;$$

although this arbitrary constant must be taken as equalling zero, otherwise the value for Y or $\int \cdot y\, du\, d\omega$ will contain $log.\theta$, which cannot be the case as y represents an ordered series using the whole and positive powers of θ. Thus $\frac{dY}{d\theta} = 0$ and as a consequence Y will be a quantity independent of θ. Thus Y is a function of h, which is designated here as ψh. However, if we consider only the (f) series, then $r\theta = \int \cdot ydh$, from which it follows that $rR = \int \cdot Ydh = \int \cdot \psi hdh$, where R represents the integral $\int \cdot \theta\, du\, d\omega$ as above; and as one must make $h = x = r - at$, then, after integration, the value of $\int \cdot \psi hd\, h$ will be a function of $r - at$, thus giving $rR = funct.\ (r - at)$. The (F) series also produces the same $rR = funct.\ (r + at)$, and consequently the complete value for rR equals:

$$rR = funct.\ (r - at) + funct.\ (r + at),$$

as already found in Section 5.

11. **THE** definite integrals which here have been substituted for the (f) and (F) series, will now be used to find a class of particular integrals for equation (2), which would be otherwise difficult to determine if only considering the series *per se*.

To do this, it should be noted that equation (3) is satisfied by taking $y = U \cdot V$, where U and V are functions of u and ω and of θ respectively, which are determined using the equations:

$$\frac{d \cdot \left(2\theta - \theta^2\right) \cdot \frac{dV}{d\theta}}{d\theta} + i \cdot i + 1 \cdot V = 0,$$

$$\frac{d\cdot\left(1-u^2\right)\cdot\frac{dU}{du}}{du}+\frac{d^2U}{d\omega^2}\cdot\frac{1}{1-u^2}+i\cdot i+1\cdot U=0,$$

in which i is a positive whole number taking whatever value is required. Any other constant coefficient may be used in place of the coefficient $i{\cdot}i{+}1$, although by using a coefficient of this form, the functions U and V are expressed in finite form. Indeed, if, using the ascending powers of the indeterminate θ , one develops the function:

$$\frac{1}{\sqrt{\left\{1-2\theta\left[uu'+\sqrt{\left(1-u^2\right)}\cdot\sqrt{\left(1-u'^2\right)}\cdot\cos.\left(\omega-\omega'\right)\right]+\theta^2\right\}}},$$

in which u' and ω' are any constants. The coefficient of the power θ^i may then be taken for U in this development, the said coefficient being a rational and whole function of u, $\sqrt{(1-u^2)}.\cos.\omega$, $\sqrt{(1-u^2)}.\ \sin.\ \omega$. (*cf. La Mécanique céleste*, Book III, Ch. II.). In addition, it is possible to solve the equation which gives V by taking:

$$V=A+A_1\cdot\theta+A_2\cdot\theta^2+A_3\cdot\theta^3+\ldots+A_i\cdot\theta^i\ ;$$

provided that the general equation:

$$2\,(n+1)^2\cdot A_{n+1}-n\cdot n+1\cdot A_n+i\cdot i+1\cdot A_{\,n}=0,$$

holds for the coefficients A, A_1, A_2, A_3, ... A_i. The values of all these coefficients may be deduced from this general equation by using the value for the first, which will remain an arbitrary function of h.

Thus this gives:

$$y=U\cdot(A+A_1\cdot\theta+A_2\cdot\theta^2+\ldots A_i\cdot\theta^i);$$

and this value for y may be accepted, given that the relation which must exist (cf. Section 9) between the values for y and for $\frac{dy}{d\theta}$ with respect to $\theta=0$, is indeed valid, as these values are $A\cdot U$ and $A_1\cdot U$. In fact, one obtains:

$$A_1=-\frac{i\cdot i+1\cdot A}{2},\ U=-\frac{\delta U}{i\cdot i+1};$$

hence it follows that:

$$A_1 \cdot U = \frac{A\delta U}{2} = \frac{\delta \cdot AU}{2}.$$

Now if one replaces θ with $\frac{x - h}{r}$ within the value for y, and multiply this by dh integrated with respect to h, and then put $h = x$ after integration, assuming that the function U is independent of h, one gets:

$$r\varphi = U \cdot \left[fx + \frac{1}{r} \cdot \int f_1 x \cdot dx + \frac{1}{r^2} \cdot \int^2 f_2 x \cdot dx^2 + \frac{1}{r^3} \int^3 f_3 x \cdot dx^3 \ldots\ldots + \frac{1}{r^i} \cdot \int^i f_i x \cdot dx^i \right],$$

where fx is an arbitrary function of x or of $r - at$, and functions $f_1x, f_2x, f_3x \ldots\ldots f_ix$, are derived from the first as per the general equation:

$$2 \cdot (n + 1)^2 \cdot f_{n+1}\, x - n\,.\, n + 1 \cdot f_n x + i \cdot i + 1 \cdot f_n x = 0.$$

As one may take for i any whole, positive number, it follows that there will be an infinite number of particular values for $r\varphi$ which will satisfy equation (2), and the sum of any number of these values will still satisfy the same equation due to its linear nature. One may also take either $r - at$ or $r + at$ for x indiscriminately, such that there will be available two sets of particular integrals from equation (2).

These various integrals are consistent with the special hypotheses on the nature of the original vibrating source which one may still assume is held within a portion of air limited from all sides, as the theory of sound requires. One could check the theorem demonstrated in general terms against each of these integrals, namely that the velocity of sound is the same along all sound radius vectors, despite its intensity varying between radius vectors.

12. **IF** i is assigned all possible whole positive numbers from $i = 0$ to $i = \infty$, and if one then takes successively $x = r - at$ and $x = r + at$, and finally produces the sum of all these particular values of $r\varphi$, arranging them in increasing order of value of i, then one obtains the complete integral for equation (2), developed in an ordered series as per the rational and whole functions of u, $\sqrt{(1-u^2)}.\ sin.\ \omega$, $\sqrt{(i-u^2)}.\ cos.\ \omega$, which has then the ability to satisfy the general equation:

$$\frac{d \cdot \left(1 - u^2\right) \cdot \frac{dU_i}{du}}{du} + \frac{d^2 U_i}{d\omega^2} \cdot \frac{1}{1 - u^2} + i \cdot i + 1 \cdot U_i = 0,$$

in which i is any subscript.

This form of development has been illustrated because of its analogy with the type of development which *Laplace* gave for the integral of the equation:

$$\frac{d \cdot \left(1 - u^2\right) \cdot \frac{dV}{du}}{du} + \frac{d^2V}{d\omega^2} \cdot \frac{1}{1-u^2} + \frac{rd^2 \cdot rV}{dr^2} = 0,$$

and which he successfully used in his theories on the attraction of spheroids and the shape of the planets.

Starting with this property of the U_i functions (*cf. la Mécanique céleste*, Book III, Ch. II), namely that the integral $\int \cdot U_i \; du \; dw$ taken from $u = 1$ to $u = -1$, and from $\omega = 0$ up to $\omega = 2\pi$, is equal to zero except when $i = 0$, then one may easily again obtain the equation:

$$rR = funct.\ (r - at) + funct.\ (r + at).$$

A Mixture of Simultaneous Sounds

13. **THE** sum of the two series (f) and (F) gives the most general value for $r\varphi$ which satisfies equation (2). However, when the air is vibrated in several places at the same time, it becomes difficult to determine the arbitrary functions which these series contain so that a representation of the state of the air at the origin of the motion can be obtained. This difficulty is avoided by noting that if the origin of the coordinates x, y, z is moved to another point, and x', y', z' are then the new coordinates, without any change in direction, then equation (1) from section 3 becomes:

$$\frac{d^2\varphi}{dt^2} = a^2\left(\frac{d^2\varphi}{dx'^2} + \frac{d^2\varphi}{dy'^2} + \frac{d^2\varphi}{dz'^2}\right);$$

In addition, if r' is used to represent the radius vector of a given molecule relative to the new origin, u' the cosine of the angle which this radius vector makes with the axis of z', and ω' the angle which the projection of the same radius vector in the place of x', y' makes with the axis of x', the above equation will then change to:

$$\frac{d^2 \cdot r'\varphi}{dt^2} = a^2\left[\frac{d^2 \cdot r'\varphi}{dr'^2} + \frac{d \cdot \left(1 - u'^2\right)\frac{d \cdot r'\varphi}{du'}}{r'^2 \cdot du'} + \frac{d^2 \cdot r'\varphi}{d\omega'^2} \cdot \frac{1}{r'^2\left(1 - u'^2\right)}\right],$$

which should be regarded as a transform of equation (2) from Section 4. One may conclude from this that this transform resembles equation (2) and that equation (2) may still be satisfied by replacing the values of u, ω, r of the (f) and (F) series with the values of u', ω', r' respectively, thus giving new series which may be designated as (f') and (F'). In addition, due to the linear form of equation (2), these new series may be added to the old, and the result will still hold.

Now, if the air was vibrated both towards the origin of the coordinates r', u', ω' and the origin of the coordinates r', u', ω', and if the first vibration ranges from $r = 0$ to $r = \alpha$, and the second from $r' = 0$ to $r' = \alpha'$, then this original state of the air may be represented assuming that the arbitrary functions of $r - at$ and $r + at$ used in the (f) and (F) series are zero for all values of these quantities which are negative or which exceed α, whilst the arbitrary functions of $r' - at$ and $r' + at$ which are contained in the (f') and (F') series are zero for all values of $r' - at$ and $r' + at$ which exceed α', and for all negative values of $r' - at$. Hence, it is obvious that the two partial vibrations will propagate independently of each other as if each existed alone; the same will also apply when the air has been vibrated at more than two different places.

In general, let r, u, ω represent the polar coordinates of an air molecule with respect to a given origin, (f) and (F) the sums of the series found in Section 6 to express the value for $r\varphi$, and then let r', u', ω', r'', u'', ω'' etc., equal the polar coordinates of the same molecule with respect to the other origins; let (f') (f''), (f'''), etc. and (F'), (F''), (F'''), etc., be what (f) and (F) become when the quantities r, u, ω are replaced by their analogues r', u', ω'; r'', u'', ω''; r''', u''', ω''' etc.: equation (2) would be satisfied by taking:

$$\varphi = \frac{(f)}{r} + \frac{(F)}{r} + \frac{(f')}{r'} + \frac{(F')}{r'} + \frac{(f'')}{r''} + \frac{(F'')}{r''} + etc.$$

However, it should be observed that the equation:

$$\varphi = \frac{(f)}{r} + \frac{(F)}{r};$$

is nonetheless the complete integral of equation (2), as this value for φ contains two arbitrary functions, hence the addition of the other terms do not in any way enhance the general nature of the value for φ, this value merely takes a more convenient form for its intended use, and this will be aptly demonstrated below using the theory of *Echoes*.

The Reflection of Sound

14. **WHEN** the mass of air in which the sound is propagated, is terminated by a surface having a fixed position, all the molecules adjacent to this surface cannot have a velocity through the normal, as obviously they are only able to slide along the said surface. However, until the sound has reached this surface, it must be able to propagate absolutely as if the surface did not exist; therefore, taking the origin of the polar coordinates r, u, ω, as the centre of the original vibrating source, then the value for φ must first contain a series of the form seen earlier, namely:

$$\varphi = \frac{f(x,u,\omega)}{r} + \frac{1}{r^2} \cdot \int \cdot f_1(x,u,\omega)\, dx + \frac{1}{r^3} \cdot \int^2 f_2(x,u,\omega)\, dx^2 + etc.,$$

in which $x = r - at$ and where f, f_1, f_2, f_3 etc., are functions of x, u, ω, linked together by a specific equation such that only the first remains an arbitrary function (*cf.* Section 6). The form of this function depends on the nature of the original vibrating source; if it occurs between $r = 0$ and $r = \alpha$, then the function f, and therefore all the other functions f_1, f_2, f_3 etc., must equal zero (*cf.* Section 7) for all values of $x < 0$ or $> \alpha$. Using this method, the previous formula may be used to determine the propagation of sound so long as it has not reached the fixed surface. However, to obtain a formula capable of describing the propagation of sound once it has reached this limit, the function φ will need to have added to it another function φ' which needs to be determined ensuring:

I) That the said function φ' has the same properties as the function φ, in other words that it complies with equation (2) in the same way as the function φ;

II) That, for each molecule adjacent to the fixed surface, the velocity along the normal and calculated from the function φ', is always equal and opposite to the velocity in the direction of the normal, as calculated from the function φ;

III) That the partial differentials of φ', with respect to the three coordinates and to time, equal zero as long as the sound has not reached the fixed surface, in order that the motion in the air is not disturbed before the sound reaches this surface.

Examples are used below to demonstrate how all these conditions may be met.

15. **ASSUMING** first of all that the air mass is delimited by an indefinite plane, then let c represent the centre of the original vibrating source, in other words the point of origin of the coordinates r, u, ω. Next, drop a perpendicular from point c to the fixed plane and extend this beyond the said plane by an equal amount until it reaches point c'. For clarity, consider that the fixed plane is horizontal, with the point c being above the plane and point c' below, the three conditions listed above are then satisfied by taking $x' = r' - at$, and:

$$\varphi' = \frac{F(x',u',\omega')}{r'} + \frac{1}{r'^2}\cdot\int\cdot F_1(x',u',\omega')dx' + \frac{1}{r'^3}\cdot\int^2\cdot F_2(x',u',\omega')dx'^2 + etc.,$$

where u', ω', r' are polar coordinates the origin of which is point c', whilst F, F_1, F_2 etc. (within u', ω', x') have the same functions as f, f_1, f_2 etc. (within u, ω, x), such that $F = f$, $F_1 = f_1$, $F_2 = f_2$ etc., when $u = u'$, $\omega = \omega'$ and $x = x'$.

Firstly, this value for φ' will satisfy equation (2); next, if this is added to the value for φ, there will be two spherical waves in the air mass which are perfectly identical, with one centred around point c, and the other around point c'. When the first arrives at the molecules located below the plane, the second will still not have arrived, from which it follows that these molecules will be vibrated by the direct sound just as if the fixed plane did not exist. However, the two waves will reach the molecules adjacent to the fixed plane at the same time, in which case if the two velocities of the same molecule are factorized both along the plane and perpendicular to the plane, these latter components will be exactly equal and opposite in sign. Consequently, the molecules adjacent to the fixed plane will never have velocities perpendicular to the said plane.

After the two waves have arrived at the fixed plane, the one produced around point c' will continue to propagate below this plane. The air molecules located below this plane will therefore be vibrated a second time, and the reflected sound will evidently be the same as if the original vibrating source which produced the direct sound had had its centre at point c'. It is in this way that light, leaving a luminous point placed in front of a mirror, is reflected by the mirror in such a way that the reflected light has the same order of magnitude and direction as if it had been emitted by the image of the light point in the mirror.

It is therefore shown that when the reflecting surface is an arbitrary plane:

I) Each sound radius vector is reflected with an angle of incidence equal to the angle of reflection;

II) The velocity of the reflected sound is the same as that of direct sound;

III) The intensity of the reflected sound at the extreme end of a zig-zag path is precisely the same as that which would occur at the extreme end of a straight path of equal length to the zig-zag path if the sound, rather than being reflected, had instead spread beyond the fixed plane.

One could easily see what would occur if the air mass was contained between two arbitrary, parallel planes. The sound would be subject to an infinite series of reflections resembling those of light between two parallel mirrors. However, rather than dwell disproportionately on the reflection of sound by flat surfaces, consider instead a special case where this occurs in a very simple manner, namely that where the air mass is terminated by two indefinite planes perpendicular to each other. The reflection in this case is determined by placing outside the air mass, fictitious wave centres such that the molecules adjacent to these two planes never take any velocities perpendicular to the said planes, either from the direct sound or from the sound reflected by one of the flat surfaces to the other. One finds that three fictitious centres are adequate; thus there will be four waves co-existing in the air mass and consequently each molecule will be vibrated at four different times.

16. **A** second example in the reflection of sound takes the case of sound produced at the focus of a surface of revolution and reflected by its surface which envelopes all parts of the air mass. c and c' represent the two foci of the ellipsoid, with point c being the origin of the coordinates r, u, ω, where r is the radius vector, u is the cosine of the angle which the radius vector makes with the axis of rotation, and ω is the angle formed by the plane of the given radius vector and the axis with an arbitrarily selected meridian. Finally, it is assumed that the focus c is the centre of the original vibration source extending from $r = 0$ to $r = \alpha$.

The formula:

$$\varphi = \frac{f(x,u,\omega)}{r} + \frac{1}{r^2} \cdot \int \cdot f_1(x,u,\omega)dx + \frac{1}{r^2} \cdot \int^2 \cdot f_2(x,u,\omega)dx^2 + etc.,$$

in which $x = r - at$, may be used to determine the propagation of sound as it would occur if the ellipsoid did not exist, and as it would continue up until the sound reaches the surface. To obtain the formula used to determine the propagation of reflected sound, let us move the origin of the coordinates to point c'. The angle ω will remain the same, but the new radius vector will be designated by r' and the cosine of the angle it makes with the axis of rotation on which the two foci are located, by u'. In addition, b is the length of the axis of rotation, which gives, for brevity, $b - r' - at = x'$. The required formula is therefore:

$$\varphi' = \frac{F(x',u',\omega)}{r'} + \frac{1}{r'^2} \cdot \int \cdot F_1(x',u',\omega)dx' + \frac{1}{r'^3} \cdot \int^2 \cdot F_2(x',u',\omega)dx' + etc.,$$

where F, F_1, F_2 etc., are a set of functions which satisfy the general equation:

$$2 \cdot (n+1) \cdot F_{n+i} = n(n+1) \cdot F_n + \frac{d \cdot \left(1-u^2\right) \cdot \frac{dF_n}{du}}{du} + \frac{d^2 F_n}{d\omega^2} \cdot \frac{1}{1-u^2};$$

such that the first remains only an arbitrary function. Unlike in the previous example, these functions are not of the same form as the functions f, f_1, f_2, f_3 etc., which enter into the value of φ. Only functions F, F_1, F_2 etc., are zero for all values of x' greater than α or negative; a specific relation exists between the two arbitrary functions f and F as will be seen later.

The aim is therefore to prove that the value for φ' satisfies the conditions given earlier (*cf.* Section 14). This value is seen to satisfy equation (2), as does that of the function φ, whilst if the value for φ' is added to that of φ, two spherical waves will exist in the air mass; one spreading continuously away from its centre c, and the second continuously approaching its centre c'. Now, given the above, consider a molecule m located within the ellipsoid. In the case of this molecule, $r + r'$ will be smaller than b and consequently, when $x = \alpha$, then at the same instant $x' > \alpha$. The first sound wave will therefore arrive at point m before the second, and the molecule m will therefore be vibrated directly by the sound just as if the fixed surface did not exist. Now, if the molecule m is adjacent to the surface of the ellipsoid, then one obtains $r + r' = b$, thus giving simultaneously $x = \alpha$ and $x' = \alpha$, and similarly $x = 0$ and $x' = 0$. Thus the molecule will be vibrated simultaneously by both sound waves, and its two motions will also cease at exactly the same time. Thus factorizing the two velocities of the molecule m, both along the tangent plane and along the normal to the ellipsoid, and representing the normal velocities by V and V', then these velocities may be obtained using the partial differentials of the functions φ and φ'. One of these will therefore be a function of x, u, ω, r, whilst the other a function of x', u', ω, r', although all these variables may be reduced to three, namely x', u', ω because as point m is at the surface of the ellipsoid, then r, r', u may be expressed as functions of u', and because also $r + r' = b$, and one obtains $x = x'$. Thus by declaring that:

$$V + V' = 0,$$

it is possible to perceive that this is used to determine the function F using the function f, as this equation does not contain any variables other than the three x', u', ω, which enter into the function F. However, with this function thus determined, the molecules adjacent to the ellipsoid's surface will never be able to take velocities perpendicular to this surface.

Once the two sound waves have reached the surface of the ellipsoid at exactly the same time, the second wave will continue to propagate by continuously moving towards focus c', and hence each of the molecules located inside the ellipsoid will therefore be vibrated a second time. The second vibration of the molecule m will begin when $x' = b - r' - at = \alpha$, in other words, at the end of $t = \dfrac{b - r' - \alpha}{a}$, and will continue up until $t = \dfrac{b - r'}{a}$. However, if the radius vector r' is extended right to the surface of the ellipsoid, if h is used to designate the distance from the meeting point to the focus c, and k the distance from the same point to the molecule m, then this gives $h + k = b - r'$. Thus the reflected sound arrives at the molecule m via the zig-zag path $h + k$ in the same amount of time as it takes for the direct sound to follow a straight line of equal length to $h + k$, or in other words, within a time period equal to $\dfrac{h + k - \alpha}{a}$.

In general, if one describes two spheres with the points c and c' as centres, the first with a radius of $at + \alpha$, and the second with a radius of $b - at - \alpha$, and if only those portions of these surfaces inside the ellipsoid are retained, then these two portions of spheres will contain all the molecules which begin to vibrate at the end of time t, under the influence of either direct or reflected sound.

17. THE velocity of air molecules, on which depends the intensity of the reflected sound, has been determined using the partial differentials of φ', and it can be clearly seen that this intensity must increase as the reflected sound moves towards points closer to the focus c', because of the decrease in r'. However, the problem remains of determining the relation between direct sound and reflected sound. To do this, one must first solve the equation $V + V' = 0$ with respect to F, in order to deduce the value of F using that of f, which it is assumed has been supplied and which depends on the nature of the original vibration. This is easily arrived at when one assumes that the distances from the foci c and c' to the various points of the surface of the ellipsoid are much larger than α, which can generally be regarded as true. Then, the value for φ is reduced to its first term (*cf.* Section 8), the velocity of the air molecules becomes:

$$\frac{d\varphi}{dr} = \frac{1}{r} \cdot \frac{d \cdot f\,(x, u, \omega)}{dx},$$

and this velocity is directed along the radius vector r. Similarly, the velocity of the air molecules from the reflected sound is directed along the radius vector r', and may be reduced to:

$$\frac{d\varphi'}{dr'} = -\frac{1}{r'} \cdot \frac{d \cdot F(x', u', \omega)}{dx'}.$$

However, as the two radius vectors r and r', which end at the same point on the surface of the ellipsoid, form equal angles with the normal at this point, then it will suffice that the velocities $\frac{d\varphi}{dr}$ and $\frac{d\varphi'}{dr'}$ are equal and of opposite sign in order that the normal velocity is zero. This thus gives, for all points of the ellipsoid:

$$\frac{d \cdot F(x', u', \omega)}{dx'} = \frac{r'}{r} \cdot \frac{d \cdot f(x, u, \omega)}{dx}.$$

But for all these points, $\frac{r'}{r}$ is a specific function of u' which will be represented by U', u is another function of u' designated this time by u_I, and x is equal to x'. Therefore assuming that $\frac{d \cdot f(x, u, \omega)}{dx} = \psi(x, u, \omega)$, then:

$$\frac{d \cdot F(x', u', \omega)}{dx'} = U' \cdot \psi(x', u_I, \omega).$$

The velocity of the air molecules from the reflected sound will therefore equal $-\frac{U'}{r'} \cdot \psi(x', u_I, \omega)$ so long as the molecules under consideration are distant from the focus c'. However, in order to determine the velocity of a molecule close to this focus, then the value of $F(x', u', \omega)$ needs to be substituted (i.e. $U' \cdot \int \cdot \psi(x', u_I, \omega)\, dx'$), in the series which provides the value of φ'. Without ignoring any terms in this series, its partial differentials with respect to x', u', ω, are taken to determine the magnitude and direction of the required velocity which will no longer be directed along the radius vector r'.

18. **AS** the sound is reflected by a concave surface, its intensity must be increased following reflection, which is to say that it must exceed that of the direct sound if that had been able to continue for the same distance beyond the reflecting surface. To determine the ratio of these two intensities, first let the two radius vectors leaving foci c and c', and which end at a specific point of the ellipsoid, equal h and $b - h$ respectively, whilst $\frac{A}{r}$ equals the velocity, relative to the direct sound, of the air molecule located on the radius vector h at a distance r

from the focus c. Using the findings from Section 17, $-\dfrac{A(b-h)}{hr'}$ gives the velocity, relative to the reflected sound, of the air molecule located on the radius vector $b - h$ at a distance r' from the focus c'. Consequently, the intensity of the reflected sound at this point is given by $\dfrac{A^2(b-h)^2}{h^2r'^2}$. The sound produced at the focus c will have travelled a zig-zag path of length equal to $b - r'$ by the time it reaches the end of the radius vector r'. However, if one imagines that the radius vector h extends beyond the surface of the ellipsoid until it equals $b - r'$, then $\dfrac{A^2}{(b-r')^2}$ gives the intensity of direct sound arriving at this distance from point c. Thus the ratio of the intensity of reflected sound to the intensity of direct sound is given by:

$$\frac{(b-r')^2 \cdot (b-h)^2}{h^2 \cdot r'^2};$$

which has a value which is always greater than 1 as $b - r' > h$ and $b - h > r'$. It may be readily checked using the catoptrics formulae (*refer to* the Note by *Malus* provided at the front of this volume), that this ratio is equal to that which would have been obtained under the same circumstances, between the intensity of reflected light and that of direct light. From this an important conclusion may be drawn, namely that, with respect to both its intensity and its direction, the sound produced at the focus of an ellipsoid and reflected by its concave surface obeys the same laws as light in the emission hypothesis.

19. **WHEN** the reflected sound reaches the focus c, as will occur after a time period equal to $\dfrac{b-\alpha}{a}$, the reflected sound wave will die away, to be succeeded by another the centre of which will be at point c', and which will spread away indefinitely from the said centre. Once it reaches the surface of the ellipsoid, it will be reflected by this surface towards the focus c, where it too will die away. This will again be followed by a new wave, the centre of which will be point c, and so this motion continues indefinitely, with each oscillation taking place over a time scale which will be the same for all and equal to $\dfrac{b-\alpha}{a}$. Indeed, if one describes a sphere having point c' as its centre, and a radius vector of α, then it is evident that at the instant where the molecule occupying focus c' is set into motion by the reflected sound, the whole air mass will be at rest with the exception of the part held within the said sphere. This is therefore analogous to having produced a partial vibration around point c', with the said vibration spreading throughout the whole mass in the same way as the initial vibration which took place around point c.

20. **GIVEN** that the two foci c and c' were separated by an unspecified distance, the point c may in fact be an infinite distance away, in which case the ellipsoid would change to a paraboloid. One may then conclude from the previous analysis that if the sound was produced at the focus of a paraboloid of revolution, it will be reflected parallel to the axis on meeting this surface. All the molecules located along the same plane perpendicular to this axis will then be vibrated at the same time by the reflected sound and will begin moving after a time period equal to that applicable to the direct sound when travelling to the apex of the paraboloid along the same plane, plus the distance from the apex to the focus. It is possible to determine directly the reflection of the sound on the paraboloid, but it is much simpler to consider it as a special case of reflection off an ellipsoid, as has been done here.

Rather than repeat the reasoning already outlined above, it will suffice here to illustrate the reflection of sound off the hyperboloid generated by one branch of a hyperbola rotating around the line joining the two foci. If the sound leaves from the external focus, it will be reflected by the convex surface of the hyperboloid. The reflected sound wave will still retain its spherical shape, but this time its centre will be the internal focus from which it will spread away indefinitely. Conversely, when the sound is produced by the internal focus, it will be reflected by the concave surface but again the reflected sound wave will retain its spherical form, whilst this time continuously moving away from its centre which will be the external focus. In either case, the sound arrives at an air molecule via a zig-zag path within the same time period taken by the direct sound to cover a straight path of length equal to that of the zig-zag path. These propositions can be seen to have been easily arrived at using the previous analysis method after appropriate modification.

Finally, if one dimension of air is ignored (which is to say that the propagation of sound is considered along one plane only), then using the principles established (*cf.* Section 14), one may determine the reflection of sound by a conical section, provided that the sound was generated at one of the foci of this curve. The results are analogous to those presented by reflection by a quadric, but the demonstrations are more complex despite their being based on the same principles. As such, they have not been presented here to prevent this paper from becoming excessively long.

Examination of the Various Physical Circumstances which may Influence the Velocity of Sound

21. IN Section 3, D and gh were used to represent respectively the density and elasticity of air before motion commenced, and $D\ (1 + \gamma)$ and $gh\ (1 + \gamma + k\gamma)$ to represent the same density and elasticity during motion, with the term $gh{\cdot}k\gamma$ representing the increase in elasticity due to the production of heat accompanying the compression of air. These may then be abbreviated to $\frac{gh}{D} \cdot (1 + k) = a^2$, where a represents the velocity of sound. Thus, to compare this with the experimentally observed value for the velocity of sound, one must first of all examine the variations which this quantity may be subjected to, then calculate its numerical value.

As density D decreases in line with an increase in temperature, one must conclude that the velocity of sound must increase with temperature. Thus, this velocity must be in general higher in summer than in winter. The following calculation hence allows one to determine the extent of this difference.

Let D equal the density of air at the melting point temperature of ice at random pressure; under the same pressure and at a temperature of n degrees centigrade, the density then equals $\frac{D}{1 + n(0.00375)}$, as the experiments conducted by *Gay-Lussac* have shown that air expands uniformly from 0 to 100°, and that its increase in volume per degree rise in temperature equals 0.00375 of its volume at zero. The velocity of sound at this temperature therefore becomes:

$$\sqrt{\frac{gh}{D}} \cdot \left[1 + n(0.00375)\right] \cdot \sqrt{(1 + k')},$$

where k' denotes the new value for k for the same temperature. If, for example, one then takes $n = 30°$, then the approximation $\sqrt{1 + n \cdot (0.00375)} = 1 + \frac{1}{16}$ is obtained, from which it follows that when the temperature rises from 0 to 30°, the velocity of sound will vary by approximately $\frac{1}{16}$th of its value at zero, ignoring the unknown variation of the quantity k. This change in velocity must be appreciable in comparative, carefully conducted experiments, while the same experiments would also offer a means of determining whether the quantity k varies significantly with air temperature.

The same is not true for the relative humidity of air, however, as experiments have shown that humidity does not influence the velocity of sound. Physicists who have measured this velocity have observed that it remains the same for both good weather conditions and wet conditions. This is in part due to water vapour density not being very different from that of air for the same temperature and pressure, and partly due to the quantity of vapour suspended in the atmosphere always being very small relative to the total air mass. Indeed, *Saussure* found (*cf. Essai sur l'Hygrométrie*) that for the same pressure and temperature, water vapour density equals $\frac{5}{7}$ths the air density. In addition, when the hygrometer registers the point of saturation, the quantity of vapour suspended in a volume of air at 15°R using the *Réaumur* scale thermometer at ordinary atmospheric pressure, may be evaluated using *Saussure*'s experimental findings, at approximately $\frac{1}{72}$ of the total mass. However, using these data, it is easy to ensure that the relative humidity of the air does not modify the velocity of sound by as much as one *five-hundredth* of its relative value when the air is at its driest.

22. **LET** us now calculate the numerical value for the velocity of sound. Taking as a unit the sexagesimal second, then g equals 9.7979 metres per second, whilst the most commonly found barometric height for our climatic region is 0.76 metres. Under this pressure and at the melting point temperature of ice, the ratio of the density of mercury to that of air equals 10475.68, according to *Biot* and *Arago* (in *Physique*, by *Haüy*, p. 299). These data then give $\sqrt{\frac{gh}{D}}$ = *279.395 metres* per second. This would equal the velocity of sound at zero temperature, if the quantity k were to be ignored. Consequently, multiplying this number of metres by $\sqrt{[1 + n\ (0.00375)]}$ then gives the velocity of sound in air at temperature n. For example, if $n = 6°$, then the velocity of sound equals 282.42 metres per second relative to this temperature.

However, the members of the Academy of Sciences concluded (*refer to* its own *Mémoires*, 1738) from a series of experiments conducted over a line of approximate length 29,000 metres at a mean thermometer temperature of 6° centigrade, that the sound travelled at 337.18 metres per second, thus exceeding by one-fifth the velocity of 282.42 metres per second calculated ignoring the quantity k. This difference is too great to attribute to errors of observation or the inaccuracy of data used for the basis of the calculation. It is therefore indispensable to take account of the heat produced by the compression of air in the calculation of the velocity of sound. The calculation may be made to agree with the observation using simply:

$$(282.42)\cdot\sqrt{1+k} = 337.18,$$

from which it can be seen that $k = 0.4254$. It now remains to determine the increase in temperature implied by the value for k.

To do this, consider a volume of air of density D, pressure gh, and temperature θ, and assume that the said air is compressed such that the density becomes $D\ (1 + \gamma)$. Let ω equal the number of degrees the said compression causes the thermometer to rise, where the mass of this instrument is not similar to that of air. Hence the problem is to determine the elasticity corresponding to the density $D\ (1 + \gamma)$ and the temperature $\theta + \omega$.

Firstly, $gh\ (1 + \gamma)$ gives the pressure which corresponds to the density $D\ (1 + \gamma)$ and the temperature θ. If the air subjected to this pressure is heated until its temperature equals $\theta + \omega$, it will expand and, according to *Gay-Lussac's* law, its density will become:

$$\frac{D \cdot (1 + \gamma) \cdot [1 + \theta \cdot (0.00375)]}{1 + (\theta + \omega) \cdot (0.00375)}.$$

Finally, with the air in this state, if one considers compressing it slowly such that it returns to density $D\ (1 + \gamma)$ without increasing its temperature, then one obtains:

$$\frac{gh \cdot (1 + \gamma)\,[1 + (\theta + \omega) \cdot (0.00375)]}{1 + \theta(0.00375)},$$

for a pressure corresponding to a density $D\ (1 + \gamma)$ and a temperature of $\theta + \omega$.

In the propagation of sound, as the compression or expansion which occurs successively throughout the air mass is assumed to be very small, one may regard an increase or a drop in temperature due to this change in density as being proportional to it. Thus by making $\omega = m\gamma$ and then ignoring the square of γ, one obtains:

$$gh \cdot \left[1 + \gamma + \frac{(0.00375) \cdot m}{1 + \theta \cdot (0.00375)} \cdot \gamma\right],$$

which gives the pressure which was given in section 3 by $gh \cdot (1 + \gamma + k\gamma)$. This then gives:

$$k = \frac{(0.00375) \cdot m}{1 + (0.00375) \cdot \theta},$$

from which $m = 115.99°$ when $\theta = 6°$ and $k = 0.4254$. Consequently, if the expansion or compression had been by $\frac{1}{116}$, the temperature would have fallen or risen by one degree centigrade. Generally, if a layer of air experiences a very small degree of compression designated by γ, between two other layers of air, the temperature of the first should increase by $(116°) \cdot \gamma$. By accepting this result, as it cannot be verified by any direct measurements, the difference noted by *Newton* between the calculated velocity of sound and the observed value is eliminated.

Motion of a Line of Air in the Case where the Velocities of the Molecules are not Assumed to be very Small

23. **BY** reducing the mass of air to a simple line, and by taking this line as the axis for the abscissa x, the quantities ρ andφ will no longer be functions of x and t. If then the gravity is ignored, equations (a) and (b) from Section 1 then become:

$$\int \cdot \frac{dp}{\rho} + \frac{d\varphi}{dt} + \frac{1}{2} \cdot \frac{d\varphi^2}{dx^2} = 0,$$

$$\frac{d\rho}{dt} + \rho \cdot \frac{d^2\varphi}{dx^2} + \frac{d\rho}{dx} \cdot \frac{d\varphi}{dx} = 0.$$

If one also assumes that the temperature is constant along the whole line of air, the pressure p will then be proportional to the density ρ, in which case $p = a^2 \cdot \rho$, a being a constant coefficient representing the square root of the ratio of air elasticity to its density before motion commences. The first of these equations then becomes:

$$a^2 \cdot log. \frac{\rho}{D} + \frac{d\varphi}{dt} + \frac{1}{2} \cdot \frac{d\varphi^2}{dx^2} = 0, \tag{1}$$

D being the density of air prior to motion. By combining this and the previous equation to eliminate ρ, one obtains:

$$\frac{d^2\varphi}{dt^2} + 2\frac{d\varphi}{dx} \cdot \frac{d^2\varphi}{dx\,dt} + \frac{d\varphi^2}{dx^2} \cdot \frac{d^2\varphi}{dx^2} = a^2 \frac{d^2\varphi}{dx^2}. \tag{2}$$

It is not possible to obtain the complete, finite integral for this equation. However, it is easy enough to ascertain this by applying the method given by *Laplace* for integrating second-order partial differential equations, after first converting to the linear form as required by the method, using a known transformation as supplied by *Legendre*. This linear transform is included amongst the equations integrated by *Laplace* using a definite integral (*cf.* '*le Calcul intégral*' by *Lacroix*, which gives all the integration methods cited in this paper). However, this type of integral is unable to determine the motion of the line of air under consideration. Thankfully, equation (2) allows for a particular, finite integral which contains an arbitrary function and which is sufficient, as will be seen later, to determine the law obeyed by this motion, provided that one assumes the velocities of the air molecules are smaller than the velocity a over the whole duration of the motion and along the whole length of the line of air.

24. IN order to find this integral, it is observed that equation (2) may be satisfied using:

$$\frac{d\varphi}{dt} - a \cdot \frac{d\varphi}{dx} + \frac{1}{2} \cdot \frac{d\varphi^2}{dx^2} = fk; \tag{3}$$

where fk is an arbitrary function of the variable k; this gives:

$$\frac{dk}{dt} + \frac{dk}{dx} \cdot \left(a + \frac{d\varphi}{dx}\right) = 0.$$

Indeed, by differentiating equation (3) successively with respect to t and to x, and multiplying the second partial differential by $a + \dfrac{d\varphi}{dx}$, and then adding it to the first, one obtains equation (2) once more. It may also be satisfied by using this second equation:

$$\frac{d\varphi}{dt} + a\frac{d\varphi}{dx} + \frac{1}{2} \cdot \frac{d\varphi^2}{dx^2} = Fk_I; \tag{4}$$

where Fk_I is an arbitrary function of the quantity k_I, giving:

$$\frac{dk_I}{dt} - \frac{dk_I}{dx} \cdot \left(a - \frac{d\varphi}{dx}\right) = 0.$$

However, as $\dfrac{d\varphi}{dx}$ and k are functions of x and t, one may regard $\dfrac{d\varphi}{dx}$ as a function of t and k; thus if we put $\dfrac{d\varphi}{dx} = \psi(k, t)$, and integrate:

$$\frac{dk}{dt} + \frac{dk}{dx} \cdot [a + \psi(k, t)] = 0,$$

using known methods, then one obtains:

$$k = funct.\left[x - at - \int \cdot \psi(k, t)\, dt\right],$$

where the integral $\int \cdot \psi(k, t)\, dt$ is used with k as a constant and, where required, equal to zero when $t = 0$. Similarly one obtains:

$$k_I = funct.\left[x + at - \int \cdot \psi_I(k_I, t)\, dt\right],$$

by assuming that $\frac{d\varphi}{dx} = \psi_I(k_I, t)$, that the integral $\int \cdot \psi(k_I, t)\, dt$ used equals zero when $t = 0$, and ensuring that the value k_I is a constant. Equations (3) and (4) therefore become:

$$\frac{d\varphi}{dt} - a \cdot \frac{d\varphi}{dx} + \frac{1}{2} \cdot \frac{d\varphi^2}{dx^2} = f\left[x - at - \int \cdot \psi(k,t)\, dt\right],$$

$$\frac{d\varphi}{dt} + a\frac{d\varphi}{dx} + \frac{1}{2} \cdot \frac{d\varphi^2}{dx^2} = F\left[x + at - \int \cdot \psi_I(k_I,t)\, dt\right],$$

from which, through equation (1), one obtains:

$$\left.\begin{aligned} 2a \cdot \frac{d\varphi}{dx} &= F\left[x + at - \int \cdot \psi_I(k_I,t)\, dt\right] - \\ &\quad f\left[x - at - \int \cdot \psi(k,t)\, dt\right], \\ 2a^2 \cdot log.\frac{\varphi}{D} &= -F\left[x + at - \int \cdot \psi_I(k_I,t)\, dt\right] - \\ &\quad f\left[x - at - \int \cdot \psi(k,t)\, dt\right]. \end{aligned}\right\} \qquad (5)$$

Now, assuming that the original vibration source for the line of air extends from $x = 0$ to $x = \alpha$, then the velocity $\frac{d\varphi}{dx}$ should equal zero and the density ρ should equal D, when $t = 0$ for all values of x greater than α or negative. From this, and because of equations (5), it follows that for all these values of x, the functions fx and Fx must equal zero. Given this, if one considers, after a given time period, a molecule corresponding to a positive abscissa x and greater than α, then the value of $x + at - \int \psi_I(k_I, t)\, dt$ must be greater than x and hence greater than α, such that one obtains:

$$F\left[x + at - \int \cdot \psi_I(k_I, t)\, dt\right] = Fk_I = 0.$$

Indeed, the integral $\int \psi_1(k, t)\, dt$, which equals zero when $t = 0$, can never become greater than Vt, ignoring the signs, as V is the largest value that the velocity $\frac{d\varphi}{dx}$ can take over the whole duration of the motion and along the whole length of the line of air. However, as this velocity has been assumed to be smaller than a, then $Vt < at$, and hence $\int \psi_1(k_1, t)\, dt < at$, from which it follows that $x + at - \int \psi_1(k_1, t)\, dt$ is a larger quantity than α, for all values of x which are positive and larger than α.

Thus, in considering the motion along a part of the line of air which corresponds to positive abscissae, the first of the equations (5) may be reduced to:

$$\frac{d\varphi}{dx} = -\frac{1}{2a} \cdot f\left[x - at - \int \psi(k, t)\, dt\right] = -\frac{1}{2a} \cdot fk;$$

from which it can be seen that $\frac{d\varphi}{dx}$, or $\psi(k, t)$, will be a function only of k. Hence this gives $\int \cdot\, \psi(k, t)\, dt = \frac{d\varphi}{dx} \cdot t$, as this integral must be used taking the value of k to be constant. Also, as f designates an arbitrary function, the previous equation becomes:

$$\frac{d\varphi}{dx} = f\left(x - at - \frac{d\varphi}{dx} \cdot t\right). \tag{a}$$

It should be noted that alone this equation cannot satisfy equation (2), hence equation (4) needs to be added to it, which reduces down to:

$$\frac{d\varphi}{dt} + a\frac{d\varphi}{dx} + \frac{1}{2} \cdot \frac{d\varphi^2}{dx^2} = 0\ ; \tag{b}$$

as $Fk_1 = 0$.

It may be easily checked that the system of these two equations satisfies equation (2), and that the values for $\frac{d\varphi}{dx}$ and $\frac{d\varphi}{dt}$ which are deduced from these calculations, that is:

$$\frac{d\varphi}{dx} = f\ , \quad \frac{d\varphi}{dt} = -af - \frac{1}{2} \cdot f^2,$$

may also be integrated, that is:

$$\frac{d \cdot \frac{d\varphi}{dx}}{dt} = \frac{d \cdot \frac{d\varphi}{dt}}{dx},$$

irrespective of the form of the $x - at - \frac{d\varphi}{dx} \cdot t$ function, which is designated by f. Therefore if the function f is assigned a determinate form, one may deduce from equations (a) and (b) the values for $\frac{d\varphi}{dx}$ and $\frac{d\varphi}{dt}$ as functions of x and t, thus providing the value for φ after integration of the formula:

$$d\varphi = \frac{d\varphi}{dx} \cdot dx + \frac{d\varphi}{dt} \cdot dt.$$

This value of φ will be a secondary integral of equation (2) with no arbitrary function. One must therefore regard all of the equations (a) and (b) as representing a secondary integral of equation (2) having a somewhat peculiar form. However, as this secondary integral only contains one single arbitrary function, it follows that it is only a particular integral.

Much work still has to be done to ascertain all the different forms in which the integrals of second and higher-order partial differential equations may occur. It would appear that nobody has as yet had the opportunity to observe that these integrals could be supplied by a system of equations between variables and the first-order differentials, in the same number as these differentials, but for which it is not possible to determine the values for these same differentials without first determining the arbitrary functions.

25. THERE now remains the task of determining from equation (a), the law which governs how the motion propagates along the positive abscissae. If the motion is considered along the negative abscissae, then similarly it could be determined using the following equations:

$$\frac{d\varphi}{dx} = F\left[x + t\left(a - \frac{d\varphi}{dx}\right)\right],$$

$$\frac{d\varphi}{dt} - a \cdot \frac{d\varphi}{dx} + \frac{1}{2} \cdot \frac{d\varphi^2}{dx^2} = 0;$$

the system of which is also a particular integral of equation (2). All that is then required is to check the value for $\frac{d\varphi}{dx}$ relative to the positive abscissae; the second will yield the same results.

As the function fx equals zero as long as $x > \alpha$ or $x < 0$, it follows that the value for $\frac{d\varphi}{dx}$ given by equation (a) will also equal zero as long as:

$$x > t\left(a + \frac{d\varphi}{dx}\right) + \alpha, \text{ or } x < t\left(a + \frac{d\varphi}{dx}\right).$$

The molecule which corresponds to the abscissa $x > \alpha$, will therefore begin to vibrate at the instant when $x = t\left(a + \frac{d\varphi}{dx}\right) + \alpha$, or more simply when $x = at + \alpha$, as the initial velocity $\frac{d\varphi}{dx}$ should be regarded as being infinitesimally small. Similarly, the same molecule will stop moving once $x = t\left(a + \frac{d\varphi}{dx}\right)$, or more simply $x = at$, as the velocity $\frac{d\varphi}{dx}$ has to be infinitesimally small at the instant motion ceases, just as with the case when motion commences.

It is therefore clear that the original vibrating source transmits uniformly, with a velocity equal to a. This velocity is independent of the nature of the original vibrating source and the magnitude of the velocities of the air molecules, while the duration of vibration will be the same for all the molecules and equal to $\frac{\alpha}{a}$. Finally, the width of the moving wave on either side of the origin, will always equal α.

Up until now, these propositions have only been demonstrated in the case where the velocities of air molecules have been assumed to be very small. Ordinarily, the powers of these velocities greater than one are ignored in the calculations, and one finds that $\frac{d\varphi}{dx} = f(x - at)$ as found from equation (a) by ignoring on the right-hand side the velocity $\frac{d\varphi}{dx}$ with respect to the velocity a. *Lagrange* took the approximation even further by retaining in the calculation the square of the velocity $\frac{d\varphi}{dx}$ (*cf. Mémoires de Turin*, volume II), thus obtaining an equation which is equivalent to:

$$\frac{d\varphi}{dx} = f\left[x - at - t \cdot f(x - at)\right];$$

from which he concluded the same as here, namely that the magnitude of the velocities of the air molecules has no influence on the velocity of sound.

Lagrange's formula may be deduced from equation (a) by substituting $\frac{d\varphi}{dx}$ on the right hand side of this equation with its approximate value $f(x - at)$; other increasingly accurate approximations may similarly be deduced for the velocity $\frac{d\varphi}{dx}$.

The Propagation of Sound when the Temperature and Density of Air are Assumed to be Variable

26. **RETAINING** the notation used in Section 1, and for the purposes of simplifying the problem, let us assume the gravity to be a force constant and parallel in direction throughout the whole expanse of the air mass in which the sound is propagated, and let the said force equal g. Using the vertical ordinates z directed from top to bottom, one then obtains $q = gz$. p' and ρ' represent the pressure and density respectively at steady state, and both quantities are functions of z. Hence, one obtains using equation (a) from Section 1:

$$gz - \int \cdot \frac{dp'}{\rho'} = 0.$$

Let D represent the value for ρ' in the air layer which satisfies $z = 0$, h the barometric height for the same layer, a^2 the value of $\frac{gh}{D}$, and finally take the density of mercury to be one. One obtains $p' = a^2 \rho'$ where consideration is first given to the special case where the temperature remains constant, and where the density varies only from one layer to another within the mass of air. Substituting this value for p' into the previous equation and differentiating with respect to z, one obtains:

$$g - a^2 \cdot \frac{d\rho'}{\rho' dz} = 0;$$

from which is obtained:

$$\rho' = De^{\frac{g}{a^2} \cdot z},$$

with e being the base of the system of logarithms with unit modulus, while ensuring that $\rho' = D$ when $z = 0$.

Now, by letting $\rho = \rho' (1 + \gamma)$, where γ is the contraction of air during motion, and by ignoring the sensible heat yielded during compression, then $p = p' + a^2 \rho'\gamma$, and consequently:

$$\int \cdot \frac{dp}{\rho} = \int \cdot \frac{dp'}{\rho'} + a^2 \cdot \gamma,$$

where the square and the powers greater than γ are ignored. Equations (a) and (b) from Section 1 then become:

$$\frac{d\varphi}{dt} + a^2 \cdot \gamma = 0,$$
$$\frac{d\gamma}{dt} + \left(\frac{d^2\varphi}{dx^2} + \frac{d^2\varphi}{dy^2} + \frac{d^2\varphi}{dz^2}\right) + \frac{d\rho'}{\rho' dz} \cdot \frac{d\varphi}{dz} = 0,$$

by eliminating the second dimension terms with respect to the velocities and the contraction. By substituting $\frac{d\rho'}{\rho' dz}$ in the second equation with $\frac{g}{a^2}$, and by eliminating γ using the first, one obtains:

$$\frac{d^2\varphi}{dt^2} = a^2\left(\frac{d^2\varphi}{dx^2} + \frac{d^2\varphi}{dy^2} + \frac{d^2\varphi}{dz^2}\right) + g\frac{d\varphi}{dz}.$$

The first-order differential contained within this equation is removed by making $\varphi = \varphi_I e^{-\frac{g}{2a^2} \cdot z}$, thus transforming it into:

$$\frac{d^2\varphi_I}{dt^2} = a^2\left(\frac{d^2\varphi_I}{dx^2} + \frac{d^2\varphi_I}{dy^2} + \frac{d^2\varphi_I}{dz^2}\right) - \frac{g^2}{4a^2} \cdot \varphi_I,$$

and consequently:

$$\frac{d^2 \cdot r\varphi_I}{dt^2} = a^2\left[\frac{d^2 \cdot r\varphi_I}{dr^2} + \frac{d \cdot \left(1 - u^2\right)\frac{d \cdot r\varphi_I}{du}}{rd^2\, u} + \frac{d^2 \cdot r\varphi_I}{d\omega} \cdot \frac{1}{r^2\left(1 - u^2\right)}\right] - \frac{g^2}{4a^2} \cdot r\varphi_I,$$

by transforming coordinates x, y, z into polar coordinates r, u, ω as in Section 4.

So as not to become entangled in lengthy calculations, let us assume that the function φ is independent of u and ω; the previous equation would then reduce to simply:

$$\frac{d^2 \cdot r\varphi_I}{dt^2} = a^2 \cdot \frac{d^2 \cdot r\varphi_I}{dr^2} - \frac{g^2}{4a^2} \cdot r\varphi_I; \qquad (1)$$

but which still does not allow finite integration.

27. **IN** order to obtain its series integral, let $r - at = s$, $r + at = s_I$ and by shortening to $\frac{g^2}{16a^4} = m$, we have:

$$\frac{d^2 \cdot r\varphi_I}{dsds_I} = m \cdot r\varphi_I;$$

which is an equation the complete integral of which is the sum of two infinite series, i.e.:

$$r\varphi_I = fs + ms_I \cdot \int fsds + \frac{m^2 s_I^2}{2} \cdot \int^2 fsds^2 + \frac{m^3 s_I^3}{2 \cdot 3} \cdot \int^3 fsds^3 + etc.$$
$$+ Fs_I + ms \cdot \int Fs_I ds_I + \frac{m^2 s^2}{2} \cdot \int^2 Fs_I ds_I^2 + \frac{m^3 s^3}{2 \cdot 3} \int^3 Fs_I ds_I^3 + etc.$$

where fs and Fs_I are two arbitrary functions.

Each of these two series may be transformed into a definite integral relative to a new variable h. Indeed, this is generally given as:

$$\int^n \cdot fsds^n = \frac{1}{1 \cdot 2 \cdot 3 ... n - 1} \cdot \int \cdot (s - h)^{n-1} \cdot fhdh$$

provided h is replaced by s after integration. If therefore $\int \cdot$fsds and $\int \cdot Fs_I ds_I$ are used instead of fs and Fs_I (which is considered reasonable), and if one makes:

$$H = 1 + (s - h) \cdot hs_I + \frac{(s - h)^2 \cdot h^2 s_I^2}{2 \cdot 2} + \frac{(s - h)^2 \cdot h^3 s_I^3}{2 \cdot 3 \cdot 2 \cdot 3} + etc.$$
$$H_I = 1 + (\mathrm{s} - \mathrm{h}) \cdot hs + \frac{(s_I - h)^2 \cdot h^2 s^2}{2 \cdot 2} + \frac{(s_I - h)^3 \cdot h^3 s^3}{2 \cdot 3 \cdot 2 \cdot 3} + etc.$$

then one obtains:

$$r\varphi_I = \int \cdot Hfhdh + \int \cdot H_I Fhdh;$$

with the first integral being taken from an arbitrary value of h up to $h = s$, and the second from an arbitrary value of h up to $h = s$.

From this value of $r\varphi_I$, it is possible to deduce the partial differentials $\frac{d \cdot r\varphi_I}{dr}$ and $\frac{d \cdot r\varphi_I}{dt}$, although as it is sufficient simply to know their form, it can be seen that this is the same as that of the function $r\varphi_I$, as by differentiating equation (1) with respect to t or to r, one obtains an equation with the same form as equation (1) for determining $\frac{d \cdot r\varphi_I}{dt}$ or $\frac{d \cdot r\varphi_I}{dr}$. From here it may be easily concluded that these must give:

$$\frac{d\varphi_I}{dr} = \frac{1}{r} \cdot \int \cdot Hf_I hdh + \frac{1}{r^2} \cdot \int \cdot Hf_2 hdh +; \quad \frac{1}{r} \cdot \int \cdot H_I F_I hdh + \frac{1}{r^2} \cdot \int \cdot H_I F_2 hdh \ ,$$

$$\frac{d\varphi_I}{dt} = \frac{1}{r} \cdot \int \cdot Hf' hdh + \frac{1}{r} \cdot \int \cdot H_I F' hdh;$$

provided that, following the integrations, h is made to equal s in all three integrals $\int \cdot Hf_I hdh$, $\int \cdot Hf_2 hdh$ and $\int \cdot Hf' hdh$, and $h = s_I$ in the case of the three others. The functions $f_I h$, $f_2 h$ and $f'h$ are derived from the arbitrary function fh, whilst the functions $F_I h$, $F_2 h$ and $F'h$ are derived from Fh.

The H series is dependent on a second-order differential equation. However, if one makes $(h - s) \cdot ms_I = \theta$, then it may be easily seen that, according to the form of the series, one then obtains:

$$\frac{d \cdot \theta \frac{dH}{d\theta}}{d\theta} + H = 0;$$

from which can be obtained a value for H containing two arbitrary constants resolved such that one obtains $H = 1$, $\frac{dH}{d\theta} = -1$ when $\theta = 0$. This equation cannot be finitely integrated using any known methods. However, it is included amongst those which *Euler* has integrated using definite integrals (*cf. Lacroix's Traité des différences et des séries*, p. 491). This is satisfied by taking:

$$H = A \cdot \int \cdot \frac{\cos.h' \sqrt{\theta} \cdot dh'}{\sqrt{4 - h'^2}},$$

where A is an arbitrary constant, and the integral relative to h' is taken from $h' = 0$ to $h' = 2$. In fact, by substituting this value of H in the left hand side of the previous equation, and by carrying out the integrations with respect to h', which become possible after making the substitution, one finds that:

$$\frac{d\cdot\theta\frac{dH}{d\theta}}{d\theta} + H = \frac{A}{4\sqrt{\theta}}\cdot\sqrt{4-h'^2}\cdot sin.h'\sqrt{\theta},$$

which is a zero quantity at the two limits $h' = 0$, $h' = 2$. In addition, the conditions $H = 1$ and $\frac{dH}{d\theta} = -1$ when $\theta = 0$, are met by taking $A = \frac{2}{\pi}$. As $H = \frac{A\pi}{2}$ and $\frac{dH}{d\theta} = -\frac{A\pi}{2}$ for this value of θ, then:

$$H = 1 - \theta + \frac{\theta^2}{2\cdot 2} - \frac{\theta^3}{2\cdot 3\cdot 2\cdot 3} + etc. = \frac{2}{\pi}\cdot\int\cdot\frac{\cos.h'\sqrt{\theta}\cdot dh'}{\sqrt{4-h'^2}}.$$

This formula will also provide the value of the H_I series by making $\theta = (h - s_I)\cdot ms$, and in this way the value for φ_I will be wholly expressed by definite integrals.

28. BECAUSE $\varphi = \varphi_I\cdot e^{-\frac{g}{2a^2}\cdot z}$, and $a^2\gamma + \frac{d\varphi}{dt} = 0$, then:

$$\frac{d\varphi}{dz} = -\frac{g}{2a^2}\cdot\frac{e^{-\frac{g}{2a^2}\cdot z}}{r}\cdot\left(\int\cdot Hfhdh + \int\cdot H_IFhdh\right),$$

$$\gamma = -\frac{e^{-\frac{g}{2a^2}\cdot z}}{a^2 r}\cdot\left(\int\cdot Hf'hdh + \int\cdot H_IF'hdh\right),$$

$$\frac{d\varphi}{dr} = \frac{e^{-\frac{g}{2a^2}\cdot z}}{r}\cdot\left(\int\cdot Hf_Ihdh + \int\cdot H_IF_Ihdh + \frac{\int\cdot Hf_2hdh}{r} + \frac{\int\cdot H_IF_2hdh}{r}\right);$$

leaving only the requirement to deduce from these formulae the law which determines the propagation of sound.

If one assumes for this that the original vibration source extends from $r = 0$ to $r = \alpha$ about the origin of the coordinates, then the partial differentials $\frac{d\varphi}{dz}$ and $\frac{d\varphi}{dr}$, and the contraction γ will then have to equal zero when $t = 0$, for all values of $r > \alpha$. In fact, this condition may be met by assuming that the arbitrary functions fh, Fh, $f'h$ etc., are zero for all values of $h > \alpha$, and that the integrals $\int \cdot Hfh$, $\int \cdot H_I Fhdh$ and $\int \cdot Hf'hdh$ etc., begin at $h = \alpha$. As these integrals need to terminate at $h = r$ when $t = 0$, then if r is greater than α, the quantities Hfh, $H_I Fh$ and $Hf'h$ etc., will equal zero over the whole range of the integrals, and as a consequence, these integrals will then also equal zero. Given this, at the end of any given time period, these same integrals will still equal zero as long as $s > \alpha$ and $s_I > \alpha$, i.e. $r - at > \alpha$ and $r + at > \alpha$. The motion will therefore reach the molecules located at a distance r from the centre of the original vibration source as soon as $r - at = \alpha$. Thus the velocity of sound will be constant along all sound radius vectors, and equal to a, just as if the mass of air was composed of layers all having the same density.

The velocity of the air molecules on which the intensity of the sound depends, shall be equal to $\sqrt{\left(\frac{d\varphi^2}{dz^2} + \frac{d\varphi^2}{dr^2}\right)}$. However, if the molecules far away from the origin of the motion are considered, then one obtains:

$$\int \cdot H_I Fhdh = 0, \int \cdot H_I F_I hdh = 0, \int \cdot H_I F_2 hdh = 0;$$

whilst if the terms divided by r^2 are ignored, then one obtains:

$$\sqrt{\left(\frac{d\varphi^2}{dz^2} + \frac{d\varphi^2}{dr^2}\right)} = \frac{e^{-\frac{g}{2a^2} \cdot z}}{r} \cdot A;$$

which simplified is:

$$\frac{g^2}{4a^2} \cdot \left[\left(\int \cdot Hfhdh\right)^2 + \left(\int \cdot Hf_I hdh\right)^2\right] = A^2.$$

The intensity of sound has been considered as being proportional to the square of the velocity of the air molecules impacting on the ear, while the density of the fluid medium remains unchanged. However, if this intensity was to be compared between different layers, one needs to assume that it is proportional to the density of the layer of air in which the sound is heard: thus the intensity would be represented by $\rho' \left(\frac{d\varphi^2}{dz^2} + \frac{d\varphi^2}{dr^2} \right)$, otherwise expressed as $\frac{D \cdot A^2}{r^2}$, since $\rho' = D \cdot e^{\frac{g}{a^2} \cdot z}$ (*cf.* Section 26). However, θ is an extremely small value between the limits of the integrals $\int \cdot Hfhdh$ and $\int \cdot Hf_l hdh$. These integrals will therefore take almost exactly the same values for equal values of *s* or of *r - at*, from which it follows that the sound intensity, or the value of $\frac{D \cdot A^2}{r^2}$, will decrease inversely to the square of the distance *r*, as in the case where the density is constant throughout the air mass. In addition, it may be seen that this intensity will be proportional to the density *D* of the layer of air in which the sound was generated, over an equal distance. Indeed, the value of *A* is independent of this density as firstly the velocity *a* is equal to the square root of the ratio of elasticity to density, and this ratio remains the same for all layers of the air mass, as the assumption has already been made that the temperature is constant. In addition, if the need arose to determine the arbitrary functions contained in *A* using the velocities and the contraction at the origin, then this is possible using those equations which do not contain *D*. Thus the intensity of the sound produced by the same source, depends only on the density of the air at the point where the sound was generated and on the distance it has travelled. Also, for the same distance travelled, the intensity falls as the height of the sound source in the atmosphere increases.

29. REFRACTION theory, and comparative experimental results, show that the temperature of the atmosphere decreases with height above the earth's surface in proportion to the vertical height (*cf. Mécanique céleste*, Book X). Thus, at steady state, rather than simply having $p' = a^2 \rho'$ as presented in Section 26, the expression should instead be $p' = a^2 \rho' \cdot (1 + bz)$, where *b* is a constant, positive coefficient. Substituting this value into the equation $\int \cdot \frac{dp'}{\rho'} - gz = 0$, and differentiating this with respect to *z*, one obtains:

$$\left(g - a^2 \cdot b\right) \cdot dz - a^2 (1 + bz) \cdot \frac{d\rho'}{\rho'} = 0;$$

from which one obtains:

$$\rho' = D (1 + bz)^{\frac{g - a^2 b}{a^2 b}},$$

ensuring that $\rho' = D$ when $z = 0$. During motion, the density becomes $\rho = \rho'(1 + \gamma)$ and the pressure $p = p' + a^2 \cdot (1 + bz) \cdot \rho'\gamma$, if the sensible heat generated by compression is ignored. Also ignored are whether the quantity of heat depends on the density and the initial temperature of the air under compression, and consequently whether the value of k in Section 1 varies or remains constant when moving from one layer of the atmosphere to the next. In any case, the changes in k have little influence on the velocity of sound as the correction due to the production of heat constitutes a minimal part of this velocity. For this reason, it is possible to ignore the effect of k in the problem at hand.

By ignoring the second dimension terms with respect to the velocities of the air molecules and their contraction, one obtains:

$$\int \cdot \frac{dp}{\rho} = \int \cdot \frac{dp'}{\rho'} + a^2 \cdot \int \cdot (1 + bz) dg,$$

and equations (a) and (b) in Section 1 become:

$$\frac{d\varphi}{dt} + a^2 \cdot \int \cdot (1 + bz) d\gamma = 0,$$

$$\frac{d\gamma}{dt} + \frac{d^2\varphi}{dx^2} + \frac{d^2\varphi}{dy^2} + \frac{d^2\varphi}{dz^2} + \frac{dr'}{\rho' \cdot dz} \cdot \frac{d\varphi}{dz} = 0.$$

By differentiating the first with respect to z, and by substituting $\frac{dr'}{\rho' \cdot dz}$ in the second equation with its value, one obtains:

$$\frac{d^2\varphi}{dtdz} + a^2 \cdot (1 + bz) \frac{d\gamma}{dz} = 0,$$

$$\frac{d\gamma}{dt} + \frac{d^2\varphi}{dx^2} + \frac{d^2\varphi}{dy^2} + \frac{d^2\varphi}{dz^2} + \frac{1}{a^2} \cdot \frac{g - a^2 b}{1 + bz} \cdot \frac{d\varphi}{dz} = 0 .$$

Eliminating γ from the two equations and then making $\frac{d\varphi}{dz} = \psi$, one finds:

$$\frac{d^2\psi}{dt^2} = a^2 (1 + bz) \cdot \left(\frac{d^2\psi}{dx^2} + \frac{d^2\psi}{dy^2} + \frac{d^2\psi}{dz^2} \right) + \left(g - a^2 b\right) \cdot \frac{d\psi}{dz} - \frac{\left(g - a^2 b\right)}{1 + bz} \cdot b\psi .$$

The first-order differential contained within this equation may be removed by taking $\psi = \psi_I \cdot (1 + bz)\,\frac{a^2b - g}{2a^2b}$. If at the same time the coordinates x, y, z are transformed into polar coordinates r, u, ω, as per section 4, and if, to simplify the problem, one assumes that the function ψ is independent of the angle ω, then one obtains:

$$\frac{d^2 \cdot r\psi_I}{dt^2} = a^2(1 + bru) \cdot \left(\frac{d^2 \cdot r\psi_I}{d^2 r} + \frac{d \cdot \left(1 - u^2\right) \frac{d \cdot r\psi_I}{du}}{r^2 du} \right) - \frac{B \cdot r\psi_I}{1 + bru},$$

B being a constant coefficient, and u the cosine of the angle made by the radius r with the vertical z.

It is not possible to assume that the function ψ is independent of the variable u because if this assumption was made, then the previous equation would reduce to:

$$\frac{d^2 \cdot r\psi_I}{dt^2} = a^2(1 + bru) \cdot \frac{d^2 \cdot r\psi_I}{dr^2} - \frac{B \cdot r\psi_I}{1 + bru};$$

such that it would still contain the variable u, which would amount to a contradiction. It is not possible to finitely integrate this equation, as its integral either in series or expressed using definite integrals, would appear to be too complex to allow the deduction of the velocity of sound in the event of variable temperature; its determination under three dimensions in air will therefore not be attempted here. Instead, consideration will be given to the special case of a mass of air reduced down to a single infinitesimally narrow channel.

30. CONSIDER therefore such a channel of any shape, filled with a fluid the temperature and density of which vary along the length of the channel. Let X equal the width of the channel at a distance x from a fixed origin chosen arbitrarily along the channel, the distance x being measured along its length. Under steady state, ρ' represents the density of the fluid at distance x from the origin, p' is the pressure at this point given per unit surface area, and q is the weight at the same point factorized along the path of the channel. The mass of the molecule occurring at distance x equals $X\rho' \cdot dx$, whilst its weight factorized along the path of the channel, shall therefore equal $X\rho'q \cdot dx$. It shall exert on its base X a pressure $p'X$, and the said pressure, transmitted to the other base $X + dX$, shall equal $p' \cdot (X + dX)$. Adding to this the weight $X\rho'q \cdot dx$, this gives the total pressure exerted at the base $X + dX$, namely $p' \cdot (X + dX) + X\rho' q \cdot dx$. However, at the distance $x + dx$, the function p' becomes $p' + dp'$, whilst the pressure exerted at the base $X + dX$ must therefore be given by $(X + dX) \cdot (p' + dp')$. Thus one obtains:

$$(X + dX) \cdot (p' + dp') = p' \cdot (X + dX) + X\rho' q \cdot dx,$$

which may be shortened to:

$$\frac{dp'}{dx} = \rho' q .$$

This is the equilibrium equation for the fluid: so as $p' = \rho' \cdot K$, where K is a constant coefficient when the temperature is constant along the whole length of the channel, and a given function of x when the temperature varies as per a known law, then this equation becomes:

$$K \cdot \frac{d\rho'}{\rho' dx} + \frac{dK}{dx} - q = 0.$$

Now let ρ, p, and v represent the density, pressure and velocity of the fluid molecule at a distance x after a given time t has elapsed since the origin of the motion. During the following instant, the same molecule will respond at distance $x + vdt$, and consequently its velocity will have changed to $v + \frac{dv}{dt} \cdot dt + \frac{dv}{dx} \cdot vdt$; the driving force will therefore, after time t, be equal to $\rho X dx \cdot \left(\frac{dv}{dt} + \frac{dv}{dx} \cdot v\right)$. However, according to *d'Alembert*'s principle, an equilibrium must exist within the fluid between the driving forces animating the molecules (with a direction opposite to their direction of motion), the weight of the molecules and the pressures exerted on them. Therefore, by substituting ρ' with ρ, p' with p, and q with $q - \frac{dv}{dt} - \frac{dv}{dx} \cdot v$ in the equation $\frac{dp'}{dx} = \rho' q$, one obtains:

$$\frac{dp}{dx} = \rho\left(q - \frac{dv}{dt} - v \cdot \frac{dv}{dx}\right),$$

as one of the motion equations.

If then $\rho = \rho' \cdot (1 + \gamma)$, and $p = p' + \rho' K\gamma$, and if one ignores the second dimension terms relating to contraction γ and velocity v, this equation becomes:

$$K \cdot \frac{dg}{dx} + \frac{dv}{dt} = 0 \quad . \tag{a}$$

As there are two unknowns γ and v to determine, a second equation is required. This is found by observing that the mass of the molecule after a time period t, at distance x, should not vary with time. However, this mass is equal to $X\rho \cdot (x' - x)$, where x' represents the distance $x + dx$. In the instant that follows, x, x', X, and ρ become $x + vdt$, $x' + vdt + \frac{dv}{dx} \cdot dxdt$, $X + \frac{dX}{dx} vdt$, and $\rho + \frac{d\rho}{dt} \cdot dt + \frac{d\rho}{dx} \cdot vdt$. These then give:

$$X \cdot \rho\,(x'-x) = \left(X + \frac{dX}{dx} \cdot vdt\right) \cdot \left(\rho + \frac{d\rho}{dt} + \frac{d\rho}{dx} \cdot vdt\right) \cdot \left(x' - x + \frac{dv}{dx} \cdot dxdt\right) ,$$

which may then be shortened to:

$$\frac{dX}{dx} \cdot \rho v + \frac{d\rho}{dx} \cdot Xv + X \cdot \frac{d\rho}{dt} + X\rho \cdot \frac{dv}{dx} = 0 .$$

By replacing ρ in this equation with ρ' $(1 + \gamma)$, and by ignoring the second dimension terms with respect to γ and v, it then becomes:

$$\frac{d \cdot log.X\rho'}{dx} \cdot v + \frac{d\gamma}{dt} + \frac{dv}{dx} = 0;$$

then by cancelling out γ between this and equation (a) above, one obtains:

$$\frac{d^2 \cdot v}{dt^2} = K\left(\frac{d^2 v}{dx^2} + \frac{dv}{dx} \cdot \frac{d \cdot log.\rho' X}{dx} + v \cdot \frac{d^2 \cdot \log.\rho' X}{dx^2}\right). \tag{b}$$

31. **MANY** hypotheses may be made regarding the form of the functions K, ρ', and X which would allow the finite integration of this equation. The most interesting case is the one where the density ρ' is assumed to be inversely proportional to the width X and the temperature constant: equation (b) would then reduce to:

$$\frac{d^2 v}{dt^2} = K \cdot \frac{d^2 v}{dx^2} ,$$

where K is a constant coefficient. From this, one may conclude that, in this case, the propagation of sound has occurred just as if the density was constant and the channel was cylindrical.

If one imagines that the channel forms part of the atmosphere, that it has a constant width along its whole length, and that it forms an angle with the vertical the cosine of which is equal to u, then (*cf.* Section 29) $K = a^2 \cdot (1+bxu)$, $q = gu$, and $X = const.$, assuming that the origin of x is above the surface of the earth and the positive values of this variable are directed towards this surface. The steady state equation then becomes:

$$\frac{d\rho'}{\rho'\, dx} = \frac{\left(g - a^2 b\right) \cdot u}{a^2 \cdot (1 + bu \cdot x)},$$

while equation (b) becomes

$$\frac{d^2 \cdot v}{dt^2} = a^2(1 + bu \cdot x) \cdot \frac{d^2 v}{dx^2} + \left(g - a^2 b\right) u \cdot \frac{dv}{dx} - \frac{\left(g - a^2 b\right) \cdot bu^2}{1 + bu \cdot x} v \,.$$

In order to integrate this equation, let $\sqrt{(1 + bu \cdot x)} = y$, thus producing:

$$\frac{d^2 v}{dt^2} = \frac{a^2 b^2 u^2}{4} \cdot \frac{d^2 v}{dy^2} + \frac{\left(2g - 3a^2 b\right) \cdot bu^2}{4y} \cdot \frac{dv}{dy} - \frac{\left(g - a^2 b\right) \cdot bu^2}{1 + bu \cdot x} \cdot v \,,$$

and hence let $v = v_1 \cdot y^{\frac{3a^2 b - 2g}{2a^2 b}}$, thus removing the first-order differential and giving:

$$\frac{d^2 v_1}{dt^2} = \frac{a^2 b^2 u^2}{4} \cdot \left(\frac{d^2 v_1}{dy^2} - \frac{m v_1}{y^2}\right),$$

where m is a constant coefficient. Finally, let $y - \frac{abu}{2} t = s$ and $y + \frac{abu}{2} \cdot t = s_1$. The equation above then becomes:

$$\frac{d^2 v_I}{ds ds_I} = \frac{m v_I}{\left(s + s_I\right)^2}.$$

The complete integral for this equation is the sum of two infinite series, namely:

$$v_I = fs - \frac{m}{s + s_I} \cdot \int \cdot fsds + \frac{m_I}{2 \cdot \left(s + s_I\right)} \cdot \int^2 \cdot fsds^2 -$$

$$\frac{m_{II}}{2 \cdot 3 \cdot \left(s + s_I\right)^2} \cdot \int^3 \cdot fsds^3 + etc. + Fs_I - \frac{m}{s + s_I} \cdot \int \cdot Fs_I ds_I +$$

$$\frac{m_I}{2 \cdot \left(s + s_I\right)^2} \cdot \int^2 \cdot Fs_I ds_I^2 - \frac{m_{II}}{2 \cdot 3 \cdot \left(s + s_I\right)^3} \cdot \int^3 \cdot Fs_I ds_I^3 + etc;$$

where *fs* and *Fs* are two arbitrary functions, and m_I, m_{II} etc., designate constant and determinate coefficients. Each of these series may be transformed into a definite integral relative to a new variable h, as done in Section 28. Thus if:

$$H = 1 + \frac{(h - s) \cdot m}{s + s_I} + \frac{(h - s)^2 \cdot m_I}{2 \cdot 2 \cdot \left(s + s_I\right)^2} +$$

$$\frac{(h - s)^3 \cdot m_{II}}{2 \cdot 3 \cdot 2 \cdot 3 \cdot \left(s + s_I\right)^3} + etc.$$

$$H_I = 1 + \frac{\left(h - s_I\right) \cdot m}{s + s_I} + \frac{\left(h - s_I\right)^2 \cdot m_I}{2 \cdot 2 \cdot \left(s + s_I\right)^2} +$$

$$\frac{\left(h - s_I\right)^3 \cdot m_{II}}{2 \cdot 3 \cdot 2 \cdot 3 \cdot \left(s + s_I\right)^3} + etc.$$

then one obtains:

$$v_I = \int \cdot Hfhdh + \int \cdot H_I Fhdh;$$

provided that following the integrations, s is used to replace h in $\int \cdot Hfhdh$, and s_I to replace h in $\int \cdot H_I Fhdh$. The H and H_I series depend on second-order differential equations which it is not possible to finitely differentiate.

Thus one obtains:

$$v = (1 + bu \cdot x)^{\frac{3a^2b - 2g}{2a^2b}} \cdot \left(\int \cdot Hfhdh + \int \cdot H_1Fhdh\right);$$

and hence if one assumes that the original vibration occurs from $x = 0$ up to $x = \alpha$, it may then be readily concluded from this formula, without having to repeat that outlined above in Section 28, that the air molecule which corresponds to distance x which is positive and larger than α, will begin to vibrate the moment that:

$$\sqrt{(1 + bu \cdot x)} = \frac{abu}{2} \cdot t + \sqrt{(1 + bu \cdot \alpha)};$$

which, by raising to the square and then dividing by bu, yields:

$$x = \alpha + at \cdot \sqrt{(1 + bu \cdot \alpha)} + \frac{a^2bu}{4} \cdot t^2.$$

Thus the sound propagates in the direction of the positive abscissae in a uniform accelerated motion, with the velocity at any instant being the same as that of a body subjected to a constant accelerating force equal to $\frac{a^2b}{2}$ and directed vertically. The said body then slides along the channel under the influence of this factorized acceleration force, with an initial velocity equal to $a \cdot \sqrt{(1 + bu \cdot \alpha)}$, or more simply a as α may be ignored. It may be noted that this initial velocity is independent of the direction of the channel, and that it equals the constant velocity of sound when the channel is horizontal.

From the formula yielding the value for v, it may be seen that the sound also propagates in the direction of the negative abscissae. The molecule which corresponds to a distance $-x$, begins to vibrate at the instant when:

$$\sqrt{(1 + bu \cdot x)} + \frac{abu}{2} \cdot t = 1,$$

which is to say:

$$x = at - \frac{a^2bu}{4} \cdot t^2;$$

such that when sound moves away from the earth's surface or in the direction of the negative abscissae, its propagation is uniformly delayed, with the initial velocity and the delaying force being identical (but for the sign) to those of sound moving in the direction of the positive abscissae. These results confirm the points made in the introduction to this paper.

32. MOREOVER, it should be noted that the velocity of sound varies very slowly due to the small magnitude of b. This may be exemplified by considering a vertical, cylindrical channel with the sound being produced at the earth's surface. Its velocity at this point is then compared with the velocity of sound at a height of 6980 metres (as achieved by *Gay-Lussac* in a hot-air balloon). Thus, by letting the said velocity equal a' and the number of sexagesimal seconds taken by the sound to reach this height as n, then:

$$a' = a - \frac{a^2 b}{2} \cdot n, \quad 6980 \; metres = an - \frac{a^2 b}{4} \cdot n^2;$$

hence:

$$a'^2 = a^2\left[1 - b \cdot (6980 \; metres)\right] \; .$$

However, at this same height, the temperature observed by *Gay-Lussac* was -9.5° centigrade, whereas it equalled 30.75°C at the earth's surface. Therefore the ratio of air elasticity to density at this height is given by (*cf.* Section 22):

$$a^2 \cdot \left(\frac{1 - (40.25) \cdot (0.00375)}{1 + (30.75) \cdot (0.00375)}\right);$$

and as this relation to a given height x has been designated as $a^2 \cdot (1 + bx)$, one finds that by using x = -6980 *metres*, then:

$$b \cdot (6980 \; metres) = \frac{71 \cdot (0.00375)}{-1 + (30.75) \cdot (0.00375)} = 0.23852.$$

This then gives:

$$a' = a\sqrt{(1 - 0.23852)} = a \cdot (0.87262),$$

from which it follows that the velocity of sound, at a height of 6980 metres in the atmosphere, is only reduced by *one-eighth* of its value at the earth's surface.

It is also noted that, irrespective of the pitch of the channel, the sound rises up to a point in the atmosphere where its elasticity falls to zero due to the drop in temperature, and therefore where $1 - bu \cdot x = 0$. This is the case as it must rise up until the variable velocity at which it propogates falls to zero. If the height at which this occurs is calculated, one finds that $x = \frac{1}{bu}$, as is the case at the atmosphere's limit.

Motion of a Heavy Chain

33. **CONSIDER** a heavy chain suspended vertically by one end, and assume that a portion of this chain is pulled to form a figure which deviates only slightly from the vertical, with the remainder of the chain both above and below this portion remaining in the vertical. Now if this portion of the chain is then released from its position, it will move back to the vertical by virtue of the chain's weight, and the motion will spread by degrees along the whole length of the chain. All those who have had the opportunity to witness this motion know that the motion spreads in a wave form which is readily visible. The problem is therefore to determine the velocity at which this wave moves, in the particular case where the heavy chain is uniform and of equal thickness along its whole length.

Let l equal the length of the total chain, g the gravity, and x the distance from a given point along this chain to the point of suspension. Also, let y and z represent the coordinates for the same point, at a time t lapsed since the motion was initiated, and assume that all these coordinates remain much smaller than x, for the whole duration of the motion. The equations describing this motion, as presented in *'la Mécanique analytique'*, p. 335, are thus:

$$\frac{d^2y}{dt^2} = g \cdot \frac{d \cdot (l - x) \cdot \frac{dy}{dx}}{dx};$$

$$\frac{d^2z}{dt^2} = g \cdot \frac{d \cdot (l - x) \cdot \frac{dz}{dx}}{dx}.$$

As these two equations are very similar, only one needs to be considered; hence the first is used below as an example but it should be noted that the values of z will follow the same law as those of y.

Placing the first equation into the form:

$$\frac{d^2y}{dt^2} = g \cdot (l - x) \cdot \frac{d^2y}{dx^2} - g \cdot \frac{dy}{dx}; \qquad (1)$$

and to integrate:

$$\sqrt{(l - x)} + \frac{t \cdot \sqrt{g}}{2} = s, \quad \sqrt{(l - x)} - \frac{t \cdot \sqrt{g}}{2} = s_I,$$

$$y = \frac{y_I}{\sqrt[4]{(l - x)}};$$

then one obtains:

$$\frac{d^2 y_I}{ds ds_I} = - \frac{y_I}{4(s + s_I)}.$$

This equation has already been integrated (*cf.* Section 31), the complete integral being:

$$y_I = \int \cdot Hfhdh + \int \cdot H_I Fhdh;$$

where fh and Fh are two arbitrary functions, the integral $\int \cdot Hfhdh$ commencing with an arbitrary value for h and ending at $h = s$, while the integral $\int \cdot H_I Fhdh$ similarly starts with an arbitrary value h and ends with $h = s_I$. Finally H and H_I are used to represent the same series as presented in Section 31, namely:

$$H = 1 + \frac{(h - s) \cdot m}{s + s_I} + \frac{(h - s)^2 \cdot m_I}{2 \cdot 2 \cdot (s + s_I)^2} + \frac{(h - s)^3 \cdot m_{II}}{2 \cdot 3 \cdot 2 \cdot 3 \cdot (s + s_I)^3} + etc.$$

$$H_I = 1 + \frac{(h - s_I) \cdot m}{s + s_I} + \frac{(h - s_I)^2 \cdot m_I}{2 \cdot 2 \cdot (s + s_I)^2} + \frac{(h - s)^3 \cdot m_{II}}{2 \cdot 3 \cdot 2 \cdot 3 \cdot (s + s_I)^3} + etc.$$

in which m must be taken to equal to $-\frac{1}{4}$.

This therefore produces:

$$y = \frac{1}{\sqrt[4]{(l - x)}} \cdot \left(\int \cdot Hfhdh + \int \cdot H_I Fhdh\right).$$

In addition, the value of $\frac{dy}{dt}$ must have the same form as that of y, as when differentiating equation (1) with respect to t, the same equation is used to determine $\frac{dy}{dt}$ as to determine y. Thus one also obtains:

$$\frac{dy}{dt} = \frac{1}{\sqrt[4]{(l - x)}} \cdot \left(\int \cdot H f_I h dh + \int \cdot H_I F_I h dh \right) ;$$

where $f_I h$ is a function which is derived from fh, and $F_I h$ from Fh.

Given the above, if the original vibration occurs from $x = c$ up to $x = c + \alpha$, then the ordinate y and the velocity $\frac{dy}{dt}$ must equal zero for all values of $x > c + \alpha$ or $< c$. This condition may now be satisfied by assuming that :

i) the functions fh, $f_I h$, Fh and $F_I h$ equal zero for all values of $h > \sqrt{(l - c)}$ or $< \sqrt{(l - c - \alpha)}$;

ii) the integrals $\int \cdot Hfhdh$ and $\int \cdot Hf_I hdh$ begin at $h = \sqrt{(l - c - \alpha)}$, while integrals $\int \cdot H_I Fhdh$ and $\int \cdot H_I F_I hdh$ begin at $h = \sqrt{(l - c)}$.

Indeed, as these four integrals must terminate at $h = \sqrt{(l - x)}$ when $t = 0$, then it follows that the functions fh and $f_I h$ will equal zero between the limits of the first two integrals for all values of $x > c + \alpha$. Consequently, these two integrals will equal zero for all values of x. The same will apply for the two other integrals for all values of $x < c$. In addition, the integrals $\int \cdot Hfhdh$ and $\int \cdot Hf_I hdh$ will also equal zero when $t = 0$ for all values of $x < c$. These integrals either have to be taken from $h = \sqrt{(l - c - \alpha)}$ up to $\sqrt{(l - x)}$, or they may be considered in two parts (effectively arriving at the same result): the first from $h = \sqrt{(l - c - \alpha)}$ up to $h = \sqrt{(l - c)}$, and the second from $h = \sqrt{(l - c)}$ up to $\sqrt{(l - x)}$. This second part evidently equals zero as the functions fh and $f_I h$ are zero between the two limits $h = \sqrt{(l - c)}$ and $h = \sqrt{(l - x)}$. The first part also equals zero as the integrals $\int \cdot Hfhdh$ and $\int \cdot Hf_I hdh$, taken from $h = \sqrt{(l - c - \alpha)}$ up to $h = \sqrt{(l - c)}$ and then divided by $\sqrt[4]{(l - c)}$, represent, at the origin of the motion, the values of y and $\frac{dy}{dt}$ at the point of the chain corresponding to $x = c$. Now irrespective of the original vibrating source, as it terminates at this point, the difference between y and the velocity $\frac{dy}{dt}$ for this point must equal zero or be infinitesimally small when $t = 0$. Similarly, it can be proven that the integrals $\int \cdot H_I Fhdh$ and $\int \cdot H_I F_I hdh$ equal zero when $t = 0$, for all values of $x > c + \alpha$. Thus the four integrals $\int \cdot Hfhdh$, $\int \cdot Hf_I hdh$, $\int \cdot H_I Fhdh$ and $\int \cdot H_I F_I hdh$ do not take any values when $t = 0$ for all values of x which fall outside the limits of the original vibration source. Consequently, these integrals may be considered as suitable for representing the original state of the heavy chain.

The integrals $\int \cdot Hfhdh$ and $\int \cdot Hf_I hdh$ only take values the instant one obtains:

$$s = \sqrt{(l - x)} + \frac{t\sqrt{g}}{2} = \sqrt{(l - c - \alpha)};$$

and the integrals $\int \cdot H_I Fhdh$ and $\int \cdot H_I F_I hdh$ will continue to equal zero up until:

$$s_I = \sqrt{(l - x)} - \frac{t\sqrt{g}}{2} = \sqrt{(l - c)}.$$

The first condition can only occur for values of $x > c + \alpha$, and the second for values of $x < c$. It therefore follows that the propagation of motion will occur below the original vibration as per $s = \sqrt{(l - c - \alpha)}$, and above the vibration as per $s_I = \sqrt{(l - c)}$. The first equation then gives:

$$x = c + \alpha + t \cdot \sqrt{g(l - c - \alpha)} - \frac{t^2 \cdot g}{4},$$

and the second gives:

$$c - x = t \cdot \sqrt{g(l - c)} + \frac{t^2 \cdot g}{4};$$

which indicate that two waves will be produced, one descending and the other ascending. The first will descend along the length of the chain in a uniformly delayed manner where the delay force equals one half the chain's weight. The second will move up the length of the chain, accelerating uniformly, and the acceleration force applied will also equal half the natural weight.

34. **THE** initial velocity of the descending wave is equal to $\sqrt{g(l - c - \alpha)}$. Now it can be seen that this velocity is precisely that necessary for the wave to travel the line $l - c - \alpha$, the distance from the original vibration to the lower end of the chain, and hence the wave reaches that end. Once it arrives, its velocity will equal zero irrespective of the location of its starting point. The initial velocity of the ascending wave is equal to $\sqrt{g(l - c)}$. This is the velocity it would have acquired by virtue of its motion having been accelerated if it had travelled from the lower end of the chain. Once it reaches the upper end, its velocity will be equal to that it would have acquired if it had travelled the whole length of the chain, irrespective of where it actually started.

From this it is possible to make the transmission of motion through a heavy chain as rapidly as required, by suspending a large weight from its lower end, as it is evident that, until the descending wave reaches this end, the motion of the chain will be the same as if it had been extended by a certain amount, such that if l' is used to designate the length of a portion of this same chain, the weight being equal to that already used, then the initial velocities of the ascending and descending waves become $\sqrt{g(l + l' - c)}$ and $\sqrt{g(l + l' - c - \alpha)}$. Consequently, these waves will reach the two ends of the chain after a smaller time period.

The limits of the integral $\int \cdot Hfhdh$ then become $h = \sqrt{(l + l' - c - \alpha)}$ and $h = \sqrt{(l + l' - x)} + \frac{t\sqrt{g}}{2}$, and as the latter limit must be smaller than $\sqrt{(l + l' - c)}$ for fh not to equal zero, it follows that the difference for these limits will be smaller than $\sqrt{(l + l' - c)} - \sqrt{(l + l' - c - \alpha)}$. The value of $\frac{(h - s) \cdot m}{s + s'}$ will therefore be smaller than $\frac{\left[\sqrt{(l + l' - c)} - \sqrt{(l + l' - c - \alpha)}\right] \cdot m}{2\sqrt{(l + l' - x)}}$ throughout the whole range of the integral $\int \cdot Hfhdh$. Thus if one assumes that the lines α and x are much smaller than $l + l'$, the H series will be highly convergent and therefore, taking only the first term, will give:

$$\int \cdot Hfhdh = funct.\left[\sqrt{(l + l' - x)} + \frac{t\sqrt{g}}{2}\right].$$

Similarly, one obtains:

$$\int \cdot H_I Fhdh = funct.\left[\sqrt{(l + l' - x)} - \frac{t\sqrt{g}}{2}\right],$$

and therefore:

$$y = \frac{funct.\left[\sqrt{(l + l' - x)} + \frac{t\sqrt{g}}{2}\right]}{\sqrt[4]{(l + l' - x)}} + \frac{funct.\left[\sqrt{(l + l' - x)} - \frac{t\sqrt{g}}{2}\right]}{\sqrt[4]{(l + l' - x)}}.$$

It would thus be easy to construct the figure of the heavy chain at any given instant, using its figure at the origin of the motion. One can also see that the length of the parts of the chain in motion at any given instant on either side of the original vibration, will be constant and equal to α during the whole duration of the motion. This result could not have been found if the lower end of the chain had not had a heavy weight (in comparison with the chain's own weight) attached to it, as, without this addition, the H series would be divergent for values of x which differ very little from l; the figure of the portion of the chain under motion would change towards the lower end, and the length of the wave would no longer be the same along the whole length of the chain.

35. **WITH** the lower end of the chain free, a point located at this end may receive the velocity communicated to it and no new conditions need to be filled. However, with the upper end fixed, the ordinate y of the point corresponding to $x = 0$ will equal zero for all values of t. To satisfy this condition, it suffices to add to the two terms yielding the value for y, a third term of the same form which only acquires a value at the instant the ascending wave arrives at the point of suspension, and for which the values relating to this point , where $x = 0$, will always be equal and opposite to the values of the term $\frac{\int \cdot H_I F h dh}{\sqrt[4]{(l - x)}}$. This addition is admissible as the value for y will continue to satisfy equation (1) due to the linear form of this equation. The added term will result in a descending wave the motion of which will be the same as that of the first wave of this type, and will follow the ascending wave as an echo follows on from the direct sound, within a line of air terminated by a fixed point. In general, the transmission of the motion within a heavy, uniform chain of standard thickness along its whole length, and suspended from one of its ends, is analogous to the propagation of sound inside a cylindrical channel where the temperature and density of the air vary, as assumed in Section 31, and where the channel is closed off at its lower end. In addition, if the thickness of the chain varied according to a given law, one would still find a law governing temperature and density along the length of the channel which would still ensure that the motion of the chain and that of the air remained similar.

Stokes
courtesy of AIP Emilio Segrè Visual Archives

GEORGE GABRIEL STOKES

Stokes was Lucasian professor of mathematics at Cambridge from 1849 until his death in 1903. This was the chair held by Newton some 150 years previously, and is held today by Stephen Hawking.

Stokes was born on 13 August, 1819, in Skreen, Sligo County, Ireland. He was the son of a rector in the established Church of Ireland, and received much of his early education from his father. Stokes entered Pembroke College, Cambridge University in 1837. He was granted a fellowship at Pembroke College following his graduation in 1841. He was knighted in 1889. He died in Cambridge on 1 February 1903, having been active throughout the Victorian era almost exclusively.

Stokes' contributions to science included the entire realm of natural philosophy, his most familiar contributions being in the field of hydrodynamics. His analysis of internal friction in fluids was published in 1845. Stokes was unfamiliar with the French scientific literature of his time, and was thus unaware of the work done previously by Navier on this subject. Hence, the present treatment of viscous behavior of fluids carries the names of *Navier-Stokes*. An interesting component of these investigations has to do with the view of the ultimate nature of material. Navier held the view that molecules existed. Stokes held the view that a fluid was infinitely divisible and was careful not to commit himself to the ideas of atoms and molecules. Stokes was also able to use his mathematical formulation of fluid motion to find the equations of motion of an elastic solid. Poisson had expressed the behavior of isotropic solids in terms of a single elastic constant. Stokes based his equations on two independent moduli, those of compression and rigidity, the viewpoint held today. This variation grew out of arguments related to whether or not purely central forces applied on an atomic scale.

The paper on the theory of sound that is presented here was published shortly before Stokes' election to the Lucasian chair. The second footnote was added when he was compiling a collection of his mathematical and physical papers (published in 1883). In the original paper Stokes had responded to James Challis' claim that there existed a contradiction in the commonly accepted theory of sound (in gases). Stokes introduced the concept of surfaces of discontinuity to account for the special situation which develops when waves steepen beyond the point of infinite gradient. Later discussions with Sir William Thomson and Lord Rayleigh convinced him that such discontinuities could not exist, since conservation of mass, momentum, and energy could not be satisfied simultaneously. The problem here lay in the assumption that the same pressure-density relationship that applied to continuous motion must also apply to discontinuous motion, and that the shock would thus be isentropic; basic concepts of irreversibility were likely being developed and evolving at this time. Taylor and Rayleigh later sorted out the problem after reading the works of Rankine and Hugoniot.

The statement in question is the one where Stokes points out (in his *Collected Papers*, 1883) that conservation of energy across the discontinuity, combined with the assumption that the pressure $p \propto \rho$, led to the following relationship between the initial density ρ' and the final density ρ:

$$2\rho\rho' \log \rho / \rho' = \rho^2 - \rho'^2 \quad .$$

This expression is a direct result of a combination of the energy jump condition

$$\Delta e = \tfrac{1}{2}(p + p')\left(\frac{1}{\rho'} - \frac{1}{\rho}\right)$$

and the adiabatic relationship

$$\Delta e_{ad} = -\int_{v'}^{v} p\, dv = \int_{\rho'}^{\rho} a^2 \rho \frac{d\rho}{\rho^2} \quad .$$

By equating these two quantities, Stokes failed to recognize the irreversible nature of the shock (discontinuity) and the need to include a more general equation of state. This would allow the discontinuous compression curve (Hugoniot) to depart from the isentropic compression curve. If Stokes had considered an equation of state of the form (ideal gas)

$$e = \frac{pv}{\gamma - 1} \quad ,$$

where γ is the ratio of the specific heats, he would have obtained the discontinuous compression curve (Hugoniot) as

$$p = p' \frac{\dfrac{\rho}{\rho'} + \left(\dfrac{\gamma - 1}{2}\right)\left(\dfrac{\rho}{\rho'} - 1\right)}{1 - \left(\dfrac{\gamma - 1}{2}\right)\left(\dfrac{\rho}{\rho'} - 1\right)}$$

which does differ from the isentropic compression curve $p = p'\left(\frac{\rho}{\rho'}\right)^{\gamma}$.

Reading the paper of Stokes we see the difficulty that even the greats had with concepts whose physical bases had not reached sufficient maturity to warrant proper treatment, and the presence of a natural time scale related to scientific investigation; that is, the necessity of waiting until the proper questions are asked before a particular problem can be viewed more completely.

On a Difficulty in the Theory of Sound

G. G. Stokes, M. A., Fellow of Pembroke College, Cambridge, *Philosophical Magazine*, Vol. XXXIII, pp. 349-356 (November, 1848) and *Mathematical and Physical Papers*, Vol. II, Cambridge University Press (1883)

Eds. note: Text in [] appears in Math. and Phys. Papers only.

The theoretical determination of the velocity of sound has recently been the occasion of a discussion between Professor Challis and the Astronomer Royal. It is not my intention to enter into the controversy, but merely to consider a very remarkable difficulty which Professor Challis has noticed in connection with a known first integral of the accurate equations of motion for the case of plane waves.

*Eds. note: The following text (up to ***) appears only in the original Phil. Mag. article.*

I would first however observe, that I do not think that we are *obliged*, in treating the subject to a first approximation, to enter into the consideration of any difficulty which may arise when we come to employ exact equations. In neglecting the squares of small quantities, we adopt a consistent system of approximation, and we arrive at a precise result, namely, the laws which the actual motion tends indefinitely to obey when the magnitude of the velocity is indefinitely diminished. And if, moreover, we have every reason to believe that the first order of small quantities contains all that are sensible in practical cases, we have arrived, not merely at abstract mathematical results, but at important physical laws. Nevertheless it will be admitted on all hands, that any apparent contradiction arrived at by employing exact equations is deserving of serious consideration.

***

The difficulty already alluded to is to be found at page 496 of the preceding volume of this Magazine.* In what follows I shall use Professor Challis's notation.

Without entering into the consideration of the mode in which Poisson obtained the particular integral

$$w = f\{z-(a+w)t\} \quad . \quad . \quad . \quad . \quad . \quad . \quad . \quad . \quad . \quad . \quad . \quad . \quad . \quad (1.)$$

it may easily be shown, by actual differentiation and substitution, that the integral does satisfy our equations. The function f being arbitrary, we may assign to it any form we please, as representing a particular possible motion, and may employ the result, *so long as no step tacitly assumed in the course of our reasoning fails.* The interpretation of the integral (1.) will be rendered more easy by the consideration of a curve. In fig. 1 let oz be the axis of z, and let the ordinate of the curve represent the values of w for $t=0$. The equation (1.)

---

* [The following quotation will suffice to put the reader in possession of the apparent contradiction discovered by Professor Challis. It should be stated that the investigation relates to plane waves, propagated in the direction of z, and that the pressure is supposed to vary as the density.

"The function f being quite arbitrary, we may give it a particular form. Let, therefore,

$$w = m\sin\frac{2\pi}{\lambda}\{z-(a+w)t\} \ .$$

This equation shows that at any time t_1 we shall have $w=0$ at points on the axis of z, for which

$$z-(a+w)t_1 = \frac{n\lambda}{2} \ ,$$

or

$$z = at_1 + \frac{n\lambda}{2} \ .$$

At the same time w will have the value $\pm m$ at points of the axis for which

$$z-(a+m)t_1 = \left(\frac{n}{2}-\frac{1}{4}\right)\lambda \ ,$$

or

$$z = at_1 + \frac{n\lambda}{2} + mt_1 - \frac{\lambda}{4} \ .$$

Here it is observable that no relation exists between the points of no velocity and the points of maximum velocity. As m, t_1 and λ are arbitrary constants, we may even have

$$mt_1 - \frac{\lambda}{4} = 0 \ ,$$

in which case the points of no velocity are also points of maximum velocity."]

merely asserts that whatever value the velocity w may have at any particular point when $t=0$, the same value will it have at the time t at a point in advance of the former by the space $(a+w)t$. Take any point P in the curve of Fig. 1, and from it draw, in the positive direction, the right line PP' parallel to the axis of z, and equal to $(a+w)t$. The locus of all the points P' will be the velocity-curve for the time t. This curve is represented in fig. 2, except that the displacement *at* common to all points of the original curve is omitted, in order that the modification in the form of the curve may be more easily perceived. This comes to the same thing as drawing PP' equal to wt instead of $(a+w)t$. Of course in this way P' will lie on the positive or negative side of P, according as P lies above or below the axis of z. It is evident that in the neighborhood of the points a, c the curve becomes more and more steep as t increases, while in the neighbourhood of the points o, b, z its inclination becomes more and more gentle.

Fig. 1.
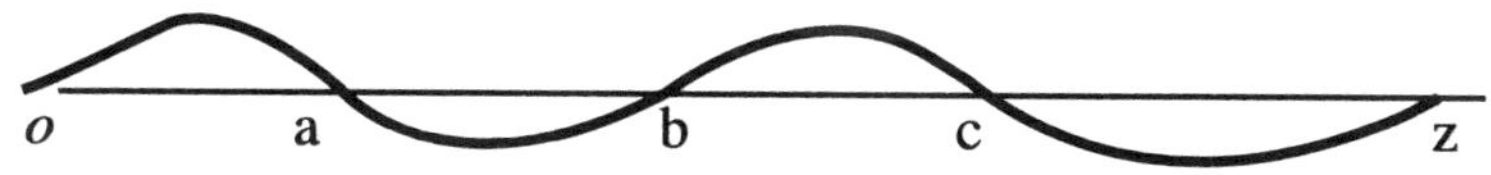

Fig. 2.

The same result may easily be obtained analytically. In fig. 1, take two points, infinitely close to each other, whose abscissae are z and $z+dz$; the ordinates will be w and

$$w+\frac{dw}{dz}dz \ .$$

After the time t these same ordinates will belong to points whose abscissae will have become (in fig. 2) $z+wt$ and

$$z+dz+\left(w+\frac{dw}{dz}dz\right)t \ .$$

Hence the horizontal distance between the points, which was dz, will have become

$$\left(1+\frac{dw}{dz}t\right)dz \ ;$$

and therefore the tangent of the inclination, which was dw/dz, will have become

$$\frac{\frac{dw}{dz}}{1+\frac{dw}{dz}t} \quad \ldots\ldots\ldots\ldots\ldots\ldots\ldots \text{(A.)}$$

At those points of the original curve at which the tangent is horizontal, $dw / dz = 0$, and therefore the tangent will constantly remain horizontal at the corresponding points of the altered curve. For the points for which dw / dz is positive, the denominator of the expression (A.) increases with t, and therefore the inclination of the curve continually decreases. But when dw / dz is negative the denominator of (A.) decreases as t increases, so that the curve becomes steeper and steeper. At last, for a sufficiently large value of t, the denominator of (A.) becomes infinite for some value of z. Now the very formation of the differential equations of motion with which we start, tacitly supposes that we have to deal with finite and continuous functions; and therefore in the case under consideration we must not, without limitation, push our results beyond the least value of t which renders (A.) infinite. This value is evidently the reciprocal, taken positively, of the greatest negative value of dw / dz; w here, as in the whole of this paragraph, denoting the velocity when $t = 0$.

By the term *continuous function*, I here understand a function whose value does not alter *per saltum*, and not (as the term is sometimes used) a function which preserves the same algebraical expression. Indeed, it seems to me to be of the utmost importance, in considering the application of partial differential equations to physical, and even to geometrical problems, to contemplate functions apart from all idea of algebraical expression.

In the example considered by Professor Challis,

$$w = m \sin \frac{2\pi}{\lambda}\{z - (a + w)t\} \ ,$$

where m may be supposed positive; and we get by differentiating and putting $t = 0$,

$$\frac{dw}{dz} = \frac{2\pi m}{\lambda} \cos \frac{2\pi z}{\lambda} \ ,$$

the greatest negative value of which is $-2\pi m / \lambda$; so that the greatest value of t for which we are at liberty to use our results without limitation is $\lambda / 2\pi m$, whereas the contradiction arrived at by Professor Challis is obtained by extending the result to a larger value of t, namely $\lambda / 4m$.

Of course, after the instant at which the expression (A.) becomes infinite, some motion or other will go on, and we might wish to know what the nature of that motion was. Perhaps the most natural supposition to make for trial is, that a surface of discontinuity is formed, in passing across which there is an abrupt change of density and velocity. The existence of such a surface will presently be shown to be possible,* on the two suppositions that the pressure is equal in all directions about the same point, and that it varies as the density. I have however convinced myself, by a train of reasoning which I do not think it worth while to give, inasmuch as the result is merely negative, that even on the supposition of the existence of a surface of discontinuity, it is not possible to satisfy all the conditions of the problem by means of a single function of the form $f\{z-(a+w)t\}$. Apparently, something like reflexion must take place. Be that as it may, it is evident that the change which now takes place in the nature of the motion, beginning with the particle (or rather plane of particles) for which (A.) first becomes infinite, cannot influence a particle at a finite distance from the former until after the expiration of a finite time. Consequently even after the change in the nature of the motion, our original expressions are applicable, at least for a certain time, to a certain portion of the fluid. It was for this reason that I inserted the words "without limitation," in saying that we are not at liberty to use our original results without limitation beyond a certain value of t. The full discussion of the motion which would take place after the change above alluded to, if possible at all, would probably require more pains than the result would be worth.

*Eds. note: The following text (up to ***) appears only in the original Phil. Mag. article.*

I proceed now to consider the possibility of the existence of a surface of discontinuity, and the conditions which must be satisfied at such a surface. Although I was led to the subject by considering the interpretation of the integral (1.), the consideration of a discontinuous motion is not here introduced in connection with that interpretation, but simply for its own sake; and I wish the two subjects to be considered as quite distinct.

Suppose that in passing across a point Q in the axis of z the velocity and density change suddenly from w, ρ to w', ρ', and let γ be the velocity of propagation of the surface of discontinuity. Let us first investigate the equation which expresses that there is no generation or destruction of mass at the surface of discontinuity, the equation in fact which takes the place of the equation of continuity, which has to be satisfied elsewhere.

Take two points A, B in the axis of z, the first on the negative and the second on the positive side of Q, and let $QA = h$, $QB = h'$. Take also $QQ' = \gamma dt$, so that Q' is the point where the surface of discontinuity cuts the axis of z at the time $t + dt$. The quantities h, h' are supposed to be very small, and will be made

* [Not so: see the substituted paragraph at the end.]

to vanish after QQ'. Consider the portion of space comprised within a cylinder whose ends consist of two planes, of area unity, drawn through the points A, B perpendicular to the axis. In the time dt, the mass of fluid which flows in at the plane A is ultimately $\rho w dt$, and that which flows out at the plane B is ultimately $\rho' w' dt$. Hence the gain of mass within the cylinder is ultimately $(\rho w - \rho' w')dt$. Now the mass at the time t is ultimately $\rho h + \rho' h'$, and that at the time $t + dt$ is

$$\left(\rho + \frac{d\rho}{dt}dt\right)(h + \gamma dt) + \left(\rho' + \frac{d\rho'}{dt}dt\right)(h' - \gamma dt);$$

and therefore the gain of mass is $\gamma(\rho - \rho')dt, h$ and h' being omitted, since they are to be made to vanish in the end. Equating the two expressions for the gain of mass, we get

$$\rho w - \rho' w' = (\rho - \rho')\gamma \qquad (2)$$

It remains to form the equation of motion. Now we have

moving force on plane $A = a^2\rho$, ultimately;
moving force on plane $B = a^2\rho'$, ultimately;
$\therefore$ resultant moving force $= a^2(\rho - \rho')$,
$\therefore$ momentum generated in time $dt = a^2(\rho - \rho')dt$.

Now the momentum of the mass contained within the cylinder at the time t is ultimately $\rho w h + \rho' w' h'$, and the momentum of the same set of particles at the time $t + dt$ is

$$\left(\rho w + \frac{d\rho w}{dt}dt\right)(h + [\gamma - w]dt) + \left(\rho' w' + \frac{d\rho' w'}{dt}dt\right)(h' + [w' - \gamma]dt);$$

and therefore the gain of momentum is ultimately

$$\left\{(\rho w - \rho' w')\gamma - \rho w^2 + \rho' w'^2\right\}dt;$$

whence we have

$$(\rho w - \rho' w')\gamma - \left(\rho w^2 - \rho' w'^2\right) = a^2(\rho - \rho'). \ldots\ldots\ldots\ldots (3)$$

By eliminating γ between (2.) and (3.), we get

$$(w'-w)^2 \rho\rho' = a^2(\rho-\rho')^2, \ldots\ldots\ldots\ldots\ldots\ldots\ldots\ldots \quad (4)$$

an equation which we may if we please employ instead of (3.).

The equations (2.), (3.) being satisfied, it appears that the discontinuous motion is dynamically possible. This result, however, is so strange, that it may be well to consider more in detail the simplest possible case of such a motion.

Conceive then an infinitely long straight tube filled with air, of which the portion to the left of a certain section s is of a uniform density ρ, and at rest, while the portion to the right is of a uniform smaller density ρ', and is moving in the positive direction with a uniform velocity w', the surface of separation s at the same time traveling backwards into the first portion with the uniform velocity γ. The conception of such a motion having been formed, consider next whether the motion is possible or impossible; that is to say, not whether it is possible or impossible in the actual state of elastic fluids, but whether it would or would not be consistent with dynamical principles in the case of an ideal elastic fluid, in which the pressure was equal in all directions about the same point, and varied as the density.

In the case under consideration, the fluid to the left of s is in equilibrium in the simplest way. The fluid to the right is of uniform pressure, and there is no generation or destruction of velocity. The only question, then, can be as to the possibility of the passage from the one state into the other taking place in the way supposed. In the first place, it is evident that, independently of any consideration of force, there must be a relation between ρ, ρ', w', and γ; for a length γt of condensed air comes to occupy, in the rarefied state, a length $(\gamma + w')t$, so that we must have $\rho\gamma = \rho'(\gamma + w')$. Next, if we take two sections s_1, s_2, the first to the left and the second to the right of s, and suppose the first to remain at rest, with the fluid in which it is situated, while the second moves, along with the fluid in which it is situated, with the velocity w' since the pressure on s_1 exceeds that on s_2 by $a^2(\rho-\rho')$, the surface of s_1 or s_2 being for simplicity's sake supposed equal to unity, there must in the time t be generated a momentum $a^2(\rho-\rho')t$ in the fluid lying between s_1 and s_2. But this will be the case in consequence of the velocity w' being communicated to a volume γt of air which was previously at rest and of density ρ, while the state of rest or motion of the remainder of the air between s_1 and s_2 has been unaltered, provided $a^2(\rho-\rho') = w'\gamma\rho$. These two relations being satisfied, it appears that the motion is dynamically possible. The two equations might have been obtained at once from (2.) and (3.) by writing - γ for γ and putting $w = 0$, but I have preferred deducing them afresh from first principles in consequence of the novelty of the subject, and the reluctance with which the conclusions that I have arrived at are likely to be received by mathematicians.

These conclusions certainly seem sufficiently startling; yet a result still more extraordinary remains behind. By solving the two equations of the preceding paragraph with respect to w' and γ, we get

$$w' = \frac{\rho - \rho'}{\sqrt{\rho\rho'}} a, \quad \gamma = \sqrt{\frac{\rho'}{\rho}} a^{*} \quad .$$

Now let ρ' vanish; then w' becomes infinite and γ vanishes. Hence the rate at which the condensed air (which remains packed like the combustible matter in a rocket) is discharged decreases indefinitely as the space into which the discharge takes place approaches indefinitely to a vacuum. Of course the velocity of discharge becomes infinite, without which the requisite momentum could not be furnished. The quantity of air which passes in a unit of time across a plane, of area unity, taken at the positive side of the tube, is $w'p'$, which is easily seen to be a maximum, for a given value of ρ, when $\rho' = \frac{1}{3}\rho$.

A similar paradox is fully considered by MM. Barré de Saint-Venant and Wantzel, in the 27th *Cahier du Journal de l' Ecole Polytechnique.*

The strange results at which I have arrived appear to be fairly deducible from the two hypotheses already mentioned. It does not follow that the discontinuous motion considered can ever take place in nature, for we have all along been reasoning on an ideal elastic fluid which does not exist in nature. In the first place, it is not true that the pressure varies as the density, in consequence of the heat and cold produced by condensation and rarefaction respectively. But it will be easily seen that the discontinuous motion remains possible when we take account of the variation of temperature due to condensation and rarefaction, neglecting, however, the communication of heat from one part of the fluid to another. Indeed, so far as the possibility of discontinuity is concerned, it is immaterial according to what law the pressure may increase with the density.

Of course the communication of heat from one particle of the fluid to another would affect the result, though whether to the extent of preventing the possibility of discontinuity I am unable to say. But there is another supposition that we have made which is at variance with the actual state of elastic fluids. It is not true that one portion of an elastic fluid is incapable of exerting any tangential

---

* It is worthy of remark, that when ρ' is very nearly equal to ρ, and consequently w' very small, the velocity of propagation γ is very nearly equal to a, to which it approaches indefinitely when w' is indefinitely diminished. Thus even this discontinuous motion offers no exception to the theorem, at once proved by neglecting the squares of small quantities, that in very small motions any disturbance is propagated in the fluid with the velocity a.

force on another portion on which it slides, even though the variation of velocity from the one portion to the other be not abrupt but continuous. In consequence of this tangential force, analogous in some respects to friction in the case of solids, the mutual pressure of two adjacent elements of a fluid is not accurately normal to the surface of separation, nor equal in all directions about the same point. In many cases the influence of this internal friction is insensible, while in other cases it is very important. Its general effect is to check the relative motion of the parts of a fluid. Suppose now that a surface of discontinuity is very nearly formed, that is to say, that in the neighborhood of a certain surface there is a very rapid change of density and velocity. It may be easily shown, that in such a case the rapid condensation or rarefaction implies a rapid sliding motion of the fluid; and this rapid sliding motion would call into play a considerable tangential force, the effect of which would be to check the relative motion of the parts of the fluid. It appears, then, almost certain that the internal friction would effectually prevent the formation of a surface of discontinuity, and even render the motion continuous again if it were for an instant discontinuous.

***

[So long as the motion is continuous, and none of the differential coefficients involved become infinite, the two principles of the conservation of mass and what may be called the conservation of momentum, applied to each infinitesimal slice of the fluid, are not only necessary but also sufficient for the complete determination of the motion, the functional relation existing between the pressure and density being of course supposed known. Hence any other principle known to be true, such for example as that of the conservation of energy, must be virtually contained in the former. It was accordingly a not unnatural mistake to make to suppose that in the limit, when we imagine the motion to become discontinuous the same two principles of conservation of mass and of momentum applied to each infinitesimal slice of the fluid should still be sufficient, even though one such slice might contain a surface of discontinuity. It was however pointed out to me by Sir William Thomson, and afterwards independently by Lord Rayleigh, that the discontinuous motion supposed above involves a violation of the principle of the conservation of energy. In fact, the equation of energy, applied to the fluid in the immediate neighbourhood of the surface of discontinuity, and combined with the two equations deduced from the two principles first mentioned, leads in the case of $p \propto \rho$ to

$$2\rho\rho' \log \rho / \rho' = \rho^2 - \rho'^2 ,$$

where ρ, ρ' are the densities at the two sides of the supposed surface of discontinuity; but this equation has no real root except $\rho = \rho'$.]

Earnshaw
courtesy of Sheffield City Council

The REVEREND SAMUEL EARNSHAW

Earnshaw's paper "On the Mathematical Theory of Sound," published in 1860, gives the author an affiliation with Sheffield the town, not the university. Earnshaw did eventually have an important connection to the University of Sheffield, but not at this time.

Samuel Earnshaw was born in Sheffield in 1805, the son of a file-cutter. When Samuel was about 13 years old, his father became master of a Sheffield school. Samuel was educated at home until the age of 8 when he entered the Carver Street National School. The custom of the time was that students were expected to teach each other; with those more advanced taking greater responsibility for tutoring. From the age of 10 years Samuel was paid for doing this. It was apparent that Samuel had an outstanding mathematical ability and in this way attracted the attention of a number of influential people. Thus, at the age of 22, he was able to enter St. John's College, Cambridge, where he was later awarded a scholarship. In 1831 he became Senior Wrangler, and was the first Smith's Prizeman. After receiving his degree he remained in Cambridge until 1847. During this period he earned his living as an academic coach, making a considerable income. He also published a number of mathematical papers and took holy orders. Upon returning to Sheffield, in order to recover from strain on his health, he took a position as Assistant Minister in the Parish Church and showed intense interest in all educational matters.

The second half of the 19th Century was a period of pronounced growth in higher education in England. The University Extension Movement reached Sheffield in 1874. This movement involved the University of Cambridge placing its resources at the disposal of the towns of the country through extension courses. Samuel Earnshaw was instrumental in establishing the Extension Lectures and played a role in the founding of Firth College (the forerunner to the University of Sheffield). Within the University his name is commemorated with the Earnshaw Scholarship, founded in 1885. Earnshaw died in 1888.

On The Mathematical Theory of Sound

Rev. S. Earnshaw, M.A., Sheffield, *Philosophical Transactions,* Vol. 150 , pp. 133-148 (1860)

Summary

In making certain investigations on the properties of the sound-wave, transmitted through a small horizontal tube of uniform bore, I found reason for thinking that the equation

$$\frac{dy}{dt} = F\left(\frac{dy}{dx}\right) \tag{1}$$

must always be satisfied; F being a function of a form to be determined. Differentiating this equation with regard to *t*, we find

$$\frac{d^2y}{dt^2} = \left\{F'\left(\frac{dy}{dx}\right)\right\}^2 \cdot \frac{d^2y}{dx^2}\ , \tag{2}$$

which by means of the arbitrary function F can be made to coincide, not only with the ordinary dynamical equation of sound, but with any dynamical equation in which the ratio of $\frac{d^2y}{dt^2}$ and $\frac{d^2y}{dx^2}$ can be expressed in terms of $\frac{dy}{dx}$.

Equation (1) is a partial first integral of (2), and by means of it we shall be able to obtain a final integral of (2), which will be shown to be the general integral of (2) for wave-motion, propagated in one direction only in such a tube as we have supposed, by its satisfying all the conditions of such wave-motion.

It will be convenient to begin with the simplest case of sound,—that in which the development of heat and cold is neglected.

I. Wave-Motion when Change of Temperature Is Neglected

1. The equations for this case of motion are, the dynamical equation

$$\left(\frac{dy}{dx}\right)^2 \cdot \frac{d^2y}{dt^2} = \mu \frac{d^2y}{dx^2} \ , \tag{3}$$

and the equation of continuity,

$$\frac{dy}{dx} = \frac{\rho_0}{\rho} \ ; \tag{4}$$

ρ_0, p_0 are the equilibrium density and pressure at any point of the fluid; ρ, p the same for a particle in motion; x the equilibrium distance of the same particle from a fixed plane cutting the tube at right angles; and t is the time when the same particle, being in motion, is at the distance y from the same plane; μ is the constant which connects ρ and p by BOYLE'S law $p = \mu\rho$.

On comparing (3) with (2), we find $\frac{dy}{dx} \cdot F'\left(\frac{dy}{dx}\right) = \pm\sqrt{\mu}$; or, for brevity, writing α for $\frac{dy}{dx}$, we have

$$F'(\alpha) = \pm\frac{\sqrt{\mu}}{\alpha} \ ,$$

and $\qquad \therefore \qquad F(\alpha) = C \pm \sqrt{\mu}\ \ln(\alpha)$.

But as $\frac{dy}{dt} = F(\alpha)$ from (1), it follows that

$$\begin{aligned} dy &= \frac{dy}{dx} \cdot dx + \frac{dy}{dt} \cdot dt \\ &= \alpha dx + F(\alpha) \cdot dt \ , \end{aligned}$$

which being integrated in the usual manner, substituting at the same time for $F(\alpha)$ its value, gives

$$\left.\begin{aligned} y &= \alpha x + \left(C \pm \sqrt{\mu}\ \ln\alpha\right)t + \varphi(\alpha) \\ 0 &= \alpha x \pm \sqrt{\mu t} + x\varphi'(\alpha) \end{aligned}\right\} \tag{5}$$

Between these equations, if we eliminate α, we have then the integral of equation (3).

2. From equation (4) we see that $\alpha = \frac{\rho_0}{\rho}$; and if we represent by u the velocity of the particle whose place is y, we find

$$u=\frac{dy}{dt}=C\pm\sqrt{\mu}\ln\alpha\ ,$$
$$=C\pm\sqrt{\mu}\ln\left(\frac{\rho_0}{\rho}\right)\ ;$$

and $$\therefore\quad \rho=\rho_0 e^{\mp\frac{u-c}{\sqrt{\mu}}}\ .$$

3. To determine the arbitrary constant C, we observe that $\rho=\rho_0$ and $u=C$ are always simultaneous equations. But the former belongs to the confines of the wave, where in fact $u=0$; and therefore C = 0. Hence for a wave transmitted through a medium which is itself at rest beyond the limits of the wave, we have these equations*:—

$$\rho=\rho_0 e^{\mp\frac{u}{\sqrt{\mu}}} \tag{6}$$

$$\left.\begin{aligned} y&=\alpha x\pm\sqrt{\mu}\ln\alpha\cdot t+\varphi(\alpha) \\ 0&=\alpha x\pm\sqrt{\mu}t+\alpha\varphi'(\alpha) \end{aligned}\right\} \tag{7}$$

4. We have now to express these results in terms of the original genesis of the motion. Let us suppose the motion generated by a piston pushed forwards in the

---

* If x and α be eliminated between the equations (7) and $u=\pm\sqrt{\mu}\ln\alpha$, we shall obtain the equation

$$u=f\left\{y-\left(u\mp\sqrt{\mu}\right)t\right\}\ ,$$

which was first obtained, though in a very different manner from that employed in this paper, by M. Poisson, and printed in the Journal of the Polytechnique School, tome vii. It seems not to have occurred to him, however, that by means of this equation he might effect another integration of the equations of fluid motion, and thus discover the relation between ρ and u, whereby his solution would have been completed.

Several of the properties of wave-motion, depending on the gradual change of type, which are included in this equation of M. Poisson's, were first brought forward and discussed by Professor Stokes in the Philosophical Magazine for November 1848, and by the Astronomer Royal in June 1849. In the latter Number of the Magazine it also appears that Professor De Morgan had discovered and communicated to the Astronomer Royal two particular forms of the function F; without perceiving, however, that a slight generalization of his results would put him in the way to the integral expressed by the equations (5).

tube in a given manner. Let the piston at the time T (having the same origin as t) be at the distance Y from the plane of reference, and moving forwards with the velocity U; and by R denote the density of the air in contact with the piston at the moment. For all particles in contact with the piston $x = 0$ (we suppose the piston to commence its motion at the origin of x). Then since at the time T the particles in contact with the piston are within the limits of the wave, equations (6) and (7) must be satisfied;

$$\therefore \quad R = \rho_0 e^{\mp\frac{U}{\sqrt{\mu}}}$$

$$\left.\begin{aligned} Y &= \pm\sqrt{\mu}\ln\alpha'\cdot T + \varphi(\alpha') \\ 0 &= \pm\sqrt{\mu}T + \alpha'\varphi'(\alpha') \end{aligned}\right\} \tag{8}$$

In these equations $\alpha' = \dfrac{\rho_0}{R}$, and at present we have not sufficiently connected the two systems of equations (7) and (8). We shall further connect them by assuming $R = \rho$, which gives $\alpha' = \alpha$; the effect of which assumption is to limit the meaning of T, Y, U as follows:-

> T is the time of genesis of the density ρ which at the time t has been transmitted to the place denoted by y;
> Y is the place where the density ρ was generated;
> U is the velocity of the piston when ρ was generated by it.

We may now write α for α', and then eliminate α, $\varphi(\alpha)$, and $\varphi'(\alpha)$ between the four equations (7), (8). By this means we obtain

$$y = Y + \left(U \mp \sqrt{\mu}\right)(t - T) \tag{9}$$

$$x = \mp\sqrt{\mu}e^{\mp\frac{U}{\sqrt{\mu}}}(t - T) \tag{10}$$

$$\rho = \rho_0 e^{\mp\frac{U}{\sqrt{\mu}}} = \rho_0 e^{\mp\frac{u}{\sqrt{\mu}}} . \tag{11}$$

5. By these equations the state of a wave at any moment is connected with its genesis; and they contain in fact the complete solution of the problem of every kind of motion, in a tube, which can be generated by a piston.

6. From (11) it appears that u = U; that is, that the particle-velocity generated by the piston is transmitted through the medium without suffering any alteration. The same equation (11) shows that between the density and the velocity there is an invariable relation, which is independent of the law of original genesis of the

motion; so that in the same wave, or in different waves, wherever there is the same density, there will also be the same velocity.

7. One of the most obvious facts on looking at the equations just found is, that for the same genesis there are *two* values of x, two of y, and two of ρ. The signification of this is, that a single disturbance generates *two* waves; and (11) shows that for one of them ρ is *greater*, and for the other *less* than ρ_0. Equation (10) shows that they are propagated in contrary directions on opposite sides of the piston, and are therefore not parts of the *same* wave.

8. In the genesis of the wave we have supposed the piston pushed forwards, that is, in the direction of $+x$. Hence for the wave generated on that side of the piston we must, as appears from (10), take the lower sign, which in (11) gives ρ greater than ρ_0. This wave we call the *positive* wave, and the wave of *condensation.* For the wave generated on the other side of the piston we must take the upper sign, which gives ρ less than ρ_0; and this wave we call the *negative* wave, and the wave of *rarefaction.*

9. As it will be useful to have a definition of these two waves, which shall be independent of their position with regard to the generating piston, we may state that in general,—

a *positive* wave is one in which the motions of the particles are in the direction of wave-transmission: and

a *negative* wave is one in which the motions of the particles are in a direction opposite to that of wave transmission.

10. If ρ_1 and ρ_2 be the densities of the air in contact with the piston before and behind at any moment, and if p_1 and p_2 be the corresponding pressures; then from (11) we have

$$\rho_1 = \rho_0 e^{\frac{U}{\sqrt{\mu}}}, \text{ and } \rho_2 = \rho_0 e^{-\frac{U}{\sqrt{\mu}}},$$

and
$$\therefore \quad \rho_1\rho_2 = \rho_0^2 \; ;$$

which may be thus expressed in words:—if a piston moves in a tube, filled with air, in any manner whatever, the densities of the air in contact with it at its front and back are such that the equilibrium density is a mean proportional between them. And since $p = \mu\rho$, we have $p_1 p_2 = p_0^2$, which furnishes us with a similar property for the *pressures* on the piston.

11. Since $p_1 = p_0 e^{\frac{U}{\sqrt{\mu}}}$ and $p_2 = p_0 e^{-\frac{U}{\sqrt{\mu}}}$, it follows that the resistance to the motion of the piston (calling S its area) is

$$(p_1 - p_2)S = \left(e^{\frac{U}{\sqrt{\mu}}} - e^{-\frac{U}{\sqrt{\mu}}} \right) p_0 S \quad .$$

Hence in different gases, if p_0 be the same in all, those will offer the greatest resistance to the piston for which μ is the least.

It will be convenient from this point to consider the two kinds of waves separately.

1. The Wave of Condensation

12. The equations for this wave are

$$y = Y + \left(\sqrt{\mu} + U\right)(t - T)$$

$$x = \sqrt{\mu} e^{\frac{U}{\sqrt{\mu}}} (t - T)$$

$$\rho = \rho_0 e^{\frac{U}{\sqrt{\mu}}} \quad .$$

13. Now with respect to the genesis of this wave, we have seen that U must satisfy the same conditions as u, and Y as y. But $u = \frac{dy}{dt}$, therefore $U = \frac{dY}{dT}$: and again, as one of the equations of the general integral (7) was obtained from the other by differentiation with regard to α, it follows that both α and U must vary continuously; and that $\frac{dU}{dT}$ must not pass through infinity; in other words, if the velocity of the piston vary it must vary continuously. Neither Y nor U must be discontinuous with regard to T. Hence there must be no discontinuity of pressure within the limits of the wave at its genesis: and if discontinuity should afterwards occur in the wave during its transmission, our equations will cease to be applicable for that part of the wave where the discontinuity has occurred. For the wave in any one position may be supposed to generate its next position; and a piston or diaphragm may at any time be supposed to act the part of the generating wave. What is necessary for the diaphragm to observe as a law of genesis must be necessary for the wave considered as the generator of its next position; and therefore the part of the wave (if any) where discontinuity occurs will be beyond the reach of our equations.

14. It has been shown that the density ρ, which at the time t is at the distance y from the plane of reference, was generated at the time T when its distance from the same plane was Y. Hence it has been transmitted through the space y —Y in the time t—T, and consequently the velocity of its transmission (as appears from the first equation of (12)) is

$\sqrt{\mu}+U$.

15. The wave as a whole is included between two points of it for each of which U=0, and consequently for each of those points the velocity of transmission is $\sqrt{\mu}$. Hence the wave as a whole is transmitted with this uniform velocity. But all the parts of the wave, with the exception of its front and rear, are transmitted with velocities greater than this,—with velocities dependent on their respective densities. Hence every part of the wave, with the exception of its rear, is perpetually gaining on the front, and the result is a *constant change of type*, –the more condensed parts hurrying towards the front, with velocities greater as their densities are greater. This cannot go on perpetually without its happening at length that a *bore* (or tendency to a discontinuity of pressure) will be formed in front; which will force its way, in violation of our equations, faster than at the rate of $\sqrt{\mu}$ feet per second; and consequently in experiments, made on sound at long distances from the origin of the sound-wave, we should expect the actual velocity observed to be greater than $\sqrt{\mu}$, especially if the sound be a violent one, generated with extreme force (see art. 17).

16. We have seen that the velocity of transmission of the density ρ is $\sqrt{\mu}+U$. Now the velocity of the particles where the density is ρ and u, which we have shown to be equal to U. In a certain sense we may consider the velocity u to be a wind-velocity in that part of the medium, and then we have an indefinitely small disturbance at that point transmitted in that wind with the velocity $\sqrt{\mu}$ imposed upon the wind. In other words, transmission-velocity is superimposed on particle-velocity, and in this sense transmission-velocity is everywhere the same, and equal to $\sqrt{\mu}$. A wave passes by every particle with this velocity, whatever be the particular and varying density of the medium where the particle is situated.

17. Since a wave's front cannot move faster than with the velocity $\sqrt{\mu}$, if the generating piston moves faster than with this velocity, it will generate a bore; and from this we infer that a bore always moves with a velocity greater than $\sqrt{\mu}$; for wherever a bore may be situated at *any time*, we may suppose it to be *just* generated by a piston. If we write $\sqrt{\mu}$ for U, we find $p=\mu\rho=\mu\rho_0 e=e p_0$. Consequently if the piston press upon the resisting air with a pressure exceeding e atmospheres, a bore will be instantly formed.

I have defined a bore to be *a tendency to discontinuity of pressure*; and it has been shown that as a wave progresses such a tendency necessarily arises. As, however, discontinuity of pressure is a physical impossibility, it is certain Nature has a way of avoiding its actual occurrence. To examine in what way she does this, let us suppose a discontinuity to have actually occurred at the point A, in a wave which is moving forwards. Imagine a film of fluid at A forming a section at right angles to the tube. Then on the back of this film there is a certain pressure which is discontinuous with respect to the pressure on its front. To restore continuity of pressure, the film at A will rush forward with a *sudden* increase of velocity, the pressure in the front of the film not being sufficient to preserve continuity of velocity. In so doing the film will play the part of a piston generating a bit of wave in front, and a small regressive wave behind. The result be a prolongation of the wave's front, thereby increasing the original length of the wave, and producing simultaneously a feeble regressive wave of a negative character.

Now all this supposes the discontinuity to have actually occurred, which, as has been said, is a physical impossibility. For actual occurrence we must therefore substitute a tendency to occur, and modify the preceding reasoning thus:—

Nature so contrives, that as the discontinuity is in its initial stage of beginning to take place, its actual occurrence is prevented by a *gradual* (not *sudden*) prolongation of the wave's front, and by the *constant* casting off, from its front in a retrogressive direction, of a long continuous wave of a negative character, which will be of greater or less intensity according as the tendency to discontinuity is more or less intense in the original wave .

The casting off of this long wave will probably manifest itself audibly as a continuous *hiss* or *rushing* around.

Hence a sound-wave, from the moment that a tendency to discontinuity begins in its front, has the property of constantly prolonging itself in front, and by this means its front travels faster than at the rate $\sqrt{\mu}$. Those sounds also will travel most rapidly whose genesis was most violent; and gentlest sounds travel with velocities not much differing from $\sqrt{\mu}$. I should expect, therefore, that in circumstances where the human voice can be heard at a sufficiently great distance, the *command* to fire a gun, if instantly obeyed, and the *report* of the gun, might be heard at a long distance in an inverse order; i.e., *first* the report of the gun, and *then* the word "fire*." In a slight degree, therefore, the experimental velocity of sound will depend on its intensity, and the violence of its genesis. I consider this article as tending to account for the *discrepancy* between the *calculated* and *observed* velocities of sound (which most experimentalists have remarked and wondered at), when allowance is made (as will be done in a future part of this paper) for change of temperature.

* See Supplment to Appendix of PARRY'S Voyage in 1819-20, Art. "Abstract of Experiments to determine the Velocity of Sound."

18. It seems reasonable to suppose that the audible character of a wave is in some way dependent on its type; and consequently, if this be the case, the sound undergoes a perpetual modification as the distance of transmission increases. One modification of the sound-wave is, as we have seen, the formation of a bore in front; but there is another which cannot but have some influence on its audible properties, as it corresponds to a remarkable change of type; and this takes place when the greater densities begin to overtake the less.

Now when one degree of density overtakes another, the values of y corresponding to those two densities are equal; and hence at the time t the equation

$$y = Y + \left(\sqrt{\mu} + U\right)(t - T) \tag{12}$$

will give two equal values of y for two consecutive values of T. Hence differentiating it with regard to T, remembering that t is constant, or the same for both, as is also y, we have

$$0 = U - \left(\sqrt{\mu} + U\right) + (t - T)\frac{dU}{dT} \quad ,$$

or

$$t = T + \frac{\mu}{\frac{dU}{dT}} \quad . \tag{13}$$

The right-hand member of this equation is of course a continuous expression, and therefore its least or minimum value will be the value of t when the modification of type, of which we are speaking, *first begins* to take place; and because of the continuity of (13), this modification once begun will gradually spread itself over the fore-part of the wave. Now t will be a minimum when

$$\left(\frac{dU}{dT}\right)^2 = \mu \cdot \frac{d^2U}{dT^2} \quad .$$

From this equation we may find T, the time of genesis of that part of the wave where this modification begins. Then (13) will give t, the actual time when the modification begins; and (12) will give the place in the tube where it begins.

19. It is perhaps impossible to say what is the audible characteristic corresponding to the wave-modification just investigated; but whatever it be, we perceive from (13) that those sound-waves soonest begin to be affected by it for which $\frac{dU}{dT}$ is largest; i.e. those whose genesis is most violent. And we may also

consider it as proved that those sounds will retain their original characteristics the longest which are the most gently generated.

It is also quite evident from (13) if the same cause generate sound-waves in different tubes filled with different gases, the wave will be soonest affected by the above modification in that tube which contains the gas for which μ is least.

We come now to speak of

2. *The Wave of Rarefaction*

20. We shall obtain the equations for this kind of wave by writing -U for + U in the equations of art. 12, which is equivalent to supposing a negative wave generated on the +y side of the piston. Hence the equations of a negative wave are

$$y = Y + \left(\sqrt{\mu} - U\right)(t - T) \ ,$$

$$x = \sqrt{\mu}\, e^{-\frac{U}{\sqrt{\mu}}} (t - T) \ ,$$

$$\rho = \rho_0\, e^{-\frac{U}{\sqrt{\mu}}} \ .$$

21. Reasoning in the same manner as in art. 14, it appears that the velocity with which the density ρ is transmitted is

$$\sqrt{\mu} - U \ .$$

From this it appears that, speaking generally, the velocity of transmission of every part of a negative wave is less than of every part of a positive wave. The exceptions to this statement are the front and rear, which in both kinds of waves move with the same velocity $\sqrt{\mu}$, because for those points U=0. It is evident also that the most rarefied parts of a wave will be transmitted the most slowly, and will consequently drop continually towards the rear. Hence in this species of wave, as in the former, a constant change of type takes place; and in the end also a *negative or rarefied bore* will be formed in the *rear* of the wave.

By a process of reasoning analogous to that of art. 17, we infer that a negative sound-wave, from the moment that a tendency to discontinuity begins in its rear, has the property of constantly shortening its rear, and by this means its rear travels faster than a the rate $\sqrt{\mu}$; and also as it progresses it is constantly casting off from its rear in a regressive direction a long continuous wave of a negative character. Art. 18 also admits of easy modification to this kind of wave.

22. The velocity of transmission of a negative wave being $\sqrt{\mu} - U$, and the last term of this expression admitting of arbitrary increase, it is evident that $U = \sqrt{\mu}$ is a critical value, and that the part of the wave corresponding to that value of U is stationary.

The corresponding value of ρ is $\frac{\rho_0}{e}$.

Every part of the wave where the density exceeds this travels forwards; but the parts where the density is less than this are regressive; hence a wave, as a whole, in which ρ begins at ρ_0, and after twice passing through $\frac{\rho_0}{e}$ ends at ρ_0, will have two stationary points in its type, viz. those where $\rho = \frac{\rho_0}{e}$. Between these points the wave will be stationary though constantly changing type; beyond them progressive.

23. But instead of supposing the piston to generate such a wave as this, let us suppose it to begin from the velocity zero, and according to any proposed law (continuous of course) increase its velocity till it becomes infinite; and let us consider the state of the medium at this moment.

Denote by A and B the places of the piston where its velocity became respectively $\sqrt{\mu}$ and infinite. Then whatever was the law of motion from A to B, and whether AB be great or small, provided it remains of finite length, the density at A will remain unchanged and equal to $\frac{\rho_0}{e}$, and the velocity of every particle as it passes by A will be equal to $\sqrt{\mu}$. The mass of air also which will rush through the section of the tube at A will be $\frac{S\rho_0\sqrt{\mu}}{e}$; and this, be it observed, cannot be made either more or less by causing the piston to move in a different manner from A to B. It is also equally independent of the law of the piston's motion before it reached A. Hence the mass of air that flows through the section at A is altogether independent of the law of the piston's motion throughout its whole course.

24. Now let us inquire what quantity of air rushes through any other section of the tube. In every part where there is motion the same relation between density and velocity obtains, viz. $\rho = \rho_0 e^{-\frac{u}{\sqrt{\mu}}}$; and consequently the quantity which rushes through any section is at the rate of

$$S\rho_0 u e^{-\frac{u}{\sqrt{\mu}}} \text{ per second.}$$

It is obvious this admits of a maximum value, which in the usual manner we find to be

$$\frac{S\rho_0 e\sqrt{\mu}}{e},$$

at which value $u=\sqrt{\mu}$ and $\rho=\frac{\rho_0}{e}$.

25. Hence one part of the tube cannot supply air to another part faster than at this rate; and consequently the greatest possible mass of air passes through the section at A: and it may be stated as a general property of motion through a tube, that a gas cannot be conveyed through a tube faster than at the rate $\frac{S\sqrt{\mu}}{e}$ cubic feet per second of gas of the density ρ_0.

Hence the escaping powers of different gases through equal tubes are proportional to the velocities with which they respectively transmit sound.

26. Since this result is independent of the law of velocity of the air, both before and after passing the section A, we are entitled to say that air cannot rush through a pipe of finite length, even into a vacuum, faster than at the rate of $\frac{S\sqrt{\mu}}{e}$ cubic feet per second. The length of the pipe seems to be a matter of perfect indifference, and may be nothing more than a hole through a partition of finite thickness.

27. Since one part of a tube cannot supply air to, nor convey air away from, another part, A, faster than at the maximum rate, it is easy to see that if the pipe be supposed of finite length, which conveys air into a vacuum, the velocity in every part of the pipe will soon be the same throughout, and equal to $\sqrt{\mu}$, and density everywhere equal to $\frac{\rho_0}{e}$.

From this it would appear that the rate of discharge into a vacuum, which has generally been supposed to be that which is due to the height of the homogeneous atmosphere, is in reality that which is due to the $\left(\frac{1}{2e^2}\right)$th part only; that is, to little more than the fifteenth part of it; but this requires correction for change of temperature.

28. If the generating piston moves forward and then backward, so as to generate a positive wave followed continuously by a negative wave, they will not separate; for, as we have seen, they are each transmitted, as wholes, at the same rate $\sqrt{\mu}$. But the main body of the positive wave will gradually advance in the type towards its front, and that of the negative wave fall back towards its rear; and consequently for the purposes of audibility the central part of the compound wave, between the front of the positive and the rear of the negative wave, will become so attenuated that it may be considered of little audible effect, after the waves have been in existence a sufficient length of time to allow the formation of bores. The compound wave will therefore have a tendency to produce the audible effect of two separate waves, separated by an interval of space nearly equal to its whole length. If therefore the length of such a compound wave be sufficiently great, it will ultimately produce *two* distinct sounds separated by a very brief interval of time.

29. If the generating piston moves backward and then forward, so as to generate a negative wave followed continuously by a positive wave, the positive and negative bores will destroy each other as rapidly as they are formed. This, however, supposes the positive and negative portions of the original compound wave to the equal. If one exceed the other in quantity of motion, the result will be a little modified. A compound wave of the kind supposed in this article will therefore be entirely devoid of bores, and the sound corresponding to it will be free from that harshness which is probably the audible character of a bore.

30. If there be a continuous succession of positive and negative waves, constituting one long compound wave, such a wave will produce a continuous even sound, called a musical note, probably owing its sweetness in some degree to the property just mentioned; and as every negative portion is succeeded by a positive portion, and every positive by a negative, the length of each portion will remain unchangeable, whatever be the distance through which the compound wave travels. Hence the pitch of a musical note cannot change by distance of transmission.

31. Suppose a portion of the tube to be filled with air of a different kind from that which fills the first part. Let $p_0'\ \rho_0'\ \mu'$ be the quantities for this air which correspond to $p_0\ \rho_0\ \mu$ of the former; and to prevent the two airs or gases from mixing, let them be supposed to be separated by an impenetrable film without weight and inertia. Then as there is equilibrium in the tube before the wave is generated, we have

$$p_0 = p_0'.$$

Let now a wave be generated in the first medium and transmitted towards the second; then when it has reached the common boundary of the media, the velocities of the particles in contact with the film on both sides will always be equal. Let U' be this velocity at any moment, and U the velocity which the film would have had at that moment, if the second medium had been the same as the first. Then $U-U'$ is the velocity lost by the particles of the first medium by the resistance due to their contact with the film. In other words, this velocity has been impressed on the particles of the first medium by the resistance of the film, in the reflex direction. This gives rise to a reflex wave in the first medium, which we may consider superimposed on the wind of the original wave. And consequently if p be the pressure at the film due to the original wave, the pressure when this reflex wave has been superimposed, i.e. the actual pressure at the film, is $= pe^{\frac{U-U'}{\sqrt{\mu}}}$, which $= p_0 e^{\frac{2U-U'}{\sqrt{\mu}}}$, $\because p = p_0 e^{\frac{U}{\sqrt{\mu}}}$.

But if we now turn to the other side of the film, the velocity U' has been impressed upon the particles of the second medium in contact with the film; and hence the pressure of those particles on the film

$$= p_0' e^{\frac{U'}{\sqrt{\mu'}}} = p_0 e^{\frac{U'}{\sqrt{\mu'}}} ;$$

and consequently, as the pressures on the two sides are equal, we have

$$\frac{U'}{\sqrt{\mu'}} = \frac{2U-U'}{\sqrt{\mu}} .$$

Hence the velocities of the particles at the film, for the *incident*, *reflected* and *refracted* waves, are respectively proportional to

$$\sqrt{\mu}+\sqrt{\mu'},\ \sqrt{\mu}-\sqrt{\mu'},\ \text{and}\ 2\sqrt{\mu'} .$$

There is nothing new in these formulae, except that they are here deduced without supposing the motions small.

II. Wave Motion when Change of Temperature Is not Neglected

32. The heat developed by that change of temperature which is produced by the sudden alteration of density due to the passage of a wave, is probably taken account of by using the following equation as that which connects pressure and density,

$$\frac{p}{p_0} = \left(\frac{\rho}{\rho_0}\right)^k ;$$

k being the ratio of the specific heat of the gas under a constant pressure, to its specific heat under a constant volume. The dynamical equation takes for this case the following form to be used instead of that in art. 1,

$$\left(\frac{dy}{dx}\right)^{k+1}\cdot\frac{d^2y}{dt^2}=k\mu\cdot\frac{d^2y}{dx^2}.$$

This equation being integrated as explained in art. 1, gives

$$\left.\begin{aligned} y&=\alpha x+\left(C\mp\frac{2\sqrt{k\mu}}{k-1}\alpha^{-\frac{k-1}{2}}\right)t+\varphi(\alpha)\\ 0&=\alpha x\pm\sqrt{k\mu}\alpha^{-\frac{k-1}{2}}t+\alpha\varphi'(\alpha)\end{aligned}\right\}. \tag{14}$$

33. From these we obtain

$$u=\frac{dy}{dt}=C\mp\frac{2\sqrt{k\mu}}{k-1}\alpha^{-\frac{k-1}{2}},$$

$$=C\mp\frac{2\sqrt{k\mu}}{k-1}\left(\frac{\rho}{\rho_0}\right)^{\frac{k-1}{2}}.$$

For the same reasons as before we shall suppose $u=0$ and $\rho=\rho_0$ to be simultaneous equations; which gives

$$C=\pm\frac{2\sqrt{k\mu}}{k-1},$$

and

$$\therefore\left(\frac{\rho}{\rho_0}\right)^{\frac{k-1}{2}}=1\mp\frac{(k-1)u}{2\sqrt{k\mu}}. \tag{15}$$

This equation gives the relation between density and velocity; from which that between pressure and velocity is easily found.

34. The general integral (14) may be expressed in terms of the original genesis precisely in the same manner as was employed in art. 4; and the result is

$$y=Y+\left(\frac{k+1}{2}U\mp\sqrt{k\mu}\right)(t-T) \tag{16}$$

$$x = \mp\sqrt{k\mu}\left(1 \mp \frac{k-1}{2\sqrt{k\mu}}U\right)^{\frac{k+1}{k-1}}(t-T) \tag{17}$$

$$u - U, \text{ and } p = p_0\left(\frac{\rho}{\rho_0}\right)^k . \tag{18}$$

These equations, with (15), are those from which the properties of the motion are to be deduced. The degree of modification of former results required by these formulae will be in most cases sufficiently evident, and need not therefore to be particularly pointed out.

35. The result of art. 10 takes the following form—

$$\rho_1^{\frac{k-1}{2}} + \rho_2^{\frac{k-1}{2}} = 2\rho_0^{\frac{k-1}{2}} \;;$$

and that of art. 11 the following—

$$(p_1 - p_2)S = Sp_0\left\{\left(1 + \frac{k-1}{2\sqrt{k\mu}}U\right)^{\frac{2k}{k-1}} - \left(1 - \frac{k-1}{2\sqrt{k\mu}}U\right)^{\frac{2k}{k-1}}\right\} .$$

36. From (16) it appears that the velocity of transmission of the front and rear of either a positive or negative wave is $\sqrt{k\mu}$; but the velocity of transmission of that part of the wave of which the density is ρ, is, for a positive wave,

$$\sqrt{k\mu} + \frac{k+1}{2}U \;;$$

and for a negative wave,

$$\sqrt{k\mu} - \frac{k+1}{2}U .$$

The part of these expressions to which the bore is due is the term $\frac{k+1}{2}U$; and as k is known to be greater than unity, this is greater than U; and consequently change of temperature hastens the formation of a bore, and also renders the property of art. 16 inapplicable here.

37. As in the case of a negative wave the equation (15) involves a negative term, it is manifestly possible for the piston, in generating a negative wave, to

move so quickly as to leave a vacuum behind it. The least velocity with which this can happen is

$$\frac{2\sqrt{k\mu}}{k-1} ,$$

which for common air is about 5722 feet per second. But it is necessary to notice, that in this and similar extreme results, we are hardly justified in supposing k to be constant up to such high velocities.

38. The expression $S\rho u$ is a maximum (see art. 24) when

$$u = \frac{2\sqrt{k\mu}}{k+1} ,$$

which in the case of common air is equal to about 904 feet per second; and the corresponding density is

$$\rho = \left(\frac{2}{k+1}\right)^{\frac{2}{k-1}} \rho_0 ;$$

or, for common air, about $\frac{2}{5}\rho_0$.

Hence, no gas can rush through a pipe faster than at the rate of

$$\sqrt{k\mu}\left(\frac{2}{k+1}\right)^{\frac{k+1}{k-1}} S$$

cubic feet per second.

39. The change of temperature due to the transmission of a wave through an elastic medium has been taken account of, by assuming a law different from that of BOYLE, to connect pressure with density (art. 32).

If we generalize the law by assuming

$$p = \varphi\left(\frac{\rho_0}{\rho}\right) ,$$

the dynamical equation takes the form

$$\frac{d^2y}{dt^2} = -\frac{1}{\rho_0} \cdot \varphi'\left(\frac{dy}{dx}\right) \cdot \frac{d^2y}{dx^2} .$$

If now we assume

$$(F'\alpha)^2 = -\frac{\varphi'\alpha}{\rho_0} ,$$

the integral of the dynamical equation will be

$$\begin{cases} y = \alpha x + (C \pm F\alpha)t + f\alpha, \\ x = \mp F'\alpha \cdot t - f'\alpha \end{cases} ,$$

with $\alpha = \frac{\rho_0}{\rho}$, and $u = C \pm F\alpha = C \pm F\left(\frac{\rho_0}{\rho}\right)$.

40. These equations are true of any motion which can be generated by a piston moving subject to the laws of continuity. See art. 13. The last shows that the relation between velocity and density is independent of the law of genesis of the motion. The medium may be, as a whole, in the motion with the uniform velocity $C \pm F(1)$, and the motion of the particles caused by the motion of a piston will be superimposed on this. For convenience we shall suppose the medium as a whole at rest, and $\therefore C \pm F(1) = 0$.

If there be a point, or any number of points, within that part of the medium which is in motion for which $\rho = \rho_0$, for all such points $\alpha = 1$, and the equation

$$x = \mp F'(1) \cdot t - f'(1) ,$$

which is always true for all such points, shows that at those points x changes its value at the rate of $F'(1)$ feet per second, *i.e.* the front of the wave travels at the rate of

$$\left\{-\frac{\varphi'(1)}{\rho_0}\right\}^{\frac{1}{2}} \text{ feet per second,}$$

which is constant, and depends not at all on the law of genesis, but only on the assumed relation between pressure and density, and not on the *general* value of even that, but only on its limiting value when $\rho = \rho_0$. Now many different forms of the function φ may give the same limiting value; and consequently all the media corresponding to these various forms of φ will transmit a wave, as a whole, with the same velocity. Hence if the relation between pressure and density be given, the wave-velocity may be instantly deduced from the expression $\left\{-\dfrac{\varphi'(1)}{\rho_0}\right\}^{\frac{1}{2}}$, or from its equal,

$$\left\{\frac{dp}{d\rho}\right\}_0^{\frac{1}{2}} \;;$$

using the subscript 0 to signify that after the differentiation has been performed ρ_0 is to be written for ρ.

41. Since $u = C \pm F\left(\dfrac{\rho_0}{\rho}\right)$, by differentiation we obtain

$$\frac{du}{d\rho} = \mp\frac{1}{\rho}\left(\frac{dp}{d\rho}\right)^{\frac{1}{2}} .$$

42. And if c denote the velocity of transmission of the density ρ, then we have

$$c = C \pm F\left(\frac{\rho_0}{\rho}\right) \mp \frac{\rho_0}{\rho} F'\left(\frac{\rho_0}{\rho}\right) = u \mp \left(\frac{dp}{d\rho}\right)^{\frac{1}{2}} \;;$$

consequently

$$\frac{dc}{d\rho} = \mp\frac{1}{\rho}\cdot\frac{d}{d\rho}\left(\rho^2\frac{dp}{d\rho}\right)^{\frac{1}{2}} .$$

Now the former of these equations shows that unless the term $\left(\dfrac{dp}{d\rho}\right)^{\frac{1}{2}}$ be constant, the property of the superposition of wave-transmission on particle-velocity, proved in art. 16, does not hold good. But if it be constant, then $p = \mu\rho + \mu'$; which is the general relation between pressure and density when that principle of superposition holds good. Hence as mentioned in art. 36, the development of heat puts an end to this property in all known gases.

43. In the case of negative waves we may institute a method of reasoning similar to that employed in arts. 23 *et seq.*, and arrive at analogous results. We

shall also find that, taking $c = \left(\frac{dp}{d\rho}\right)^{\frac{1}{2}} - u$ for this case, the maximum value of $S\rho u$ will occur in that section of the tube where $c=0$; from which it follows that at that section

$$u = \left(\frac{dp}{d\rho}\right)^{\frac{1}{2}} ;$$

which is always positive and finite. Hence may be determined the limit to the quantity of a gas that can pass through a pipe in a given time, even into a vacuum.

44. The expression in art. 42 for $\frac{dc}{d\rho}$ shows that c is in general a function of ρ, so that in general there will be a constant change of type. In one case, however, there will be no change of type. This will take place when $\frac{dc}{d\rho} = 0$, that is when $\rho^2 \cdot \frac{dp}{d\rho}$ is constant. Assume for this case

$$\rho^2 \frac{dp}{d\rho} = B ;$$

$$\therefore p = A - \frac{B}{\rho} .$$

This equation expresses the nature of the medium which is distinguished by the property, that it transmits waves without change of type. And if we pass from this to the dynamical equation, we find

$$\frac{d^2 y}{dt^2} = \frac{B}{\rho_0^2} \cdot \frac{d^2 y}{dx^2} .$$

Now it has been usual to reduce the equation (3) to this form for the purposes of approximation; but the process appears to be allowable only so far as the equation $p = A - \frac{B}{\rho}$ may be taken to be a physical approximation to BOYLE'S law $P = \mu\rho$. To me it does not seem to be an allowable approximation; and consequently I do not consider the solution of the dynamical equation, which has been obtained by this means, to be applicable to the problem of sound at all. Many *analytically* approximate forms might be invented and used for BOYLE'S law, and each would have its peculiar physical attributive effects on the sound-wave; and we might thus, by adopting first one and then another of these analytically approximate laws, invent *ad libitum* an inexhaustible list of properties of the sound-wave which have no real existence where BOYLE'S law is strictly true. From which therefore it would seem to be a necessary consequence, that an equation between p and ρ must not only be *analytically* but also *physically* approximative, in order that the results deduced from it may be accepted as real approximations to the true laws of nature.

45. By means of the expressions for $\frac{du}{d\rho}$ and $\frac{dc}{d\rho}$, we may not only discover the properties of motion in a tube without having recourse to the usual equations, when the relation between ρ and p is known, but we may also solve many inverse problems.

Also, if the tube be filled with a medium of such a nature that the relation between p and ρ changes continuously from point to point, or is different in different parts, yet if $\left(\frac{dp}{d\rho}\right)_0$ has the same value everywhere, waves will travel through the tube with a uniform velocity.

If the nature of the medium should vary slowly and continuously, the velocity of the wave-transmission would be known, from the equations given above, by integration.

46. If, through the partial radiation of heat, or from any other cause, the dynamical equation should take the form

$$\frac{d^2y}{dt^2} = f\left(\frac{dy}{dx}, \frac{dy}{dt}\right) \cdot \frac{d^2y}{dx^2} \quad ,$$

we must integrate it as before by assuming

$$\frac{dy}{dt} = F\left(\frac{dy}{dx}\right) \quad ;$$

which gives

$$(F'\alpha)^2 = f(\alpha, F\alpha) \quad .$$

This equation being integrated will furnish the form of F; and then the integral of the proposed dynamical equation will be

$$\begin{cases} y = \alpha x + (C \pm F\alpha)t + \varphi\alpha, \\ x = \mp F'\alpha \cdot t - \varphi'\alpha \ , \end{cases}$$

which does not present any new difficulty.

Riemann
courtesy of AIP Emilio Segrè Visual Archives

GEORG FRIEDRICH BERNHARD RIEMANN

Riemann was a German mathematician born on 17 September, 1826, in Breselenz. His father was a Protestant minister and served as Riemann's teacher for his elementary education. Riemann showed remarkable skill in mathematics at an early age.

Riemann entered Göttingen University in 1846 to study theology and philology. He attended mathematical lectures and it soon became apparent that this was where his real interests lay. He later transferred to Berlin University for a brief time. He returned to Göttingen and received his Ph.D. in 1851. When Gauss died in 1855, his chair was given to Dirichlet. After Dirichlet's death in 1857, Riemann became full professor.

In 1862 Riemann suffered an attack of tuberculosis, from which he never fully recovered. In his profession his evolution was slow and his life was short; he died in 1866 (20 July, in Selasca, Italy). What his work lacks in quantity is fully compensated for in its extremely high quality. Most of his work was in the area of pure mathematics and differential geometry. Probably his most important contribution to physics was the paper on the propagation of planar air waves of finite amplitude (1860). By this time the theory of sound waves of infinitesimal amplitude was well established. Riemann studied finite-amplitude waves in a fluid in which the pressure was an arbitrary function of density. The method of characteristics has proved to be a very powerful tool in the formulation and understanding of wave motion, as well as in terms of numerical solution of flow problems.

Although somewhat out of style at the present time, because of large-scale computer codes that solve transient fluid flow and wave-propagation problems in multidimensions, the method of characteristics still has some application, particularly with regard to rate-dependent solid material behavior in one spatial dimension. The major development that eventually led to the neglect of this approach to wave-propagation solution is the work of Von Neumann and Richtmyer ["A Method for the Numerical Calculation of Hydrodynamic Shocks," J. Appl. Phys. 21, pp. 232-237, 1950] in which an artificial viscous stress is included in the equations of motion to limit the formation of shock fronts. To quote the abstract of Von Neumann and Richtmyer's paper,

"The equations of hydrodynamics are modified by the inclusion of additional terms which greatly simplify the procedures needed for stepwise numerical

solution of the equations in problems involving shocks. The quantitative influence of these terms can be made as small as one wishes by choice of a sufficiently fine mesh for the numerical integrations. A set of difference equations suitable for the numerical work is given, and the condition that must be satisfied to insure their stability is derived."

This was indeed a great step forward for numerical calculation, but has had an obvious negative effect on the understanding of wave interactions responsible for observed physical behavior. Riemann's method of characteristics, while being elegant from a mathematical perspective, also possesses physical interpretation with regard to the forward and backward propagation of waves in one dimension.

The Propagation of Planar Air Waves of Finite Amplitude

B. Riemann, Translated from *Abhandlungen der Gesellschaft der Wissenschaften zu Göttingen*, **8**, pp. 43-65 (1860)

Even though the differential equations for determining the motion of gases have long been known, their integration has been carried out almost exclusively for the case when the pressure differences can be viewed as infinitesimally small fractions of the entire pressure, and until recently one had to be satisfied with taking account of only the first powers of these fractions. Until quite recently, Helmholtz was able to bring the second order terms into the calculation and in this way he clarified the objective generation of combination tones. But still, the precise differential equations can be completely integrated for the case where the initial motion takes place in the same direction and where the velocity and pressure are constant in every plane perpendicular to this direction. And even though the former treatment is entirely sufficient even for explaining the experimental phenomena found to date, still in all, it is possible--given the tremendous progress made most recently by Helmholtz in the experimental treatment of acoustic problems--that the results of this more precise calculation will provide additional insight into experimental research in the not too distant future. Regardless of its purely theoretical interest pertaining to the treatment of non-linear, partial differential equations, this might contribute toward a further elucidation of the problem.

Boyle's law could be used as the basis for the dependence of pressure on the density, when the pressure changes caused by temperature differences are compensated so quickly that the temperature of the gas might be viewed as constant. But probably the thermal exchange should be neglected entirely, and thus for this dependence we must use the law that states that the pressure of the gas changes with its density when it is not absorbing or releasing any heat.

According to Boyle's law and the Gay-Lussac law, when v is the volume of the unit weight, p the pressure and T the temperature calculated from -273°C, we have the expression:

$$\log p + \log v = \log T + const.$$

Let us look at T as a function of p and v, and call the specific heat at constant pressure c, and that at constant volume c', both relative to the unit weight, and then if p and v change by dp and dv, the amount of heat absorbed by this unit weight will be

$$c\frac{dT}{dv}dv + c'\frac{dT}{dp}dp$$

or, since $$\frac{d\log T}{d\log v} = \frac{d\log T}{d\log p} = 1,$$

$$T(c\,d\log v + c'\,d\log p)$$

Now if no heat absorption takes place, then $d\log p = -\frac{c}{c'}d\log v$, and thus, if we make the Poisson assumption that the ratio of the two specific heats $c/c' = k$ is independent of temperature and pressure, then

$$\log p = -k\log v + const.$$

According to more recent tests by Regnault, Joule and W. Thomson, these laws are quite probably very closely applicable for oxygen, nitrogen and hydrogen and their mixtures under all feasible pressures and temperatures.

Now Regnault ascertained a very close relationship to Boyle's law and to the Gay-Lussac law for these gases and determined that the specific heat c is independent of temperature and pressure.

For atmospheric air, Regnault found:

between	- 30°C and + 10°C	c = 0.2377
"	+ 10°C and + 100°C	c = 0.2379
"	+100°C and + 215°C	c = 0.2376.

Thus for pressures from 1 to 10 atmospheres, no notable difference in specific heat was found.

According to experiments by Regnault and Joule furthermore, for these gases Mayer's assumption as adopted by Clausius seems to be very nearly correct, that is, that a gas expanding at constant temperature will absorb only enough heat as needed to produce the external work. If the volume of the gas changes by dv, while the temperature stays constant, then $d\log p = -d\log v$, the absorbed quantity of heat $T(c - c')\,d\log v$, the performed work is $p\,dv$. This hypothesis will yield the following expression when A is the mechanical equivalent of heat:

$$AT\,(c - c')\,d \log v = p\,dv$$

or

$$c - c' = AT/pv \quad ,$$

thus independent of pressure and temperature.

Accordingly, $k = c/c'$ is independent of pressure and temperature and when $c = 0.237733$, and we have A according to Joule = 424.55 kg, and for the temperature of 0°C or T = 100 °C/0.3665, pv according to Regnault = 7990.267, the factor k becomes equal to 1.4101. The distance of sound in dry air of 0°C in one second amounts to $\sqrt{7990.267 \times 9.8088\, k}$ and with this value of k it was found to be equal to 332.440 m/s, while the two most complete test series by Moll and van Beek calculated separately yield 332.528 m/s and 331.867 m/s, for an average of 332.271 m/s and the experiments by Martins and A. Bravais according to their own calculation yield 332.37 m/s.

1. Initially it is not necessary to make any particular assumption about the dependence of pressure on density; we thus assume that at a density ρ the pressure will be $\varphi(\rho)$ and we will initially leave the function φ arbitrary.

Now let us introduce rectangular coordinates x, y, z with the x-axis in the direction of motion and let ρ be the density, p the pressure and u the velocity for coordinate x at time t and let ω be an element in the plane whose coordinate is x.

The content of the straight cylinder standing on the element ω with height dx is then ωdx and the mass held by it is $\omega\rho dx$. The change in this mass during the time element dt, or the quantity $\omega \frac{d\rho}{dt} dt\, dx$, is determined by the mass flowing into it, which is found to be $=-\omega \frac{d\rho u}{dx} dx\, dt$. Its acceleration is $\frac{du}{dt} + u\frac{du}{dx}$ and the force driving it in the direction of the positive x-axis is $-\frac{dp}{dx}\omega dx = -\varphi'(\rho)\frac{d\rho}{dx}\omega dx$, when $\varphi'(\rho)$ denotes the derivative of $\varphi(\rho)$. Thus for ρ and u we have the two differential equations:

$$\frac{d\rho}{dt} = -\frac{d\rho u}{dx} \qquad \text{and} \qquad \rho\left(\frac{du}{dt} + \frac{du}{dx}\right) = -\varphi'(\rho)\frac{d\rho}{dx} \text{ or}$$

$$\frac{du}{dt} + u\frac{du}{dx} = -\varphi'(\rho)\frac{d \log \rho}{dx}$$

and

$$\frac{d \log \rho}{dt} + u\frac{d \log \rho}{dx} = -\frac{du}{dx}.$$

If we multiply the second equation with $\pm\sqrt{\varphi'(\rho)}$ and add it to the first one, and use the abbreviations:

$$(1) \qquad \int\sqrt{\varphi'(\rho)}d\log\rho = f(\rho) \quad \text{and}$$
$$(2) \qquad f(\rho)+u=2r, f(\rho)-u=2s$$

then these equations will obtain the more simple form

$$(3) \qquad \frac{dr}{dt}=-\left(u+\sqrt{\varphi'(\rho)}\right)\frac{dr}{dx}, \quad \frac{ds}{dt}=-\left(u-\sqrt{\varphi'(\rho)}\right)\frac{ds}{dx}$$

where u and ρ are particular functions of r and s due to equation (2). From this follows

$$(4) \qquad dr=\frac{dr}{dx}\left(dx-\left(u+\sqrt{\varphi'(\rho)}\right)dt\right)$$

$$(5) \qquad ds=\frac{ds}{dx}\left(dx-\left(u-\sqrt{\varphi'(\rho)}\right)dt\right)$$

Under the assumption (generally applicable in real situations) that $\varphi'(\rho)$ is positive, these equations state that r remains constant when x changes with t so that $dx=\left(u+\sqrt{\varphi'(\rho)}\right)dt$ and s remain constant when x changes with t so that $dx=\left(u-\sqrt{\varphi'(\rho)}\right)dt$

One particular value of r, or $f(\rho)+u$, , moves toward larger values of x with the velocity $\sqrt{\varphi'(\rho)}+u$, a particular value of s,or $f(\rho)-u$, moves toward smaller values of x with the velocity $\sqrt{\varphi'(\rho)}-u$.

One particular value of r will thus coincide gradually with each occurring value of s, and the velocity of its propagation will depend at every moment on the value of s with which it coincides.

2. Now the analysis offers the means of answering the question of where and when a value r' of r encounters a value s' of s, that is to determine x and t as functions of r and s. In fact, if we introduce r and s as independent variables in equation (3), then these equations go over into linear differential equations for x and t and thus they can be integrated according to known methods. In order to transform the differential equations to a linear one, it is best to set equations (4) and (5) of the previous type into the following form:

(1)

$$dr = \frac{dr}{dx}\left(d\left(x-\left(u+\sqrt{\varphi'(\rho)}\right)t\right)+\left(dr\left(\frac{d\log\sqrt{\varphi'(\rho)}}{d\log\rho}+1\right)+ds\left(\frac{d\log\sqrt{\varphi'(\rho)}}{d\log\rho}-1\right)\right)t\right)$$

(2)

$$ds = \frac{ds}{dx}\left(d\left(x-\left(u-\sqrt{\varphi'(\rho)}\right)t\right)+\left(ds\left(\frac{d\log\sqrt{\varphi'(\rho)}}{d\log\rho}+1\right)-dr\left(\frac{d\log\sqrt{\varphi'(\rho)}}{d\log\rho}-1\right)\right)t\right)$$

If we view r and s as independent variables, then for x and t we obtain the two linear differential equations:

$$\frac{d\left(x-\left(u+\sqrt{\varphi'(\rho)}\right)t\right)}{ds} = -t\left(\frac{d\log\sqrt{\varphi'(\rho)}}{d\log\rho}-1\right)$$

$$\frac{d\left(x-\left(u-\sqrt{\varphi'(\rho)}\right)t\right)}{dr} = t\left(\frac{d\log\sqrt{\varphi'(\rho)}}{d\log\rho}-1\right)$$

Consequently:

(3) $$\left(x-\left(u+\sqrt{\varphi'(\rho)}\right)t\right)dr-\left(x-\left(u-\sqrt{\varphi'(\rho)}\right)t\right)ds$$

is a complete differential whose integral, w, satisfies the equation

$$\frac{d^2w}{drds} = t\left(\frac{d\log\sqrt{\varphi'(\rho)}}{d\log\rho}-1\right) = m\left(\frac{dw}{dr}+\frac{dw}{ds}\right)$$

where $m = \frac{1}{2\sqrt{\varphi'(\rho)}}\left(\frac{d\log\sqrt{\varphi'(\rho)}}{d\log\rho}-1\right)$, is a function of $r + s$. If we set

$f(\rho) = r + s = \sigma$ then we have $\sqrt{\varphi'(\rho)} = \frac{d\sigma}{d\log\rho}$, and consequently also

$$m = -\frac{1}{2}\frac{d\log\frac{d\rho}{d\sigma}}{d\sigma}.$$

Under the Poisson assumption $\varphi(\rho) = aa\rho^k$ we have $f(\rho) = \frac{2a\sqrt{k}}{k-1}\rho^{\frac{k-1}{2}}$ + constant, and if we select the value zero for the arbitrary constant, then

$$\sqrt{\varphi'(\rho)} + u = \frac{k+1}{2}r + \frac{k-3}{2}s, \qquad \sqrt{\varphi'(\rho)} - u = \frac{k-3}{2}r + \frac{k+1}{2}s$$

$$m = \left(\frac{1}{2} - \frac{1}{k-1}\right)\frac{1}{\sigma} = \frac{k-3}{2(k-1)(r+s)}$$

Under the assumption of Boyle's law $\varphi(\rho) = aa\rho$ we obtain

$$f(\rho) = a\log\rho$$

$$\sqrt{\varphi'(\rho)} + u = r - s + a, \quad \sqrt{\varphi'(\rho)} - u = s - r + a$$

$$m = -\frac{1}{2a}$$

that is, values that obtain from the previous case if we omit from $f(\rho)$ the constant $\frac{2a\sqrt{k}}{k-1}$ and from r and s the constant $\frac{a\sqrt{k}}{k-1}$, and then set $k=1$.

The introduction of r and s as independent, variable quantities is only possible when the determinants of these functions of x and t, that are equal to $2\sqrt{\varphi'(\rho)}\frac{dr}{dx}\frac{ds}{dx}$ do not disappear, that is, only when $\frac{dr}{dx}$ and $\frac{ds}{dx}$ are both different from zero.

When $\frac{dr}{dx} = 0$ then from (1) we have $dr = 0$ and from (2) $x - \left(u - \sqrt{\varphi'(\rho)}\right)t$ is a function of s. Consequently even the expression (3) is a complete differential and w will be a mere function of s.

For similar reasons, when $\frac{ds}{dx} = 0$ then s is also constant in regard to t, $x - \left(u + \sqrt{\varphi'(\rho)}\right)t$ and w are functions of r.

If finally $\frac{dr}{dx}$ and $\frac{ds}{dx}$ are both equal to zero, then as a result of the differential equations, r, s and w are constants.

3. In order to solve the problem, first we must determine w as a function of r and s so that it will satisfy the differential equation:

(1) $$\frac{d^2w}{drds} - m\left(\frac{dw}{dr} + \frac{dw}{ds}\right) = 0$$

and the initial conditions, which determine it within an arbitrary constant. Where and when some value of x encounters some value of s result from the equation

(2) $$\left(x - \left(u + \sqrt{\varphi'(\rho)}\right)t\right)dr - \left(x - \left(u - \sqrt{\varphi'(\rho)}\right)t\right)ds = dw$$

and then we find u and ρ exclusively as functions of x and t by bringing in the equations:

(3) $$f(\rho) + u = 2r, \quad f(\rho) - u = 2s$$

In fact, when dr or ds are not zero along some finite interval, and consequently r or s are variable, then from (2) we obtain the equations:

(4) $$x - \left(u + \sqrt{\varphi'(\rho)}\right)t = \frac{dw}{dr}$$

(5) $$x - \left(u - \sqrt{\varphi'(\rho)}\right)t = \frac{dw}{ds}$$

Through linkage with (3) we obtain u and ρ expressed in x and t.

But when r initially has the same value r' in a finite interval, then this path gradually moves off toward greater values of x. Within this region, where $r = r'$ from equation (2) we then cannot derive the value of $x - \left(u + \sqrt{\varphi'(\rho)}\right)t$ since $dr = 0$; and in fact the question of where and when this value of r' encounters a particular value of s cannot be definitively answered. Equation (4) then will apply only at the boundaries of this region and indicates between which values of x the constant value r' of r will appear at a particular time, and during which time interval this value of r will exist at a particular location. Between these limits, u and ρ are determined as functions of x and t from equations (3) and (5). By similar means we find these functions when s has the value s' in a finite region, while r is invariant, and also when r and s are both constants. In the latter case they assume constant values arising from (3) between certain limits determined by (4) and (5).

4. Before we proceed with the integration of equation (1) from the previous section, it seems expedient to examine how to carry out this integration. By means of the function $\varphi'(\rho)$, the only assumption that is needed is that its derivative does not decrease with increasing ρ, which in reality is certainly always the case; and let us comment right here -- as will be used in the following section--that $\frac{\varphi(\rho_1)-\varphi(\rho_2)}{\rho_1-\rho_2}=\int_0^1 \varphi'(\alpha\rho_1+(1-\alpha\rho_2))d\alpha$, either remains constant or rises and falls when only one of the quantities ρ_1 and ρ_2 changes, from which it follows that the value of this expression always rests between $\varphi'(\rho_1)$ and $\varphi'(\rho_2)$.

First let us examine the case where the initial disturbance of equilibrium is limited to a finite region bounded by the inequalities $a < x < b$, so that outside of this region u and ρ, and thus also r and s are constant. The values of these quantities for $x < a$ can be denoted by application of the index 1; for $x > b$ by attachment of the index 2. The region where r is invariant, moves gradually forward according to section 1, and of course, its rear bound recedes with the velocity $\sqrt{\varphi'(\rho_1)}+u_1$, while the front boundary of the region where s is invariant, recedes with the velocity $\sqrt{\varphi'(\rho_2)}-u_2$. After passage of time $\frac{\sqrt{\varphi'(\rho_1)}+\sqrt{\varphi'(\rho_2)}+u_1-u_2}{b-a}$ both regions move apart and a gap forms between them where $s = s_2$ and $r = r_1$, and consequently the gas particles are again at equilibrium. From the initially disturbed location, two waves will emanate going in opposite directions. In the forward direction, $s = s_2$; thus with a particular value ρ of density the velocity $u=f(\rho)-2s_2$ will always be attached, and both values move forward with a constant velocity $\sqrt{\varphi'(\rho)}+u=\sqrt{\varphi'(\rho)}+f(\rho)-2s_2$. But in the backward-moving direction, the velocity $-f(\rho)+2r_1$ is connected with the density ρ, and these two values move backward at a velocity $\sqrt{\varphi'(\rho)}+f(\rho)-2r_1$. The rate of propagation is greater for greater densities, since both $\sqrt{\varphi'(\rho)}$ and $f(\rho)$ increase simultaneously with ρ.

Now if we imagine ρ as the ordinate of a curve for the abscissa x, then each point of this curve moves parallel with the abscissa axis at a constant velocity, and of course, the velocity will be all the greater, the larger its ordinate. It is easy to see that under this law, points with larger ordinates will eventually overtake preceding points with smaller ordinates, so that more than one value of ρ would belong to one value of x. Now since in reality this cannot occur, then a circumstance would have to occur where this law will be invalid. But in fact the derivation of the differential equations is based on the fact that u and ρ are constant functions of x and have finite derivatives. But this prerequisite stops being met as soon as the density curve becomes perpendicular to the abscissa axis at any point, and from this moment on a discontinuity occurs at this curve, so that a larger value of ρ will directly follow a smaller one; this is a case that will be discussed in the following section.

The compression wave, that is, the portions of the wave where the density decreases in the direction of propagation, will accordingly become increasingly more narrow as it progresses, and finally goes over into compression shocks; but the width of the expansion or release wave grows proportional with time.

It can be easily shown--at least under the prerequisite of the Poisson (or Boyle's) law--that even when the initial disturbance of equilibrium is not limited to a finite region, that compression shocks will always form in the course of the movement, except for quite special cases. The velocity at which a value of r moves forward, under this assumption is $\frac{k+1}{2}r+\frac{k-3}{2}s$; thus larger values will on average move at greater velocity, and a greater value r' would finally have to overtake a preceding smaller value r'', unless the value of s coincident with r'' is on average smaller by $(r'-r'')\ \frac{1+k}{3-k}$ than the one that coincides simultaneously with r'. In this case s would go negative infinity for a positive, infinite x, and thus for $x = +\infty$ the velocity $u = +\infty$ (or instead of this, the density will be infinitely small according to Boyle's law). Except for special cases, the case would always have to occur that a value of r greater by a finite amount, will directly follow a smaller one; consequently, as $\frac{dr}{dx}$ goes to infinity, the differential equations will lose their validity and forward-moving compression shocks would have to appear. Likewise almost always backward moving compression shocks will form since $\frac{ds}{dx}$ goes to infinity.

To determine the times and locations for which $\frac{dr}{dx}$ or $\frac{ds}{dx}$ go to infinity and sudden compressions begin, we obtain the following expressions from equations (1) and (2) from section 2 by inserting the function w therein:

$$\frac{dr}{dx}\left(\frac{d^2w}{dr^2}+\left(\frac{d\log\sqrt{\varphi'(\rho)}}{d\log\rho}+1\right)t\right)=1$$

$$\frac{ds}{dx}\left(-\frac{d^2w}{ds^2}+\left(\frac{d\log\sqrt{\varphi'(\rho)}}{d\log\rho}+1\right)t\right)=1$$

5. Now since sudden compressions almost always set in, even when the initial density and velocity vary continuously everywhere, we must seek the laws for propagation of compression shocks.

We assume that at time t for $x = \xi$ a sudden change in u and ρ will occur, and let us call their values and the quantities dependent on them for $x = \xi - 0$ by appending the index 1, and for $x = \xi + 0$ by the index 2. The relative velocities of the gas moving toward the unsteady state $u_1 - \frac{d\xi}{dt}, u_2 - \frac{d\xi}{dt}$, will be denoted by v_1 and v_2 The mass passing through an element w of the plane, where $x = \xi$, in the positive direction in time element dt, is then $v_1 \rho_1 w dt = v_2 \rho_2 w dt$; the acting force $\left(\varphi(\rho_1) - \varphi(\rho_2)\right) w dt$ and the resultant increase in velocity $v_2 - v_1$; thus we have $\left(\varphi(\rho_1) - \varphi(\rho_2)\right) w dt = (v_2 - v_1) v_1 \rho_1 w dt$ and $v_1 \rho_1 = v_2 \rho_2$, from which follows $v_1 = \mp \sqrt{\frac{\rho_2}{\rho_1} \frac{\varphi(\rho_1) - \varphi(\rho_2)}{\rho_1 - \rho_2}}$, therefore

$$\text{(1)} \qquad \frac{d\xi}{dt} = u_1 \pm \sqrt{\frac{\rho_2}{\rho_1} \frac{\varphi(\rho_1) - \varphi(\rho_2)}{\rho_1 - \rho_2}} = u_2 \pm \sqrt{\frac{\rho_1}{\rho_2} \frac{\varphi(\rho_1) - \varphi(\rho_2)}{\rho_1 - \rho_2}}$$

For a compression shock, $\rho_2 - \rho_1$ must have the same sign as v_1 and v_2, and of course for the forward-moving shock it will be negative, for a backward-moving shock it will be positive. In the first case the upper sign will apply and ρ_l is greater than ρ_2. Therefore, under the assumption made at the beginning of the preceding article regarding the function $\varphi(\rho)$ we have

$$\text{(2)} \qquad u_1 + \sqrt{\varphi'(\rho_1)} > \frac{d\xi}{dt} > u_2 + \sqrt{\varphi'(\rho_2)}$$

and consequently the discontinuity point will move more slowly way than the following one, and will move faster than the preceding values of r. Thus r_l and r_2 are determined at any moment by means of the differential equations applicable on both sides of the discontinuity point. The same will apply also for s_2 and consequently for ρ_2 and u_2, since the values of s go backward with the velocity $\sqrt{\varphi'(\rho)} - u$, but not for s_l. The values of s_l and $\frac{d\xi}{dt}$ are determined from r_l, ρ_2 and u_2 are determined uniquely through the equations (1). In fact, the equation:

$$\text{(3)} \qquad 2(r_1 - r_2) = f(\rho_1) - f(\rho_2) + \sqrt{\frac{(\rho_1 - \rho_2)\left(\varphi(\rho_1) - \varphi(\rho_2)\right)}{\rho_1 \rho_2}}$$

will only satisfy a value of ρ_1, because when ρ_l grows from ρ_2 to infinity, the right side will take on every positive value, since both $f(\rho_1)$ and also the two factors $\sqrt{\frac{\rho_1}{\rho_2}}-\sqrt{\frac{\rho_2}{\rho_1}}$ and $\sqrt{\frac{\varphi(\rho_1)-\varphi(\rho_2)}{\rho_1-\rho_2}}$, in which the last term can be broken down, grows steadily or only the latter factor stays constant. But when ρ_l is determined, by means of equation (1), we obtain evidently completely determined values for u_l, and $\frac{d\xi}{dt}$.

A backward-moving compression shock will behave quite similarly.

6. Now we have found that in a propagating compression shock between the values of u and ρ on both sides of it, the equation $(u_1-u_2)^2=\frac{(\rho_1-\rho_2)(\varphi(\rho_1)-\varphi(\rho_2))}{\rho_1\rho_2}$ will always appear. Now the question arises of what will happen if at a given time and at a given position, random discontinuities are present. Then from this site, depending on the values of u_1,ρ_1,u_2,ρ_2, either two compression shocks will emanate in opposing directions, or a single forward-moving or a backward-moving shock, or finally no compression shock at all will occur. In the last case movement will thus take place according to the differential equations.

If we call the values that u and ρ take on after or between the compression shocks in the first moment of propagation, by adding in a prime, then in the first case $\rho' > \rho_1$ and $> \rho_2$, and we have:

(1)

$$u_1-u'=\sqrt{\frac{(\rho'-\rho_1)(\varphi(\rho')-\varphi(\rho_1))}{\rho'\rho_1}},\quad u'-u_2=\sqrt{\frac{(\rho'-\rho_2)(\varphi(\rho')-\varphi(\rho_2))}{\rho'\rho_2}}$$

$$\text{(2)}\quad u_1-u_2=\sqrt{\frac{(\rho'-\rho_1)(\varphi(\rho')-\varphi(\rho_1))}{\rho'\rho_1}}+\sqrt{\frac{(\rho'-\rho_2)(\varphi(\rho')-\varphi(\rho_2))}{\rho'\rho_2}}.$$

Thus since both terms on the right side of (2) increase with ρ' and u_l-u_2 is positive and $(u_1-u_2)^2 > \frac{(\rho_1-\rho_2)(\varphi(\rho_1)-\varphi'(\rho_2))}{\rho'\rho_2}$; and conversely, when these conditions are met, always one and only one pair of values of u' and ρ' are obtained that will satisfy equation (1).

In order for the latter case to occur, and thus the motion be determined according to the differential equations, it is necessary and sufficient that $r_l < r_2$

and $s_1 \geq s_2$, that is, $u_1 - u_2$ is negative and $(u_1 - u_2)^2 \geq (f(\rho_1) - f(\rho_2))^2$. The values r_1 and r_2 , s_1 and s_2 will move apart since the preceding value advances with a greater speed, so that the discontinuity will disappear.

If neither the first nor the last condition is met, then the initial values will satisfy one compression shock, and of course either a forward or backward-moving shock will occur, depending on whether ρ_1 is greater or smaller than ρ_2.

In fact when $\rho_1 > \rho_2$, $2(r_1 - r_2)$ or $f(\rho_1) - f(\rho_2) + u_1 - u_2$ is positive, because $(u_1 - u_2)^2 < (f(\rho_1) - f(\rho_2))^2$, and at the same time,

$$\leq f(\rho_1) - f(\rho_2) + V \frac{(\rho_1 - \rho_2)(\varphi(\rho_1) - \varphi(\rho_2))}{\rho_1 \rho_2},$$

because

$$(u_1 - u_2)^2 \leq \frac{(\rho_1 - \rho_2)(\varphi(\rho_1) - \varphi(\rho_2))}{\rho_1 \rho_2};$$

thus for the density ρ' after the compression shock a value that satisfies the conditions (3) of the previous section will be found and it is $\geq \rho_1$. Consequently, since $s' = f(\rho') - r_1$, $s_1 = f(\rho_1) - r_1$, also $s' < s_1$, so that the motion after the compression shock can occur according to the differential equations.

The other case, when $\rho_1 < \rho_2$, is evidently quite different from this.

7. In order to explain the foregoing with a simple example, where the movement can be determined with the derived equations, let us assume that the pressure and density will depend on each other according to Boyle's law and the initial density and velocity change suddenly at $x = 0$, but are constant on both sides of this location.

Now we make a distinction between the four cases above.

I. When $u_1 - u_2 > 0$, that is, when the two gas masses move toward each other and $\left(\frac{u_1 - u_2}{a}\right)^2 > \left(\frac{\rho_1 - \rho_2}{\rho_1 \rho_2}\right)^2$, then two oppositely moving compression shocks will form. According to Section 6 (1), when $\sqrt[4]{\rho_1 / \rho_2}$ is denoted by a

and the positive root of the equation $\frac{u_1 - u_2}{a\left(a + \frac{1}{a}\right)} = \theta - \frac{1}{\theta}$ is denoted by θ, the density between the compression shocks $\rho' = \theta\theta\sqrt{\rho_1\rho_2}$, and according to Section 5 (1) for the forward moving compression shock, we have:

$$\frac{d\xi}{dt} = u_2 + a\alpha\theta = u' + \frac{a}{\alpha\theta},$$

and for the backward-moving compression shock:

$$\frac{d\xi}{dt} = u_1 - a\frac{\theta}{\alpha} = u' - a\frac{\alpha}{\theta}$$

Thus the values for velocity and density with respect to time t, when $\left(u_1 - a\frac{\theta}{\alpha}\right)t < x < (u_2 + a\alpha\theta)t$, are u' and ρ'. For a smaller x they are u_1 and ρ_1 and for a larger x they are u_2 and ρ_2.

II. When $u_1 - u_2 < 0$, consequently the gas masses move apart, and at the same time $\left(\frac{u_1 - u_2}{a}\right)^2 > \left(\log\frac{\rho_1}{\rho_2}\right)^2$, then two gradually broadening expansion or release waves move out in opposite directions from the boundary. According to Section 4, between them we have $r = r_1$, $s = s_2$, $u = r_1 - s_2$. In the forward moving wave, $s = s_2$, and $x - (u + a)t$ is a function of r, whose value is equal to zero from the initial values $t = 0, x = 0$. But for the backward running wave, we have $r = r_1$ and $x - (u - a)t = 0$. The one equation to determine u and ρ, when $(r_1 - s_2 + a)t < x < (u_2 + a)t$, is $u = -a + x/t$, for smaller values of x, $r = r_1$ and for larger values $r = r_2$; the other equation, when $(u_1 - a)t < x < (r_1 - s_2 - a)t$ is $u = a + x/t$, for a smaller x, $s = s_1$ and for a larger x, $s = s_2$.

III. If neither of these two cases occurs and $\rho_1 > \rho_2$, then a backward-running dilution wave and a forward-moving compression shock occurs. For the latter from Section 5 (3), when θ is the root of the equation $\frac{2(r_1 - r_2)}{a} = 2\log\theta + \theta - \frac{1}{\theta}$, then $\rho' = \theta\theta\rho_2$ and from Section 5 (1), we have $\frac{d\xi}{dt} = u_2 + a\theta = u' + \frac{a}{\theta}$. After passage of time t, before the compression shock, that is, when $x > (u_2 + a\theta)t$, then $u = u_2$, $\rho = \rho_2$. But behind the compression shock we have $r = r_1$ and in addition, when $(u_1 - a)t < x < (u' - a)t$, we have $u = a + x/t$; for a smaller x we have $u = u_1$ and for a larger x, we have $u = u'$.

IV. Now finally, if the first two cases do not occur and $\rho_1 < \rho_2$ then the profile is exactly as in III, only the direction is changed.

8. Now to solve our problem in general, according to section 3 the function w must be determined so that it satisfies the differential equation

$$\text{(1)} \qquad \frac{d^2 w}{dr ds} - m\left(\frac{dw}{dr} + \frac{dw}{ds}\right) = 0$$

and the initial conditions.

If we eliminate the case where discontinuities occur, then evidently according to Section 1 the values of x and t for which a particular value of r' of r encounters with a particular value s' of s, are fully determined when the initial values of r and s (between r = r' and s = s') are given over the x-interval, and when the differential equations (3) of Section 1 hold everywhere in the whole region (S) which consists of all intermediate values of x between the one where r = r' and the one where s = s', for any value of t. Therefore the vlaue of w for r = r' and s = s' is also completely determined, when w satisfies thedifferential equation everywhere in (S), and when the values of $\frac{dw}{dr}$ and $\frac{dw}{ds}$ are given for the initial values of r and s — which determines the initial value of w, except for an arbitrary additive constant. Actually, these conditions are equivalent to those above. From Section 3 it also follows that $\frac{dw}{dr}$ will take on different values on both sides of a value r'' of r when this value occurs in a finite distance, but changes steadily with s; likewise $\frac{dw}{ds}$ changes with r, but the function w itself changes steadily with r and also with s.

After these preparations we can now solve our problem, to determine the value of w for two random values, r' and s' of r and s.

For illustration let us imagine x and t as the abscissa and ordinate of a point in a plane and draw the curves in this plane, where r and where s have constant values. Of these curves, the first one is designated by (r), the latter one by (s), and in them the direction in which t increases, is viewed as positive. The large region (S) will then be represented by a piece of the plane that is limited by the curve (r'), the curve (s') and the piece of the abscissa axis located between the two; we are interested in determining the value of w in the points of intersection of the two former values given in the latter line. Let us generalize the problem somewhat and assume that the large region (S) is bounded by an arbitrary curve c through this latter line, and none of the curves (r) and (s) intersects more than once, and that the pair of values of r and s belonging to this curve are given by the values of $\frac{dw}{dr}$ and $\frac{dw}{ds}$. As will be seen from the solution of the problem, these values of $\frac{dw}{dr}$ and $\frac{dw}{ds}$ are subject only to the condition of changing steadily with the location in the curve, and can be taken arbitrarily. These values would not be independent of each other when the curve c intersects one of curves (r) or (s) more than once.

In order to determine functions that are to satisfy linear partial differential equations and linear boundary conditions, we can use an entirely similar method. This is done by multiplying all equations by arbitrary factors, adding them up and then determining these factors so that from the sum all known quantities expect one will drop out.

Let us image the piece (S) of the plane as being cut by the curves (r) and (s) into an infinite number of small parallelograms and use δr and δs to denote changes that the quantities r and s undergo when the curve elements that form the sides of these parallelograms are transited in a positive direction. Furthermore, we call v an arbitrary function of r and s that is continuous and has continuous derivatives. As a result of equation (1) we then have

$$(2) \qquad 0=\int v\left(\frac{d^2w}{drds}-m\left(\frac{dw}{dr}+\frac{dw}{ds}\right)\right)\delta r\delta s$$

expanded across the entire large region (S). Now the right side of this equation must be arranged according to the unknowns; that is, the integral will be changed by partial integration so that besides the known quantities, only the desired function, and not its derivatives are contained. During execution of this operation, the integral first goes over into the integral expanded across (S):

$$\int w\left(\frac{d^2v}{drds}+\frac{dmv}{dr}+\frac{dmv}{ds}\right)\delta r\delta s$$

and a simple integral that extends only across the boundary of (S) because $\frac{dw}{dr}$ changes with s, $\frac{dw}{ds}$ changes with r, and w changes steadily with both quantities. If dr and ds mean the changes in r and s in a boundary element when the boundary is transited in the direction that is opposite the direction inward, like the positive direction in the curves (r) is opposite the positive direction in the curves (s), then this boundary integral is equal to

$$-\int\left(v\left(\frac{dw}{ds}-mw\right)ds+w\left(\frac{dv}{dr}+mv\right)dr\right).$$

The integral through the entire boundary of S is equal to the sum of the integrals through the curves c, (s'), (r') that form this boundary, that is, when their points of intersection are designated by (c, r'), (c, s'), (r', s')

$$\int_{c,r'}^{c,s'} + \int_{c,s'}^{r',s'} + \int_{s',r'}^{c,r'}$$

Of these three terms, the first one contains only known quantities besides the function v; the second contains only the unknown function w itself, since in it $ds = 0$, and not its derivatives; the third term can be converted through partial integration into

$$(vw)_{r',s}-(vw)_{c,r'}+\int_{s',r'}^{c,r'} w\left(\frac{dv}{ds^2}+mv\right)ds$$

so that it likewise will contain only the desired function w itself.

After these transformations equation (2) evidently provides the value of the function w at point (r', s') expressed by known quantities, when the function v is determined according to the following conditions:

(3)

1) Everywhere in S: $\dfrac{d^2v}{drds}+\dfrac{dmv}{dr}+\dfrac{dmv}{ds}=0$

2) For $r=r'$: $\dfrac{dv}{ds}+mv=0$

3) For $s=s'$: $\dfrac{dv}{dr}+mv=0$

4) For $r=r'$, $s=s'$: $v=1$

We then have

(4) $$w_{r',\,s'}=(vw)_{c,r'}+\int_{c,r'}^{c,s'}\left(v\left(\frac{dw}{ds}-mw\right)ds+w\left(\frac{dv}{dr}+mv\right)dr\right)$$

9. By means of the method just described, the problem becomes one of determining a function w of a linear differential equation and linear boundary conditions and applying it to the solution of a similar, but much simpler problem for another function v. Determination of this function is usually attained most easily by treatment of one special case using the Fourier method. We will have to be satisfied here merely with pointing out the calculation, but finding the result by other means.

If we substitute $\sigma = r + s$ and $u = r - s$ into equation (1) of the previous section for r and s as independent, variable quantities, and for the curve c we select a curve in which σ is constant, then the problem can be treated according to Fourier rules, and by comparison of the result with equation (4) of the previous section, when $r' + s' = \sigma$, $r' - s' = u'$ we obtain:

$$v = \frac{2}{\pi}\int_o^\infty \cos\ \mu(u - u')\frac{d\rho}{d\sigma}\left(\psi_1(\sigma')\psi_2(\sigma) - \psi_2(\sigma')\psi_1(\sigma)\right)d\mu$$

wherein $\psi_1(\sigma)$ and $\psi_2(\sigma)$are two such particular solutions of the differential equation $\psi'' - 2m\psi' + \mu\mu\psi = 0$, that

$$\psi_1\psi_2' - \psi_2\psi_1' = \frac{d\sigma}{d\rho}$$

If we assume the Poisson law, according to which $m = \left(\frac{1}{2} - \frac{1}{k-1}\right)\frac{1}{\sigma}$, then we can express ψ_1 and ψ_2 by certain integrals so that for v we obtain a three-fold integral that can be reduced to:

$$v = \left(\frac{r' + s'}{r + s}\right)^{\frac{1}{2} - \frac{1}{k-1}} \mathrm{F}\left(\frac{3}{2} - \frac{1}{k\ \text{-}1}, \frac{1}{k-1} - \frac{1}{2}, 1, -\frac{(r - r')(s - s')}{(r + s)(r' + s')}\right).$$

Now we can easily demonstrate the accuracy of this expression by showing that it actually does satisfy the conditions (3) of the previous section.

If we set $v = y\exp\left(-\int_{\sigma'}^{\sigma} mds\right)$ then for y they go over into $\frac{d^2y}{drds} + \left(\frac{dm}{d\sigma} - mm\right)y = 0$ and $y = 1$, both for $r = r'$ and also for $s = s'$. Under the Poisson assumption, these conditions can be satisfied when we assume that y is a function of $z = -\frac{(r - r')(s - s')}{(r + s)(r' + s')}$. Because if we denote $\frac{1}{2} - \frac{1}{k-1}$ by λ then $m = \frac{\lambda}{\sigma}$, and thus $\frac{dm}{d\sigma} - mm = -\frac{\lambda + \lambda^2}{\sigma^2}$ and $\frac{d^2y}{drds} = \frac{1}{\sigma^2}\left(\frac{d^2y}{d\log z^2}\left(1 - \frac{1}{z}\right) + \frac{dy}{d\log z}\right)$. Consequently we have $v = \left(\frac{\sigma}{\sigma'}\right)^\lambda y$ and y is a solution of the differential equation

$$(1 - z)\frac{d^2y}{d\log z^2} - z\frac{dy}{d\log z} + \left(\lambda + \lambda^2\right)zy = 0$$

or according to the designation used in my treatment of the Gaussian series, we have a function

$$P\begin{pmatrix} 0 & -\lambda & 0 & \\ 0 & 1+\lambda & 0 & z \end{pmatrix}$$

and of course, that particular solution is equal to 1 when $z = 0$.

According to the transformation principles developed in that paper, y cannot be expressed merely by the functions $P(0,\, 2\lambda + 1,\, 0)$ but also by the functions $P(1/2,\, 0,\, \lambda + 1/2)$, $P(0,\, \lambda + 1/2,\, \lambda + 1/2)$. Thus we obtain for y a large quantity of representations through hyper-geometric series and certain integrals of which we will comment only on the following ones:

$$y = F(1+\lambda, -\lambda, 1, z) = (1-z)^{\lambda} F\left(-\lambda, -\lambda, 1, \frac{z}{z-1}\right)$$
$$= (1-z)^{-1-\lambda} F\left(1+\lambda, 1+\lambda, 1, \frac{z}{z-1}\right)$$

that will be sufficient in all cases.

In order to derive the results applicable for Boyle's law from these results found for the Poisson law, according to section 2 the quantities r, s, r', s' must be reduced by $\dfrac{a\sqrt{k}}{k-1}$ and then we can let k = 1, so that we obtain $m = -\dfrac{1}{2a}$ and

$$v = e^{\frac{1}{2a}(r-r'+s-s')} \sum_{0}^{\infty} \frac{(r-r')^n (s-s')^n}{n!n!(2a)^{2n}}.$$

10. If we substitute the expression for v found in the previous section into equation (4) of section 8, then we obtain the value of w for $r = r'$, $s = s'$ through the value of w, $\dfrac{dw}{dr}$ and $\dfrac{dw}{ds}$ expressed in curve c; but since in our problem in this curve $\dfrac{dw}{dr}$ and $\dfrac{dw}{ds}$ are always directly given and w has to be found from a square root, then it is expedient to transform the expression for $w_{r',s'}$ so that under the integral sign only the derivative of w will appear.

Let us call the integral of the expressions $-mvds + \left(\dfrac{dv}{dr} + mv\right)dr$ and $\left(\dfrac{dv}{ds} + mv\right)ds - mvdr$, that are complete differentials as a result of the equation

$\frac{d^2v}{drds}+\frac{dmv}{dr}+\frac{dmv}{ds}=0$, as P and Σ; likewise the integral of Pdr + Σds is a complete differential due to $\frac{dP}{ds}=-mv=\frac{d\Sigma}{dr}$.

If we determine the integration constants in these integrals so that $\omega, \frac{d\omega}{dr}$ and $\frac{d\omega}{ds}$ disappear for $r=r', s=s'$, then w satisfies the equations $\frac{d\omega}{dr}+\frac{d\omega}{ds}+1=v$, $\frac{d^2\omega}{drds}=-mv$. Both for $r=r'$, and also $s=s'$ for the equation, w = *0* and it is fully determined by this boundary condition and by the differential equation $\frac{d^2\omega}{drds}+m\left(\frac{d\omega}{dr}+\frac{d\omega}{ds}+1\right)=0$.

Now if we insert the function w in the expression of $w_{r',s'}$ for v, then we can convert it by partial integration into

$$w_{r',s'}=w_{c,r'}+\int_{c,r'}^{c,s'}\left(\frac{d\omega}{ds}+1\right)\frac{dw}{ds}ds-\frac{d\omega}{dr}\frac{dw}{dr}dr \tag{1}$$

In order to determine the motion of the gas from the initial state, we have to take the curve c in which t = 0; in this curve we then have $\frac{dw}{dr}=x, \frac{dw}{ds}=-x$, and by means of repeated partial integration we have

$$w_{r',\,s'}=w_{c,\,r'}+\int_{c,r'}^{c,s'}(wdx-xds)$$

consequently from Section 3, (4) and (5) we have

$$\begin{aligned}\left(x-\left(\sqrt{\varphi'(\rho)}+u\right)t\right)_{r',s'}&=x_{r'}+\int_{x_{r'}}^{x_{s'}}\frac{d\omega}{dr'}dx\\ \left(x+\left(\sqrt{\varphi'(\rho)}-u\right)t\right)_{r',s'}&=x_{s'}-\int_{x_{r'}}^{x_{s'}}\frac{d\omega}{ds'}dx\end{aligned} \tag{2}$$

But these equations (2) only express the motion as long as $\frac{d^2w}{dr^2}+\left(\frac{d\log\sqrt{\varphi'(\rho)}}{d\log\rho}+1\right)t$ and $-\frac{d^2w}{ds^2}+\left(\frac{d\log\sqrt{\varphi'(\rho)}}{d\log\rho}+1\right)t$ are not equal to zero. As soon as one of these quantities disappears, a compression shock will occur, and equation (2) will only apply within that region that rests on one and the same side of this compression shock. The principles developed here in general will not be sufficient to determine the movement from the initial state;

but of course, by means of equation (1) and the equations obtained for the compression shock in Section 5, we can determine the movement when the site of the compression shock is known at time t, that is ξ is a function of t. We will not pursue this further and likewise will ignore a treatment of the case when the air is bounded by a fixed wall, since the calculation presents no difficulties and a comparison of the results with experiment is presently impossible.

Rankine

WILLIAM JOHN MACQUORN RANKINE

W. J. Macquorn Rankine was a Scottish engineer. He was born in Edinburgh on July 5, 1820, and was probably descended from Robert the Bruce. His father was David Rankine who was employed, later in life, in the construction of railways. Rankine, the son, repeatedly credits his father with much useful instruction as a youth. This consisted of his first introduction to arithmetic, mathematics, mechanics and physics. He attended Ayr Academy from 1828 to 1829 and then Glasgow High School in 1830. In 1834 his uncle, Archibald Grahame, gave him a copy of Newton's *Principia*. Rankine read this carefully and obtained his knowledge of higher mathematics mainly by private study.

In November of 1836 he entered Edinburgh University. During this period he studied natural philosophy and botany. He also claims to have wasted a great deal of time studying the "theory of numbers." He left Edinburgh University in 1838 without a degree. It was during the years of 1837 and 1838 that he entered the practice of engineering, working with his father in superintending the works of the Leith branch of the Edinburgh and Dalkeith Railway. After leaving Edinburgh University he was employed for four years on engineering projects in Ireland. His first published work entitled "An Experimental Inquiry into the advantage of Cylindrical Wheels on Railways" appeared in 1842.

In 1849 Rankine began to publish papers in the physical sciences on "the hypothesis of molecular vortices." Here matter was assumed to be composed of atoms; nuclei surrounded by atmospheres of innumerable vortices of matter that were elastic and automatically tended to increase their volume. This model formed the basis of several of Rankine's theories for the behavior of matter, particularly the theory of heat. Rankine is considered the founder of the "science of energetics." One of Rankine's final applications of thermodynamics appeared in 1870 (the paper that follows), when he examined the propagation of finite longitudinal disturbances, later commonly called shock waves. This theory took into account the effects of heat conduction in the propagation of finite disturbances in perfect gases. These equations were later generalized independently by Hugoniot in 1889 and became known as the Rankine-Hugoniot relations.

Thus, for Rankine, the theory of finite-amplitude wave propagation was only a very small part of his life's work. Most of this work had been in the field of engineering sciences. Rankine insisted that the application of pure science to engineering constituted a separate academic subject in its own right. Rankine published a series of highly successful engineering textbooks: *A Manual of Applied Mechanics* (1858); *A Manual of the Steam Engine and Other Prime Movers* (1859); *A Manual of Civil Engineering* (1862); *A Manual of Machinery and Millwork* (1869). Another of Rankine's major interests was naval architecture. With J.R. Napier, a shipbuilder, and others, Rankine wrote *Shipbuilding, Theoretical and Practical* (1860).

According to P.G. Tait, a contemporary, "his appearance was striking and preposterous in the extreme, and his courtesy resembled almost that of a gentleman of the old school." Tait recalls that Rankine had a highly cultivated musical taste. It was considered a great pleasure to see him take his seat at the piano to accompany himself as he sang some humorous or grotesquely plaintive song; the words and music generally being of his own composition.

It is probably clear from this brief description that Rankine was an extremely creative and energetic man. While the study of finite-amplitude sound waves was only a very small part of contribution to science, it nevertheless helped form the basis for a unique discipline approximately a century later.

Rankine died on December 24, 1872, of a general decline in health rather than of any special disease, still a relatively young man by today's standards.

On The Thermodynamic Theory of Waves of Finite Longitudinal Disturbance

W. J. Macquorn Rankine, C.E., LL.D., F.R.SS. Lond. & Edin., &c., *Philosophical Transactions,* Vol. 160, Part II, pp. 277-288 (1870)

§1. The object of the present investigation is to determine the relations which must exist between the laws of the elasticity of any substance, whether gaseous, liquid, or solid, and those of the wave-like propagation of a finite longitudinal disturbance in that substance; in other words, of a disturbance consisting in displacements of particles along the direction of propagation, the velocity of displacement of the particles being so great that it is not to be neglected in comparison with the velocity of propagation. In particular, the investigation aims at ascertaining what conditions as to the transfer of heat from particle to particle must be fulfilled in order that a finite longitudinal disturbance may be propa-gated along a prismatic or cylindrical mass without loss of energy or change of type: the word *type* being used to denote the relation between the extent of disturbance at a given instant of a set of particles, and their respective undisturbed positions. The disturbed matter in these inquiries may be conceived to be contained in a straight tube of uniform cross-section and indefinite length.

§2. *Mass-Velocity.*—A convenient quantity in the present investigation is what may be termed the *mass-velocity* or *somatic velocity*—that is to say, the mass of matter through which a disturbance is propagated in a unit of time while advancing along a prism of the sectional area unity. That mass-velocity will be denoted by m.

Let S denote the *bulkiness*, or the space filled by unity of mass, of the substance in the undisturbed state, and a the linear velocity of advance of the wave; then we have evidently

$$a = mS \tag{1}$$

§3. *Cinematical Condition of Permanency of Type.*—If it be possible for a wave of disturbance to be propagated in a uniform tube without change of type, that possibility is expressed by the uniformity of the mass velocity m for all parts of the wave.

Conceive a space in the supposed tube, of an invariable length Δx, to be contained between a pair of transverse planes, and let those planes advance with the linear velocity a in the direction of propagation. Let the values of the bulkiness of the matter at the foremost and aftermost planes respectively be denoted by s_1 and s_2, and those of the velocity of longitudinal disturbance by u_1 and u_2. Then the linear velocities with which the particles traverse the two planes respectively are as follows: for the foremost plane $u_1 - a$, for the after-most plane $u_2 - a$. The uniformity of type of the disturbance involves as a condition, that equal masses of matter traverse the two planes respectively in a given time, being each, in unity of time, expressed by the *mass-velocity*; hence we have, as the *cinematical condition* of uniformity of type, the following equation:

$$\frac{a-u_1}{s_1} = \frac{a-u_2}{s_2} = \frac{a}{S} = m \tag{2}$$

Another way of expressing the same condition is as follows:

$$\Delta u = -m\Delta s \quad . \tag{3}$$

§4. *Dynamical Condition of Permanency of Type.*—Let p_1 and p_2 be the intensities of the longitudinal pressure at the foremost and aftermost advancing planes respectively. Then in each unit of time the difference of pressure, $p_2 - p_1$, impresses on the mass m the acceleration $u_2 - u_1$, and consequently, by the second law of motion, we have the following value for the difference of pressure:

$$p_2 - p_1 = m(u_2 - u_1) \quad . \tag{4}$$

Then substituting for the acceleration $u_2 - u_1$ its value in terms of the change of bulkiness as given by equation (3), we obtain, for the *dynamical condition* of permanency of type, the following equation,

$$p_2 - p_1 = m^2(s_1 - s_2) \quad , \tag{5}$$

which may also be put in the form of an expression giving the value of the square of the mass-velocity, viz.

$$m^2 = -\frac{\Delta p}{\Delta s} = -\frac{dp}{ds} \quad . \tag{6}$$

The square of the linear velocity of advance is given by the following equation:

$$a^2 = m^2 S^2 = -S^2 \frac{dp}{ds} \quad . \tag{7}$$

The integral form of the preceding equations may be expressed as follows. Let S, as before, be the bulkiness in the undisturbed sate, and P the longitudinal pressure; then in a wave of disturbance of permanent type, we must have the following condition fulfilled:

$$p + m^2 s = P + m^2 S \quad . \tag{8}$$

§5. *Waves of Sudden Disturbance.*—The condition expressed by the equations of the preceding section holds for any type of disturbance, continuous or discontinuous, gradual or abrupt. To represent, in particular, the case of a single abrupt disturbance, we must conceive the foremost and aftermost advancing planes already mentioned to coalesce into one. Then P is the longitudinal pressure, and S the bulkiness, in front of the advancing plane; p is the longitudinal pressure, and s the bulkiness, behind the advancing plane, and the advancing plane is a wave-front of *sudden compression or of sudden rarefaction** according as *p* is greater or less than P. The squares of the *mass-velocity* and of the linear velocity of advance are respectively as follows:

$$m^2 = \frac{p-P}{S-s} \quad ; \tag{9}$$

$$a^2 = m^2 S^2 = \frac{p-P}{S-s} \cdot S^2 \quad . \tag{10}$$

The velocity of the disturbed particles is as follows:

$$u = m(S-s) = \frac{p-P}{m} = \sqrt{(p-P)\cdot(S-s)} \quad ; \tag{11}$$

and it is forward or backward according as the wave is one of compression or of rarefaction.

---

* (Note, added lst August, 1870.) Sir William Thomson has pointed out to the author, that a wave of sudden rarefaction, though mathematically possible, is an unstable condition of motion; any deviation from absolute suddeness tending to make the disturbance become more and more gradual. Hence the only wave of sudden disturbance whose permanency of type is physically possible, is one of sudden compression; and this is to be taken into account in connection with all that is stated in the paper respecting such waves.

The energy expended in unity of time, in producing any such wave, is expressed by pu; for the wave may be conceived to be produced in a tube closed at one end by a moveable piston of inappreciable mass, to which there is applied a pressure p different from the undisturbed pressure P, and which consequently moves with the velocity u. The way in which that energy is disposed of is as follows: actual energy of the disturbance $\frac{mu^2}{2}$; work done in altering bulkiness, $\frac{m(p+P)(S-s)}{2}$; and the equation of the conservation of energy is

$$pu = \frac{m}{2}\left\{u^2 + (p+P)(S-s)\right\} \ . \tag{11A}$$

§6. *Thermodynamic Conditions.*—While the equations of the two preceding sections impose the constancy of the rate of variation of pressure with bulkiness during the disturbance $\left(\frac{dp}{ds} = -m^2\right)$ as an indispensable condition of permanency of type of the wave, they leave the limits of pressure and of bulkiness, being four quantities, connected by one equation only $\left(\frac{p_2 - p_1}{s_1 - s_2} = -\frac{dp}{ds} = m^2\right)$. Two only of those quantities can be arbitrary; therefore one more equation is required, and that is to be determined by the aid of the laws of thermodynamics.

It is to be observed, in the first place, that no substance yet known fulfills the condition expressed by the equation $\frac{dp}{ds} = -m^2 =$ constant, between finite limits of disturbance, at a constant temperature, nor in a state of non-conduction of heat (called the *adiabatic* state). In order, then that permanency of type may be possible in a wave of longitudinal disturbance, there must be both change of temperature and conduction of heat during the disturbance.

The cylindrical or prismatic tube in which the disturbance is supposed to take place being ideal, is to be considered as non-conducting. Also, the foremost and aftermost transverse advancing planes, or front and back of the wave, which contain between them the particles whose pressure and bulkiness are in the act of varying, are to be considered as non-conducting, because of there being an indefinite length of matter before the foremost and behind the aftermost plane, to resist, conduction.

The transfer of heat, therefore, takes place wholly amongst the particles undergoing variation of pressure and bulkiness; and therefore for any given particle, during its passage from the front to the back of the wave, the *integral amount of heat received must be nothing*; and this is the thermodynamic condition which gives the required equation. That equation is expressed as follows:

$$\int_{\varphi_1}^{\varphi_2} \tau d\varphi = 0 \quad ; \tag{12}$$

in which τ denotes absolute temperature, and φ the "thermodynamic function." The value of that function, as explained in various papers and treatises on thermodynamics, is given by the following formula:

$$\varphi = Jc\ hyp \log \tau + \chi(\tau) + \left(\frac{dU}{d\tau}\right) , \tag{12A}$$

in which J is the dynamical value of a unit of heat; *c*, the real specific heat of the substance; $\chi(\tau)$, a function of the temperature alone, which is = 0 for all temperatures at which the substance is capable of approximating indefinitely to the perfectly gaseous state, and is introduced into the formula solely to provide for the possible existence of substances which at some temperatures are incapable of approximating to the perfectly gaseous state; and U, the work which the elastic forces in unity of mass are capable of doing at the constant temperature τ. The substitution for the integral in equation (12) of its value in terms of p and s for any particular substance, gives a relation between the limits of pressure p_1 and p_2, and the limits of bulkiness s_1 and s_2, which being combined with equation (5), or with any one of the equivalent equations (6), (8), or (9), completes the expression of the laws of the propagation of waves of finite longitudinal disturbance and permanent type in that particular substance.

§7. *Assumption as to Transfer of Heat.*—In applying the principles of the preceding section to the propagation of waves of longitudinal disturbance, it is obviously assumed that the transfer of heat takes place between the various particles which are undergoing disturbance at a given time, in such a manner as to ensure the fulfillment of the dynamical condition of permanency of type. It appears highly probable, that how great soever the resistance of the substance to the conduction of heat may be, that assumption as to the transfer is realized when the disturbance is *sudden*, as described in §5; for then particles in all the successive stages of the change of pressure and bulkiness within the limits of the disturbance are at inappreciable distances from each other; so that the resistance to the transfer of heat between them is inappreciable.

But when the disturbance is not sudden, it is probable that the assumption as to the transfer of heat is fulfilled in an approximate manner only; and if such is the case, it follows that *the only longitudinal disturbance which can be propagated with absolute permanence of type is a sudden disturbance.*

§8. *Combination of the Dynamic and Thermodynamic Equations.*—In every fluid, and probably in many solids, the quantity of heat received during an indefinitely small change of pressure dp and of bulkiness ds is capable of being expressed in either of the following forms:

$$\frac{\tau d\varphi}{J} = c_s \frac{d\tau}{dp} dp + c_p \frac{d\tau}{ds} ds \quad ;$$

in which c_s and c_p denote the specific heat at constant bulkiness and at constant pressure respectively; and the differential coefficients $\frac{d\tau}{dp}$ and $\frac{d\tau}{ds}$ of the absolute temperature are taken, the former on the supposition that the bulkiness is constant, and the latter on the supposition that the pressure is constant. Let it now be supposed that the bulkiness varies with the pressure according to some definite law; and let the actual rate of variation of the bulkiness with the pressure be denoted by $\frac{ds}{dp}$. Then equation (12) may be expressed in the following form:

$$\int_{p_1}^{p_2} dp \cdot \left\{ c_s \frac{d\tau}{dp} + c_p \frac{d\tau}{ds} \cdot \frac{ds}{dp} \right\} = 0 \ .$$

Now, according to the dynamic condition of permanence of type, we have by equation (6),

$$\frac{ds}{dp} = -\frac{1}{m^2} \quad ;$$

which, being substituted in the preceding integrals, gives the following equations from which to deduce *the square of the mass-velocity*:

$$\int_{p_1}^{p_2} dp \cdot \left\{ m^2 c_s \frac{d\tau}{dp} - c_p \frac{d\tau}{ds} \right\} = 0 \ . \tag{13}$$

It is sometimes convenient to substitute for $c_p \frac{d\tau}{ds}$ the following value, which is a known consequence of the laws of thermodynamics:

$$c_p \frac{d\tau}{ds} = c_s \frac{d\tau}{ds} + \frac{\tau dp}{J d\tau} \ , \tag{13A}$$

the differential coefficient $\frac{dp}{d\tau}$ being taken on the supposition that s is constant. The equations (13) and (13A) are applicable to all fluids, and probably to many solids also, especially those which are isotropic.

The determination of the squared mass-velocity, m^2, enables the bulkiness s for any given pressure p, and the corresponding velocity of disturbance u, to be found by means of the following formulae, which are substantially identical with equations (8) and (3) respectively:

$$s = S + \frac{P-p}{m^2} \quad ; \tag{14}$$

$$u = m(S-s) = \frac{p-P}{m} \tag{15}$$

Equation (15) also serves to calculate the pressure p corresponding to a given velocity of disturbance u. It may here be repeated that the linear velocity of advance is $a = mS$ (equation 1).

§9. *Application to a Perfect Gas.*—In a perfect gas, the specific heat at constant volume c_s, and the specific heat at constant ratio, c_p, are both constant; and consequently bear to each other a constant ratio, $\frac{c_p}{c_s}$, whose value for air, oxygen, nitrogen, and hydrogen is nearly 1·41, and for steam-gas nearly 1·3. Let this ratio be denoted by γ. Also, the differential coefficients which appear in equations (13) and (13A) have the following values:-

$$\left.\begin{aligned}
\frac{d\tau}{dp} &= \frac{\tau}{p} = \frac{s}{J(c_p - c_s)} = \frac{s}{J(\gamma-1)c_s}; \\
\frac{d\tau}{ds} &= \frac{\tau}{s} = \frac{p}{J(c_p - c_s)} = \frac{p}{J(\gamma-1)c_s}; \\
\frac{dp}{d\tau} &= \frac{p}{\tau} = \frac{J(c_p - c_s)}{s} = \frac{J(\gamma-1)c_s}{s}.
\end{aligned}\right\} \tag{16}$$

When these substitutions are made in equation (13), and constant common factors canceled, it is reduced to the following:

$$\int_{p_1}^{p_2} dp \cdot \{m^2 s - \gamma p\} = 0 \ . \tag{17}$$

But according to the dynamical condition of permanence of type, as expressed in equation (8), we have $m^2 s = m^2 S + P - p$; whence it follows that the value of the integral in equation (17) is

$$\int_{p_1}^{p_2} dp \cdot \left\{ m^2 S + P - (\gamma + 1) p \right\} = \left(m^2 S + P \right) \left(p_2 - p_1 \right) - \frac{\gamma + 1}{2} \left(p_2^2 - p_1^2 \right) = 0 \ ;$$

which, being divided by $p_2 - p_1$, gives for the square of the mass-velocity of advance the following value:

$$m^2 = \frac{1}{S} \left\{ (\gamma + 1) \cdot \frac{p_2 + p_1}{2} - P \right\} \ . \tag{18}$$

The square of the linear velocity of advance is

$$a^2 = m^2 S^2 = S \left\{ (\gamma + 1) \cdot \frac{p_2 + p_1}{2} - P \right\} \ . \tag{19}$$

The velocity of disturbance u corresponding to a given pressure p, or, conversely, the pressure p corresponding to a given velocity of disturbance, may be found by means of equation (15).

Such are the general equations of the propagation of waves of longitudinal disturbance of permanent type along a cylindrical mass of a perfect gas whose undisturbed pressure and bulkiness are respectively P and S. In the next two sections particular cases will be treated of.

§10. *Wave of Oscillation in a Perfect Gas.*—Let the mean between the two extreme pressures be equal to the undisturbed pressure; that is, let

$$\frac{p_2 + p_1}{2} = P \ ; \tag{20}$$

then equations (18) and (19) become simply

$$m^2 = \frac{\gamma P}{S} \ , \tag{21}$$

and

$$a^2 = \gamma P S \ ; \tag{22}$$

the last of which is LAPLACE's well-known law of the propagation of sound. The three equations of this section are applicable to an indefinitely long series of waves in which equal disturbances of pressure take place alternately in opposite directions.

§11. *Wave of Permanent Compression or Dilation in a Tube of Perfect Gas.*—To adapt equation (18) to the case of a wave of permanent compression or dilatation in a tube of perfect gas, the pressure at the front of the wave is to be made equal to the undisturbed pressure, and the pressure at the back of the wave to the final or permanently altered pressure. Let the final pressure be denoted simply by p; then $p_1 = P$, and $p_2 = p$; giving for the square of the mass-velocity

$$m^2 = \frac{1}{S}\left\{(\gamma+1)\frac{p}{2}+(\gamma-1)\frac{P}{2}\right\} , \tag{23}$$

for the square of the linear velocity of advance

$$a^2 = m^2 S^2 = S\left\{(\gamma+1)\frac{p}{2}+(\gamma-1)\frac{P}{2}\right\} , \tag{24}$$

and for the final velocity of disturbance

$$u = \frac{p-P}{m} = (p-P)\sqrt{\left\{\frac{S}{(\gamma+1)\frac{p}{2}+(\gamma-1)\frac{P}{2}}\right\}} . \tag{25}$$

Equations (23) and (24) show that a wave of condensation is propagated faster, and a wave of rarefaction slower, than a series of waves of oscillation. They further show that there is no upper limit to the velocity of propagation of a wave of condensation; and also that to the velocity of propagation of a wave of rarefaction there is a lower limit, found by making p=0 in equations (23) and (24). The values of that lower limit, for the squares of the mass-velocity and linear velocity respectively, are as follows:-

$$m^2(p=0) = \frac{(\gamma-1)P}{2S} ; \tag{26}$$

$$a^2(p=0) = \frac{(\gamma-1)PS}{2} ; \tag{27}$$

and the corresponding value of the velocity of disturbance, being its negative limit, is

$$u(p=0) = -\sqrt{\left\{\frac{2PS}{\gamma-1}\right\}} \quad . \tag{28}$$

It is to be borne in mind that the last three equations represent a state of matters which may be approximated to, but not absolutely realized.

Equation (25) gives the velocity with which a piston in a tube is to be moved inwards or outwards as the case may be, in order to produce a change of pressure from P to *p*, traveling along the tube from the piston towards the further end. Equation (25) may be converted into a quadratic equation, for finding *p* in terms of *u*; in other words, for finding what pressure must be applied to a piston in order to make it move at a given speed along a tube filled with a perfect gas whose undisturbed pressure and bulkiness are P and S. The quadratic equation is as follows:

$$p^2 - \left(2P + \frac{\gamma+1}{2S}u^2\right)p - \frac{\gamma-1}{2}\cdot\frac{Pu^2}{S} + P^2 = 0 \ ;$$

and its alternative roots are given by the following formula:

$$p = P + \frac{\gamma+1}{4S}u^2 \pm \sqrt{\left\{\frac{\gamma P u^2}{S} + \frac{(\gamma+1)^2 u^4}{16S^2}\right\}} \quad . \tag{29}$$

The sign + or - is to be used, according as the piston moves inwards so as to produce condensation, or outwards so as to produce rarefaction. Suppose, now, that in a tube of unit area, filled with a perfect gas whose undisturbed pressure and volume are P and S, there is a piston dividing the space within that tube into two parts, and moving at the uniform velocity *u*: condensation will be propagated from one side of the piston, and rarefaction from the other; the pressures on the two sides of the piston will be expressed by the two values of *p* in equation (29); and the force required in order to keep the piston in motion will be the difference of these values; that is to say,

$$\Delta p = 2u\cdot\sqrt{\left\{\frac{\gamma P}{S} + \frac{(\gamma+1)^2 u^2}{16S^2}\right\}} \tag{30}$$

Two limiting cases of the last equation may be noted: first, if the velocity of the piston is very small compared with the velocity of sound, that is if $\frac{Su^2}{\gamma P}$ is very small, we have

$$\Delta p \text{ nearly } = 2u \cdot \sqrt{\left(\frac{\gamma P}{S}\right)} \;; \tag{30A}$$

secondly, if the velocity of the piston is very great compared with the velocity of sound, that is if $\frac{\gamma P}{Su^2}$ is very small, we have

$$\Delta p \text{ nearly } = \frac{(\gamma+1)u^2}{2S} \;. \tag{30B}$$

§12. *Absolute Temperature.*—The absolute temperature of a given particle of a given substance, being a function of the pressure p and bulkiness s, can be calculated for a point in a wave of disturbance for which p and s are given. In particular, the absolute temperature in a perfect gas is given by the following well-known thermodynamic formula:

$$\tau = \frac{ps}{Jc_p - c_s} \;; \tag{31}$$

and if, in that formula, there be substituted the value of s in terms of p, given by equations (8) and (18) combined, we find, for the absolute temperature of a particle at which the pressure is p, in a wave of permanent type, the following value:

$$\tau = \frac{PS}{J(c_p - c_s)} \cdot \frac{(\gamma+1)(p_1+p_2)p - 2p^2}{(\gamma+1)(p_1+p_2)P - 2P^2} \;; \tag{32}$$

in which the first factor $\frac{PS}{J(c_p - c_s)}$ is obviously the *undisturbed* value of the absolute temperature. For brevity's sake, let this be denoted by T.

The following particular cases may be noted. In a wave of oscillation, as defined in §10, we have $p_1 + p_2 = 2P$; and consequently

$$\tau = T \cdot \frac{(\gamma+1)Pp - p^2}{\gamma P^2} \tag{32A}$$

In a wave of permanent condensation or rarefaction, as described in §11, let $p_1 = P$, $p_2 = P$; then the final temperature is

$$\tau = T \cdot \frac{(\gamma+1)Pp+(\gamma-1)p^2}{(\gamma+1)Pp+(\gamma-1)P^2} \ .$$

§13. *Types of Disturbance capable of Permanence.*—In order that a particular type of disturbance may be capable of permanence during its propagation, a relation must exist between the temperatures of the particles and their relative positions, such that the conduction of heat between the particles may effect the transfers of heat required by the thermodynamic conditions of permanence of type state in §6.

During the time occupied by a given phase of the disturbance in traversing a unit of mass of the cylindrical body of area unity in which the wave is traveling, the quantity of heat received by that mass, as determined by the thermodynamic conditions, is expressed in dynamical units by

$$\tau d\varphi \ .$$

The time during which that transfer of heat takes place is the reciprocal $\frac{1}{m}$ of the mass-velocity of the wave. Let $\frac{d\tau}{dx}$ be the rate at which temperature varies with longitudinal distance, and k the conductivity of the substance, in dynamical units; then the same quantity of heat, as determined by the laws of conduction, is expressed by

$$\frac{1}{m} \cdot d\left(k \frac{d\tau}{dx} \right) \ .$$

The equality of these two expressions gives the following general differential equation for the determination of the types of disturbance that are capable of permanence:

$$m\tau d \cdot \varphi = d \cdot \left(k \frac{d\tau}{dx} \right) \tag{33}$$

The following are the results of two successive integrations of that differential equation:

$$\frac{dx}{d\tau} = \frac{k}{A + m\int \tau d\varphi} \ , \tag{33A}$$

$$x = B + \int \frac{k d\tau}{A + m\int \tau d\varphi} \ , \tag{33B}$$

in which A and B are arbitrary constants. The value of A depends on the magnitude of the disturbance, and that of B upon the position of the point from which x is reckoned. In applying these general equations to particular substances, the values of τ and φ are to be expressed in terms of the pressure p, by the aid of the formulae of the preceding section; when equation (33B) will give the value of x in terms of p, and thus will show the type of disturbance required.

Our knowledge of the laws of the conduction of heat is not yet sufficient to enable us to solve such problems as these for actual substances with certainty. As a hypothetical example, however, of a simple kind, we may suppose the substance to be perfectly gaseous and of constant conductivity. The assumption of the perfectly gaseous condition gives, according to the formulae of the preceding sections,

$$\tau = \frac{PS}{(\gamma-1)Jc_s} \cdot \frac{(\gamma+1)(p_1+p_2)p-2p^2}{(\gamma+1)(p_1+p_2)P-2p^2} \ ,$$

and

$$\tau d\varphi = \frac{\gamma+1}{m^2(\gamma-1)}\left\{\frac{p_2+p_1}{2}-p\right\}dp \ .$$

It is unnecessary to occupy space by giving the whole details of the calculation; and it may be sufficient to state that the following are the results. Let

$$p-\frac{p_1+p_2}{2}=q \ ,$$

$$\frac{p_2-p_1}{2}=q_1 \ ;$$

then

$$\frac{dx}{dp}=\frac{dx}{dq}=\frac{k}{(\gamma+1)mJc_s}\cdot\frac{(\gamma-1)(p_1+p_2)-4q}{q_1^2-q^2} \tag{34}$$

$$x=\frac{k}{(\gamma+1)mJc_s}\left\{\frac{(\gamma-1)(p_1+p_2)}{2q_1}\cdot \text{hyp log}\frac{q_1+q}{q_1-q}+2\ \text{hyp log}\left(1-\frac{q^2}{q_1^2}\right)\right\} \tag{34A}$$

In equation (34A) it is obvious that x is reckoned from the point where q=0; that is, where the pressure $p=\frac{p_2+p_1}{2}$; a mean between the greatest and least pressures. The direction in which x is positive may be either the same with or contrary to that of the advance of the wave; the former case represents the type of a wave of rarefaction, the latter that of a wave of compression. For the two limiting pressures when $q=\pm q_1$, $\frac{dx}{dq}$ becomes infinite, and x becomes positively or negatively infinite; so that the wave is infinitely long. The only exception to this is the limiting case, when the conductivity k is indefinitely small; and then we have the following results: when $p=p_1$, or $p=p_2$, $\frac{dx}{dp}$ is infinite, and x is indefinite; and for all values of p between p_1 and p_2, $\frac{dx}{dp}$ and x are each indefinitely small. These conditions evidently represent the case of a wave of abrupt rarefaction or compression, already referred to in §§6 and 7.

Supplement to a Paper "On the Thermodynamic Theory of Waves of Finite Longitudinal Disturbance;" by W. J. Macquorn Rankine, C.E., LL.D., F.R.SS. Lond & Edin.

Received October 1,—Read December 16, 1869

Note as to previous investigations.-Four previous investigations on the subject of the transmission of waves of finite longitudinal disturbance may be referred to, in order to show in what respects the present investigation was anticipated by them, and in what respects its results are new.

The first is that of POISSON, in the Journal de l'Ecole Polytechnique, vol. vii. Cahier 14, p. 319. The author arrives at the following general equations for a gas fulfilling MARIOTTE'S law:-

$$\frac{d\varphi}{dx}=f\left\{x-at-\frac{d\varphi}{dx}t\right\},$$

$$\frac{d\varphi}{dt}+a\frac{d\varphi}{dx}+\frac{1}{2}\cdot\frac{d\varphi^2}{dx^2}=0\ ;$$

in which φ is the velocity-function; $\frac{d\varphi}{dx}$ the velocity of disturbance, at the time t, of a particle whose distance from the origin is x; a is the limit to which the velocity of propagation of the wave approximates when $\frac{d\varphi}{dx}$ becomes indefinitely small, viz. $\sqrt{\frac{dp_o}{d\rho_o}}$, p_o being the undisturbed pressure and ρ_o the undisturbed density; and f denotes an arbitrary function. This equation obviously indicates the quicker propagation of the parts of the wave where the disturbance is forward (that is, the compressed parts) and the slower propagation of the parts where the disturbance is backward (that is, the dilated parts).

The second is that of Mr. STOKES, in the Philosophical Magazine for November 1848, 3rd-series, vol. xxxiii, p. 349, in which that author shows how the type of a series of waves of finite longitudinal disturbance in a perfect gas alters as it advances, and tends ultimately to become a series of sudden compressions followed by gradual dilations.

The third is that of Mr. AIRY, Astronomer Royal, in the Philosophical Magazine for June 1849, 3rd series, vol. xxxiv. p. 401, in which is pointed out the analogy between the above-mentioned change of type in waves of sound, and that which takes place in sea-waves when they roll into shallow water.

The fourth, and most complete, is that of the Rev. SAMUEL EARNSHAW, received by the Royal Society in November 1858, read in January 1859, and published in the Philosophical Transactions for 1860, page 133. That author obtains exact equations for the propagation of waves of finite longitudinal disturbance in a medium in which the pressure is any function of the density; he shows what changes of type, of the kind already mentioned, must go on in such waves; and he points out, finally, that in order that the type may be permanent $\rho^2 \frac{dp}{d\rho}\left(= -\frac{dp}{ds}\text{ in the notation of the present paper}\right)$ must be a constant quantity; being the proposition which is demonstrated in an elementary way near the beginning of the present paper. Mr. EARNSHAW regards that condition as one which cannot be realized.

The *new results*, then, obtained in the present paper may be considered to be the following:—the conditions as to transformation and transfer of heat which must be fulfilled, in order that permanence of type may be realized, exactly or approximately; the types of wave which enable such conditions to be fulfilled, with a given law of the conduction of heat; and the velocity of advance of such waves.

The *method of investigation* in the present paper, by the aid of *mass-velocity* to express the speed of advance of a wave, is new, so far as I know; and it seems to me to have great advantages in point of simplicity, enabling results to be demonstrated in a very elementary manner, which otherwise would have required comparatively long and elaborate processes of investigation.

Hugoniot
courtesy of Ecole Polytechnique Archives

The Life and Work of Pierre Henri Hugoniot

R. Chéret, Commissariat A L'Energie Atomique, *Shock Compression of Condensed Matter - 1989,* Elsevier Sci. Pub. B.V., pp. 12 - 19 (1990)

Summary

None of the scientific shock wave community will deny that the name of Hugoniot has become, as it were, a password. But who was this man, whose life was as brief as W. A. Mozart's or V. Van Gogh's, and whose most famous work appeared posthumously one century ago, two years after being written? Which point did he make, that B. Riemann and W. J. M. Rankine failed to? Through which later contributions - J. Crussard, E. Jouguet, H. A. Bethe, H. Weyl - did we get to our present understanding of the jump relations? This historical perspective is a modest attempt to better the knowledge of a field which has so deeply influenced theoretical physics, and has also allowed tremendous applications which still control the history of the world.

1. Hugoniot and His Time

The register of births, marriages and deaths of Allenjoie, a small village of the Doubs department which now has 650 inhabitants, shows the birth on 5th June, 1851 at 10 p.m., of Pierre Henri, second son of Pierre Hugoniot, 35, metallist, and Suzanne Catherine Nardin, 31, unemployed.

The village has kept no memories of his childhood, his brother, his family or his studies, nor any recognition of the posthumous fame of this child, as I discovered on seeing the astonishment of the town hall clerk at my insistence on finding out about Pierre-Henri Hugoniot.

The few scraps of information are nevertheless precious as they allow one to assume that the young boy soon became familiar, due to the profession of his father, with the working and forming of metals. Nevertheless, to judge from his early success at school, it is not in this field that he rapidly came to excel, but in that of mathematics.

By one of the miracles of which the French education system is so proud, the young Pierre Henri, although of modest origin, succeeded brilliantly in his studies. In 1869, at the age of eighteen, he was already a graduate in mathematical sciences.

In 1870, he was first at l'Ecole Normale Supérieure. However, it was to l'Ecole Polytechnique, where he was admitted the same year, that his preference went. We can only imagine the reasons for his choice. Was this the consequence of the rebellious and patriotic ideas which spread among the students during the summer of 1870, during which France declared war on Germany on 19th July and Napoleon III capitulated at Sedan on the 2nd of the following September? The work and the career chosen by Pierre Henri Hugoniot on leaving l'Ecole Polytechnique offer some support for this theory, as we shall see.

2. The Three Steps of a Lightning Career

2.1 The student

The Ecole Polytechnique entrance documents constitute one of the most precious sources of information concerning the young man.

First, the photograph of Pierre Henri Hugoniot (frontispiece). What is it that both surprises and makes him different? Is it the fact that he is bare-headed or his faraway look? The photograph is of mediocre quality, and a portrait could only be made from it by using an image processing technique specially designed for the purpose.[14] Luckily, the archives of the school can help us recreate his face and figure:

> *"brown hair and eyebrows,*
> *hooked nose,*
> *brown eyes,*
> *average mouth,*
> *round chin,*
> *oval face,*
> *height 176 cm".*

The student Hugoniot went up from second to first division, 54th out of 145 students. He was declared acceptable for public service in 1872, 38th out of 141 students. Finally, he was accepted in the marine artillery service, first out of six students. From then on, it was to this service that he devoted body and soul for the next fifteen years, as he died suddenly during a trip from Paris to Nantes at the age of 36.

2.2 The gunner

There is a description of the scientific works of P. H. Hugoniot in the archives of the Ecole Polytechnique. It tells us of the ease with which the student reached the level of his teachers. It was thus that he became:

- professor of mechanics and ballistics at the Lorient Artillery School, from 1879 to 1882.
- assistant director of the Central Laboratory of Marine Artillery, from 1882 to 1884.

On 15th January 1884, he was appointed captain of the Marine Artillery.

This period was marked by the assistance he provided to Monsieur Helie, professor of the Lorient Artillery School, in the publication of a second revised and corrective edition in 1884. Here, in a few words, the author expresses his gratitude to his colleague:

"A second edition of this work was requested of me, but this required research which the weakness of my sight appeared to forbid. Luckily, I found in M. Hugoniot, Captain of the Marine Artillery, a colleague with whom I have always found myself in perfect agreement, with the result that he shares with me the responsibility for this work."

The work received an award by the Paris Academy of Sciences in 1884.

2.3 The researcher

On 3rd April 1884, Pierre Henri Hugoniot was appointed an auxiliary assistant in mechanics at the Ecole Polytechnique. Here, where the need for practical applications was less immediate, and where mathematics was the foremost discipline, he finally found the time and surroundings he needed to settle the question which he was constantly faced with as a gunner: *"the propagation of movement in bodies, particularly in perfect gases."* It only took him fifteen months to find the answer, as on 26th October 1885 he submitted a memoir to the secretariat of the Institute. This was finally published a few years later. It appeared in two parts in the proceedings of the Ecole Polytechnique, the first part in Volume 57 in 1887, the second in Volume 58 in 1889. The reason for this delay is clearly indicated in an editor's note in the proceedings *"...The author, carried off before his time, was unable to make the necessary changes and additions to his original text which it seems he had the intention of making;..."* Indeed, it is in the second part, in the fifth and last chapter, that the expression which was made Hugoniot famous is to be found:

$$e' - e = (1/2)(p + p')(\mathrm{v}' - \mathrm{v}) \qquad (1)$$

(N.B. Hugoniot only actually wrote equation (1) for an ideal gas of constant gamma. Nevertheless, the wording clearly implies the general expression.) It is for this reason that it has been decided to celebrate its centenary this year, in 1989.

In point of fact, it would be wrong to limit the findings in this memoir simply to the single correct expression which we now call the **Hugoniot relationship**. It is essential to understand the original thoughts of their inventor. Let us hear what he has to say in section 3 of the introduction.

"3. The equation with partial derivatives which governs the movement of a body is simply a general condition with which all individual movements must comply. In fact, each integral represents a possible movement which would remain substantially unmodified if the body were indefinite in the two directions.

"But the problem that it is normally required to solve is the following: given a limited body, the initial state is known or rather the integral which represents its movement at the initial instant. The boundaries are subjected to known conditions and it is required to determine the subsequent movement.

"I give the general methods which should be used to solve this problem, that is to say that I indicate with precision the analytic conditions which should satisfy the integrals which express the solutions."

"I shall now briefly describe the principles of this method.

"When the movement of a body takes place by parallel sections, it may, at a given instant, be divided into two parts by a plane ξ perpendicular to the direction of the speed, the movement being represented on one side of ξ by a certain integral and on the other side by another integral.

"Section ξ *is generally the seat of a complex phenomenon, after which new integrals appear. Nevertheless, in certain cases, one of the integrals purely and simply supplants the other, so that at the next instant in time the movements are still represented by the original integrals, but the section of separation is no longer the same.*

"In such cases, I contend that the two movements are ***compatible*** *with each other. One of the movements propagates in the other with a certain speed which is the* ***speed of propagation.***

"When the two movements are not compatible, a new movement begins in the section of separation which is compatible with the two others, and which then propagates in both directions at the expense of the original movements."

Such is the guidance - where *"analysis, geometry and mechanics mutually support each other"* - under which Hugoniot mentions in the first four chapters everything that is necessary for understanding the fifth and last chapter on *"discontinuities found in the propagation of movement."*

3. Before and After Hugoniot

3.1 Three men, two approaches, one relationship

On reading Hugoniot's memoir, one is struck by the lack of references to Riemann and Rankine who, in modern books on the history of science, are described as coming before him. This silence is somewhat surprising, especially as Hugoniot in his earlier studies makes many references to other authors, and as he takes the care in the last few lines of his introduction to cite the names of Phillips, Saint-Venant and Darboux, who are now considered to have no link with the problem. Let us consider this paradox.

Those familiar with the works of Bernhardt Riemann (1826-1866) will have read the memorable pages[1] (1860) in which he introduces the **characteristic curves method** - without naming it - when describing the forming of abrupt condensations (Verdichtungstoss) and actually comes upon the formulae giving the **relative speeds of propagation**. However, in this approach, consideration is only given to the partial derivative equations expressing the conservation of mass and momentum, the conservation of energy being - erroneously of course - considered to be equivalent to the adiabatic law of Poisson according to which pressure is uniformly proportional to the gamma power of density. This incorrect hypothesis shows that Riemann has not even thought of the jump relationship (1).

Let us now consider the work[2] of William John Macquorn Rankine (1820-1872) in 1870, two years before his death. This already famous Scottish engineer was taking an interest in the permanent movement in planar sections of a dissipative fluid (conductive but nonviscous), and demonstrated relationship (1) as being necessary between the far upstream state and the far downstream state.

What is the position of the work[3] of Hugoniot with respect to these illustrious forerunners? Clearly his ambitions lay in another direction. Let us quote a few sentences from the introduction:

(Ref. 3, § 8)

"It is generally accepted, without being sufficiently demonstrated, that in the movement of a perfect gas, which receives no heat from outside and whose own conductivity is disregarded, the decompression of each section takes place

according to the adiabatic law, whatever the variations in speed of the molecules. I rigorously establish, by means of the equivalence principle, that this hypothesis is true when the variation of speed takes place in a continuous manner.

"But the situation is very different when the movement is discontinuous. When a section is subjected to the effects of discontinuity, the original relationship between pressure and volume is suddenly changed."

(Ref. 3, § 13)

"13. In Chapter V I shall examine the phenomena which occur when discontinuities of movement are introduced."

(Ref. 3, § 15)

"When a discontinuity occurs in a perfect gas, all of the sections which are affected by it undergo sudden changes in pressure and subsequently density. I demonstrate that the relationship between the new density and the original density remains within two limits equal to (m - 1)/(m+1) and (m+1)/(m-1), in which m represents the ratio of the specific heats. This theorem, which may appear to be paradoxical, is nevertheless a direct consequence of the first law of thermodynamics."

These passages suggest, and careful reading confirms, that Hugoniot covers the movement of non-dissipative media (elastic solids and perfect fluids) without excluding discontinuities and without postulating permanent flow. His subject - and also his methods - place him far closer to Riemann than to Rankine, which may explain him being unaware of the work of the latter. On the other hand, the reason for ignoring the former's work is more probably the linguistic barrier, being particularly strong in the years which followed the French defeat in 1870. Indeed, the first publication containing all the works of Riemann appeared in 1876 in German ... and in 1897 in French.

If this was effectively the case, it would be closer to historical truth to speak not of forerunners but of three independent approaches, the deepest and most fruitful of which is undeniably that of Hugoniot.

3.2 From Hugoniot to Crussard

At the beginning of Chapter V, consecrated to "discontinuities in the propagation of movement," Hugoniot writes "the question may be raised as to whether the discontinuities encountered in movement propagation theory are simply analytical fiction or whether they correspond to a physical reality. This is a question which is very difficult to answer at the present stage of scientific knowledge." This sentence alone proves, were it still necessary, that the intention of Hugoniot was not to break new ground in the description of waves (the word **wave** appears not even once in the nearly 200 pages of the memoir) but to develop a mathematical approach.

Paradoxically, it is among the pioneers of the **explosive wave**, Marcelin Berthelot and Paul Vieille (1881, 1882), that the first heralds of the concepts and results of Hugoniot are to be found. Indeed, as emphasized by Numa Manson in his study[13] of the contributions of Paul Vieille to the understanding of detonation and shock waves, as early as 1891 in a note[4] entitled *"concerning the influence of the covolume of gas on the speed of propagation of explosive phenomena,"* Paul Vieille refers to the work of Hugoniot. In passing, we can accept the assertion of N. Manson that V. A. Michelson is the first (1894) to quote[5] Riemann, Rankine and Hugoniot together.

However, despite his thorough understanding of the state of the art, Michelson was unable to set up the representation of relationship (1) in a reactive fluid. Undeniably, the merit belongs to Jules Crussard, as is shown by careful analysis of the publications during the first few years of this century or simply by carefully reading the celebrated Mécanique des explosifs[8] by Jouguet:

(Ref. 8, p. 241):

"Let us immediately consider the case in which the fluid is the seat of chemical reaction ... The wave is then a shock and combustion wave, to use the description introduced by M. Crussard."

Similarly, Crussard is the first to make systematic use of the representation of (1) in the following form: volume per unit mass v, pressure p.

(Ref. 8, p. 267):

"We shall use a method of geometrical representation which we owe to M. Crussard (6)."

It is also to him that we owe - again according to Jouguet in the preceding document - the statement of one of the essential properties of critical waves (Chapman-Jouguet waves).

(Ref. 8, p. 278):

"The two critical waves have an important property demonstrated by M. Crussard: the critical waves of maximum and minimum velocity are such that

their velocity w is equal to the speed of sound a in the medium which follows them."

On the other hand, it is effectively to Jouguet[7] (1910) that we owe the correct analysis of variation of the ratio w/a along the curve (H), and as a corollary the nature of the extremum of entropy at each of the critical points (minimum for detonation and maximum for deflagration) which Chapman had erroneously described in 1899.

These points may seem to be merely anecdotal. This is not the case. Indeed, they clearly show that more than twenty years were necessary for the understanding of specific features of relationship (1) with chemical change; they also show the significant difference between curves (h) and (H) representative of (1) respectively in the absence and the presence of chemical reactions, in the same way as an explosive wave differs from a shock wave. It is for this reason that I am profiting from the occasion of this homage to Hugoniot to attempt to convince the *"shock wave community"* to give Crussard his due, i.e. credit for the discovery of curve (H). In other terms, historical background and accuracy both prompt us to label the representation of (1) ***"Crussard curve"*** when it refers to the detonation of an explosive, and ***"Hugoniot curve"*** when it refers to shock waves in the detonation products of the explosive.

3.3 From Hugoniot to Jouguet

As has already been made clear, Hugoniot, in dealing with discontinuities in a perfect fluid, had ignored the question of their effective occurrence. Jouguet on the other hand - although an assistant at the Ecole Polytechnique exactly like Hugoniot - was more inclined towards thermodynamics than mathematics, and was not fully satisfied with this geometric image of the phenomenon, particularly as he was well acquainted with the experimental work conducted by Marcelin Berthelot and Paul Vieille since 1882 which came at the right moment to provide a physical picture for the concept of discontinuity.

How can a link be made between the physical phenomena of shock waves and explosive waves and the work of Hugoniot and Crussard? How can it be explained that the adiabatic law of Poisson does not apply to wave flow? These are the questions that Jouguet was perhaps the first to ponder and to which he was the first to provide satisfactory answers. Although his name has come to be associated with the critical points of curve (H), it would have been fairer for the credit to have gone to Jouguet for two results:

(i) The introduction of the concept of the **quasi-wave** (8, pp. 249-250) which foreshadows the modeling of the boundary layer by Von Mises[9] in 1927, **the shock layer** by Weyl[11] in 1944 and the **detonation layer** by Chéret[12] in 1971.

(ii) The idea that whether a flow is of a dissipative nature or not depends on the scale of the work of the viscous forces an/or the heat circulating by conduction.

3.4 From Hugoniot to Bethe and Weyl

The very title of Hugniot's memoir *(concerning the propagation of movement in bodies, particularly in perfect gases)* clearly shows that the author intended to concern himself with a general problem and not to go into detail concerning specific solutions for perfect gases. The same approach is to be found in Crussard and Jouguet: they effectively mention the relationship (1) but they plot the corresponding curve in the plane (v, p) in the case of an ideal gas at constant gamma: $e'=p'v'/(\gamma - 1)$ (In passing, it should be noticed that they refer to the ratio Cp/Cv, by making use of the symbol gamma suggested by Rankine rather than k as suggested by Riemann or m as suggested by Hugoniot).

It was up to H. Bethe[10] (1942) some forty years after Crussard, to study curve (h) for an arbitrary fluid in a report entitled: "*On the theory of shock waves for an arbitrary equation of state.*" Let us quote the introductory sentences of Bethe's work:

"The theory of shock waves thus far has been developed mainly for ideal gases. Even for these, the question of stability of shock waves has received little attention. Recently the problem of shock waves in water has gained much practical importance. Therefore it seems worth while to investigate the properties of shock waves under conditions as general as possible."

It is to Bethe that we owe the representation of (h) in terms of volume per unit mass v and entropy per unit mass s, and the analysis of the situation in the vicinity of the downstream state.

It is also him that we owe a set of conditions:

$$\frac{\partial^2 p}{\partial v^2}(v,s) > 0$$

$$v\frac{\partial p}{\partial e}(v,e) > -2$$

$$\frac{\partial p}{\partial v}(v,e) < 0$$

which ensures the stability of a shock wave. As concerns their importance and usefulness, need it be reminded that the first inequality constitutes the basis of interpretation of phenomena as well-known as the elasto-plastic structure of certain shock waves and the double shock wave structure in a phase transition zone?

A review of the work directly derived from Hugoniot would be incomplete if we were not to consecrate a few lines to a note - now long forgotten - that Hermann Weyl wrote[11] concerning the same problem in 1944. Wartime circumstances explain why Weyl had no knowledge of the report written by Bethe two years earlier. Paradoxically, the fact that the research was carried out independently turned out to be an asset. Indeed, together with a first part *(thermodynamics and the shock phenomenon)* which is less rich than the wide ranging report written by Bethe, there is a capital second part *(the problem of the shock layer)* from which are derived:

(i) the concept of the **shock layer** and more generally the dissipation layer, equivalent to the Prandtl boundary layer.

(ii) the **asymptotic expansion method** which has acquired widespread recognition in hypersonic aerodynamics and whose field of application now extends to many branches of physics.

This is not the place to consider in greater detail the consequences of this concept and of this method. Anyway it would be unforgivable not to emphasize that H. Weyl's approach is the one which finally reconciles Hugoniot's and Rankine's. Indeed, H. Weyl demonstrates that relationship (1) is not only that which links the thermodynamic states of a perfect fluid on either side of a discontinuity, but also that which links the initial state to the final state of the transformation of a dissipative fluid "inside" the discontinuity.

4. Conclusion

Here it is necessary to end this review.

After describing these lofty figures and such remarkable and fruitful discoveries, I feel reluctant to draw any conclusion.

On the contrary, I cannot help pondering upon the range of talent necessary for a question, which was badly posed by Riemann, to find a satisfactory answer more than a century later; and also the great diversity of these talents: the rigorous mechanics of Hugoniot, the intuition of the physicists Crussard and Jouguet, Bethe's flair in thermodynamics and the mathematical creativeness of Weyl.

But is it not this very diversity which makes your research so difficult and which justifies the passion you devote to it?

Chronological Index of Cited References

1. B. Riemann, Uber die Fortpflanzung ebener luftwellen von endlicher Schwingungsweite, Abhandlungen der Gesellschaft der Wissenschaften zu Göttingen, 8 (1860), 43.

2. W.J.M. Rankine, On the thermodynamic theory of waves of finite amplitude, Philosophical transactions, vol. CLX, part 1 (1870), 277.

3. P.H. Hugoniot, Sur la propagation du mouvement dans les corps et plus spécialement dans les gaz parfaits, Journal de l'Ecole Polytechnique, 57e cahier (1887) 3, and 58e cahier (1889) 1.

4. P. Vieille, De l'influence du covolume de gaz sur la vitesse de propagation des phénomènes explosifs, Mémorial des poudres et salpêtres, 4 (1891) 20.

5. V.A. Michelson, Sur la vitesse normale d'inflammation des mélanges gazeux explosifs. Thèse (1894) rééditée à Moscou en 1930.

6. J. Crussard, Ondes de choc et onde explosive, Bulletin de la société de l'industie minérale, 4e série, t. VI (1907).

7. E. Jouguet, Sur la vitesse des ondes de choc et combustion, Comptes rendus Académie des Sciences Paris, t. 149 (1909), 1361, and Comptes rendus Académie des Sciences Paris, t. 150 (1910), 91.

8. E. Jouguet, Mécanique des explosifs, Octave Doin, Paris (1917).

9. R. von Mises, Zeitung ang. Math. und Mech., 7 (1927), 425.

10. H.A. Bethe, On the theory of shock waves for an arbitrary equation of state, Report OSRD No. 545 (1942).

11. H. Weyl, Shock waves in arbitrary fluids, National Defense Research Committee, Applied Mathematics Panel, note No. 12 (1944).

12. R. Chéret, Contribution à l'étude des détonations sphériques divergentes dans les explosifs solides, Thèse de doctorat es-sciences, Poitiers (1971).

13. N. Manson, Contribution de Paul Vieille à la connaissance des détonations et des ondes de choc, Mémorial Art. francaise, t. 60 (1986), 26.

14. J.M. Dinten and M. Tatard, Application des champs de Markov à la restauration d'images, Internal Report CEA (April 1989).

On the Propagation of Motion in Bodies
and in Perfect Gases in Particular - I[1]

H. Hugoniot, Translated from the *Journal de l'École Polytechnique, Vol. 57, pp. 3-97* (1887)

Introduction

1. The questions dealt with in this paper cover analysis as well as pure physics. The motions of natural bodies are governed by partial differential equations, the theory of which remains obscure even today. All progress made in the study of these types of motion provides new information on the nature of the equations, and on the relations that exist between the various integrals in particular.

Our rational minds are often tempted to make the too abstract results of analyses more concrete. This is one way of clarifying the theory of partial differential equations with two independent variables using a geometric representation, by considering them as defining a class of surfaces. However, when there are three or four independent variables, this representation becomes impossible, unless four or five dimensional space is brought into play, which would require a further degree of abstraction.

---

[1] This paper is substantially the same as that delivered to the Academy of Sciences by Mr. Hugoniot and filed with the Institute's Secretariat on 26 October 1885.

The author was unable, on account of his untimely death, to make the modifications and additions to his preliminary draft which he apparently intended to make; however, as it is, this paper will still suffice to show the reader Mr. Hugoniot's great talent and the great loss that science has suffered with his death.

With the intention of respecting the author's reasoning, we have in no way modified his paper; we restricted ourselves to dividing it into two parts for the purposes of printing: the second part will be incorporated in a future edition.

The reader will find that some of the results given in this part are not new; this is not a sign of poor erudition, merely the author's constant concern to include in his papers all the information needed to understand them in their entirety. *(Ed.)*

Despite this, many equations with three or four variables take on an exact meaning when they are considered as defining the motion of a body.

2. In this first paper I shall only cover the simplest of cases, that where the motion of a body, by strata, is governed by an equation with two independent variables. Each integral can then be shown by a surface as well as by a motion; each motion has its own corresponding surface, each property of the surface has a corresponding property of motion. These properties are often shown more clearly by geometrical representation; in other cases, they are more evident in mechanical representations. In short, analysis, geometry and mechanics are all involved in the study of integrals.

3. The partial differential equation which governs the motion of a body is a general condition which must be met by all particular types of motion. In fact, each integral represents a possible motion which would generally continue unchanged if the body was undefined in both directions.

Ordinarily, however, the problem which must be solved is the following: we are given a limited body, and know the initial state or rather the integral which represents its motion at the initial moment. The ends are subject to known conditions. We wish to ascertain the subsequent motion.

I shall set out the general method which must be used to solve this problem, in other words I shall state the precise analytical conditions which must be met by the integrals which will provide a solution to the problem.

I shall briefly state the principles behind this method.

When a body moves by parallel strata, it may, at a given moment, be divided into two parts by a profile ξ perpendicular to the velocity axis, the motion being represented on one side of ξ by one particular integral, and on the other side by another particular integral.

The stratum ξ is generally the site of a complex phenomenon, which results in the creation of new integrals. However, in certain cases, one of the integrals simply expands at the expense of the other, such that at the following moment the types of motion are again represented by the original integrals; only the separation profile is changed.

In this case, I postulate that the two motions are *mutually* compatible. One of the motions propagates into the other at a certain velocity, which is the *velocity of propagation*.

When the two motions are incompatible, a new motion begins in the separation stratum which is compatible with the other two, and then propagates in both directions at the expense of the original motions.

4. Given the initial motion of a body and the conditions imposed at its extremities, motions develop at these end points which are compatible with the primitive motion and propagate in the opposite direction at its expense, ending with the disappearance of the initial motion and the meeting of the two new motions. At the meeting point a motion is generated compatible with the previous two; it propagates unchanged in both directions until it meets one of the ends. Here a new motion is produced which propagates like the others and so on.

The analytical determination of the motion of a body, under the conditions in which we find ourselves, obliges us to calculate an infinite number of analytical functions, all different, and each of which represents the motion of one part of the body for a certain period of time. In other words, we are not determining a single motion, but an infinite number of motions which succeed each other according to a particular law.

When one of these motions is produced at one of the extremities, it must satisfy the conditions imposed at that point and be compatible with the primitive motion. This is sufficient to define it.

5. It is the *compatibility conditions* which provide the rules for determining the various integrals representing the phenomenon. In this paper I give the analytical expression of the compatibility conditions.

These take on a particularly striking appearance when both motions constantly have the same velocity, same expansion and the same pressure for the same profile, in other words when no discontinuities occur.

The motion being studied can be shown by a surface whose vertical ordinate represents the motion u of a stratum x at time t, the horizontal coordinates being the abscissa and time.

We know that these surfaces, which consist of the integrals from a single partial differential equation, can be considered as being generated by certain *characteristic* surfaces.

When there are no discontinuities, the following theorem can be postulated:

The surfaces representing two compatible motions meet along the line of intersection, which is a common characteristic of both surfaces.

Consequently, *the velocity of propagation of a motion A into a motion B is equal to the angular coefficient of the horizontal projection of the common characteristic.*

This theorem can be used to calculate the velocity of propagation when only motion B is known. In particular, if this motion is rest, the velocity of propagation can be calculated using the partial differential equation, without integrating.

6. The above method is mainly applied in relation to conductive fluids and, in particular, perfect gases contained in cylindrical tubes. I give the integrals, in finite form, representing all the types of motion that can propagate in bodies assumed to be at rest in the primitive state. *These integrals are represented in geometric form by developable surfaces.*

This makes it easy to study the phenomenon of wave deformation, up to the moment when the reflexions intervene.

The integrals I use have been known for many years; however, no one, at least to my knowledge, had thought of using them to deduce the properties of the motion of fluids.

I will then show that the second order partial differential equation normally used only covers very specific cases of gas motion. It should be modified as soon as any discontinuity is introduced. The phenomenon can be represented in its entirety only by two simultaneous equations of the fourh order which however may be reduced to a single equation of the third order by changing the variables.

7. This paper is divided into five chapters.

Those who are only interested in Mechanics or Physics are often fairly unfamiliar with the characteristics so widely used in higher geometry. Chapter I gives a brief synopsis of the theory behind these surfaces, adequate for an understanding of the following chapters.

I consider this chapter as being of only secondary importance. However, I give a precise definition of the characteristics, which is, I have reason to believe, new, and which immediately gives two equations specific to these surfaces, irrespective of the order or the form of the partial two-variable differential equation considered. This definition can be generalized without any difficulty and applies to equations with n variables as well as to systems of equations which contain several unknown functions, as I shall show in a future paper.

8. Chapter II deals with the search to find equations for motion by strata.

It is generally held, although insufficiently demonstrated, that in the motion of a perfect gas, which does not receive heat from an external source, and if its own conductivity is disregarded, each stratum expands adiabatically, irrespective of the variations in velocity undergone by the molecules. I shall irrefutably show, using the equivalence principle, that this hypothesis is correct when velocity changes in a continuous manner.

However, things are very much different when discontinuities are introduced into the motion. When a stratum meets a discontinuity, the original relation between the pressure and the volume is suddenly changed.

We are also obliged to leave an arbitrary function in the second order equation governing the motion; the initial form of this function can change as the motion progresses, and it would, if removed, produce two fourth-order simultaneous equations with a common characteristic.

I shall apply a similar method when writing equations for the motion of other non-conducting fluids; finally, I shall rewrite these equations to cover the case of motion by concentric spherical layers.

9. Chapter III covers the theory of the propagation of motion when there are no discontinuities. The general method for determining the integrals is set out in detail. There then follows a demonstration of the theorem that two compatible integrals have a common characteristic, and that consequently the velocity of propagation is represented by the angular coefficient of the tangent to the horizontal projection of the characteristic; from this I deduce the actual velocity of propagation for the various natural bodies.

Certain consequences are worth mentioning.

In a perfect gas at rest, the velocity of propagation is equal to the normal velocity of sound, irrespective of the motion which is being propagated, until discontinuities are introduced. This velocity remains the same whether the motion is by parallel strata or by spherical layers. Finally, it is in no way modified by friction or viscosity, at least if we consider viscosity as having no other effect than introducing resistance proportional to the velocity. The same applies for non-conducting fluids.

I provide an application of the method for determining the integrals of the motion of a vertical elastic rod, fixed at its top end, and with a weight at the bottom end, a problem which had never been solved in finite terms until now.

10. Chapter IV studies the motion of a non-conducting fluid, enclosed in a cylindrical heat-proof casing. External forces, friction and viscosity are not taken into account.

Given that the fluid is homogeneous and at rest, it is assumed that one of its ends is subjected to a given condition, for instance that a known motion is communicated to it. I shall show that the motion which originates from this point, irrespective of the condition applied at this end and provided it does not contain any discontinuities, can be expressed in geometrical terms as a developable surface. For a given fluid, all these developable surfaces belong to a single class defined by a linear equation, easy to deduce from the second order equation which governs the motion. The edges of regression of all these surfaces have the same spherical indicatrix.

The motion corresponding to a developable surface propagates through the fluid at a constant velocity until it reached the opposite end, where the phenomenon of reflection occurs. However, a new phenomenon occurs when the motion meets the edge of regression of the surface. A discontinuity is then formed in the furthest stratum reached by the motion. This discontinuity usually expands gradually; however, from the moment this surface comes into existence, the primitive second order equation ceases to govern all of the motion, since the relationship between the pressure and the expansion is suddenly changed for all the strata which experience this effect.

Conditions can be imposed at the extremity to make the discontinuity in question come into existence suddenly. In this case, the surface representing the motion is reduced to a cone.

11. I then go on to apply the general theory to the perfect gases and to provide fully determined formulae for these bodies which are suitable for all values of the specific heat ratio. The case of this ratio being equal to the unit must be dealt with apart, since the gas would behave according to Mariotte's Law, making the validity of many general formulae dubious.

I provide a few examples by determining the condition imposed on the end and then study the way in which the curve, which shows the expansion or the velocity of the various points of the gas at a given moment, is deformed. According to the conventional theory, the only one taught in Physics lessons, the expansion curve moves, without deforming, at a constant velocity. In actual fact, the curve is deformed and follows a law which is extremely simple.

As we already pointed out, the phenomenon ceases to be fully represented by these formulae when the motion meets the edge of regression of the representative surface. The distance at which this meeting occurs obviously depends on the condition imposed at the end. In the case of air under normal conditions of temperature and pressure, this distance is less than 9 m when the end of the tube is closed with a vibrating diaphragm performing 500 complete vibrations every second, with an amplitude of 2 mm. It is inversely proportional to the amplitude and the square of the number of vibrations.

Finally, I go on to determine the law of motion which would have to be applied to the extremity of it to suddenly produce a discontinuity in the velocity of the strata at a given distance X, together with their expansion, in which case the developable surface representing the integral is reduced to a cone.

12. This chapter contains a few additional interesting results. I thus determine the upper bound velocity for the gas flow, a velocity which, furthermore, could only be attained if the fluid retained the same properties, even when its temperature drops to absolute zero. This velocity is equal to $\dfrac{2a}{m - 1}$, where a is the normal velocity of sound in the gas and m is the specific heat ratio.

I go on to consider a problem raised by Lagrange — a subject on which Piobert has written most of his papers, by virtue of its possible applications to ballistics in a closed space. A column of gas is at rest; one of its extremities is fixed, the other limited by a piston subjected only to the pressure of the gas; the problem concerns how the piston moves. I show that, to the contrary of the ideas put forward by Piobert, the mobile part cannot, no more than the gas, be represented by a single integral, but rather depends on an infinite number of integrals, each of which represents motion at a certain time. I calculate the first of these integrals which would alone be sufficient to provide the answer to the problem if the gas column was of infinite length. Under these conditions, the velocity of the mobile part would converge to a limit, independent of its mass, which would be equal to the upper bound flow velocity $\dfrac{2a}{m - 1}$, where a is the velocity of sound in the gas and m the specific heat ratio.

13. In Chapter V, I examine the phenomena which occur when discontinuities are introduced into the motion.

Two cases should be differentiated, according to whether or not the relationship existing, in a particular stratum, between pressure and expansion depends on the transformations it has undergone.

When the relationship between pressure and expansion does not vary, the motion is always governed by the same differential second order equation. This is what happens for a fluid whose temperature is kept constant at all points; we can also assume that the same applies for an elastic solid.

In this case, a single equation expresses the compatibility conditions for the two motions, despite the discontinuities. The surfaces which represent compatible motions are often cut along curves which can be viewed as the generalisation of characteristics. When the equation for the motion is perfectly linear, these curves are again confused with the characteristics; the compatibility condition is then the same, irrespective of whether or not discontinuities are introduced.

14. However, when the relationship between pressure and expansion for a stratum depends on the transformations it has undergone, as can be the case for a non-conducting fluid which is not heated from an external source, the phenomenon becomes more complex. There are two compatibility conditions.

When two incompatible motions A and B are present simultaneously, a further motion is developed in each of the two touching strata, and the two new motions A' and B' propagate, the first in one direction, and the second in the other, at the expense of the original motions. The stratum common to the two motions A' and B' is always the same; in this stratum, the velocities and pressures are equal, but not the expansions.

15. I have restricted myself to applying these general theories to perfect gases enclosed in cylindrical tubes. Some of the results I obtained are worthy of mention.

When a discontinuity is produced in a perfect gas, each of the strata which suffers its effects experiences, as we know, a sudden change in pressure and thereafter of density. I show that the ratio of the new density to the original density still ranges between two limits equal to $\frac{m-1}{m+1}$ and $\frac{m+1}{m-1}$, where m is the specific heat ratio. This theorem, which appears paradoxical, is, however, a direct consequence of the first law of thermodynamics.

16. When two homogeneous columns of gas, each animated by a uniform translational motion, are adjacent, the motions which are generated in the touching strata are similar to the original motions, in other words, for all strata affected by a particular motion, the velocity, pressure and expansion are the same. The velocities are equal in both new motions, as are the pressures.

The pressure common to both motions is given by a sixth-power algebraic equation which always contains a real root which provides the answer to the question.

17. Let us consider a column of gas at rest and limited by a piston; this piston is suddenly driven at a finite velocity which is thereafter kept constant. Under these conditions, the motion which is generated and propagated in the gas is a uniform motion with the same velocity as that of the piston; all strata successively brought into motion instantaneously attain this velocity and undergo a sudden corresponding change in pressure and expansion.

If the piston is moved in such a way as to compress the gas, the velocity of propagation is greater than the normal velocity of sound; it increases indefinitely at the same time as the velocity of the piston increases.

However, when the piston is moved in such a way as to expand the gas, there is a velocity above which an absolute vacuum forms behind the piston. This velocity is equal to $a\sqrt{\dfrac{2}{m(m-1)}}$, where a, as before, is the normal velocity of sound, and m the specific heat ratio.

As the velocity of the piston increases from zero to $a\sqrt{\dfrac{2}{m(m-1)}}$, the rate of propagation of the motion decreases from a to $a\sqrt{\dfrac{m-1}{2m}}$.

18. The question of the resistance offered by air to the motion of the bodies is an extremely important one, especially from the point of view of ballistics in an open space; however, its theoretical determination is posing difficulties which even today we cannot claim to surmount. Newton's attempts, including a theory founded, it is true, on very dubious assumptions, gives results which are at variance with observed facts, and will not encourage research into geometry.

What makes the question so difficult, and almost impossible to calculate, is that the air driven forward by the projectile escapes sideways. However, a study of the motion of a mobile piston in a cylinder preventing lateral motion of the gas would provide some useful information on this matter.

This study can be found in Chapter V. I shall show that the resistance offered by the gas to the motion of the mobile piston is almost proportional to the velocity as it approaches zero. The coefficient of proportionality increases with velocity; when the velocity becomes very high, the resistance is nearly proportional to the square of the velocity.

It must be borne in mind that this last result is consistent with experiments carried out in free air[1]; however, the coefficient of proportionality is, as could be expected, lower than that which corresponds to motion in the cylinder. It represents about $\frac{6}{10}$ for cylindrical shot fired from a rifled barrel; the difference is even more noticeable in the case of projectiles with a hemispherical or ogival head.

[1]HÉLIE, *Balistique expérimentale*, 2e édition, t. II, p. 177 et 338. Paris, Gauthier-Villars; 1884.

19. You will forgive me for not citing the names of the geometricians who have dealt with these questions before me. It would have taken a great deal of time to provide the full historical background. Furthermore, much of this background information can be found in de Saint-Venant's foreword to his comments on the Works of Navier.

True, since this paper was published, many other works have been produced, among which mention must be made of those by de Saint-Venant on longitudinal shock, and a remarkable paper by Phillips on the motion of elastic prisms subjected to longitudinal stress. Phillips's paper was the first paper to provide finite solutions to the problems which geometricians had until then tried to solve by trigonometric means.

I must also mention Darboux, whose exemplary papers on second order partial differential equations will almost certainly soon be applied in Mechanics and Pure Physics.

Chapter I: Theory of Characteristic Curves

I. - General observations

20. Studying the motion of natural bodies leads us to define partial differential equations with four independent variables. Let us consider three axes with the co-ordinates Ox, Oy, Oz, each point of the body is defined by the three co-ordinates x, y and z, corresponding to a determined state with which all the others must be compared.

At time t, the co-ordinates x, y and z of this point have become $x + u$, $y + v$, $z + w$. The values of u, v and w are functions of the four independent variables x, y, z and t and usually satisfy three partial differential equations, whose form depends on the nature of the body.

In certain cases, the body moves *by parallel strata*. A direction MM' then exists in space, such that all the points in the same section perpendicular to MM' at all times have the same velocity and are parallel to this straight line.

If MM' is then taken to be the x axis, it can be assumed that $v = w = 0$ at all times; the only function then remaining to be solved is u, but it depends only on the two variables x and t. The motion is then governed by a single partial differential equation with only two independent variables.

In this paper I shall only study motion in parallel strata, leaving the task of drawing general conclusions from the results until later.

However, a few general considerations about partial differential equations with two independent variables must first be stated; these considerations will be of much use in helping us to understand the phenomena of motion.

This preliminary study is dealt with in this chapter. The theories set out herein are not entirely new; however, they are currently being expounded in the works of Monge and Ampère and have not, to my knowledge, been brought together to form a body of policy.

21. In the second order partial differential equations which govern motion by parallel strata, u is normally used to designate the function, and x and t the independent variables which designate abcissa of the stratum and time respectively. However, when considering partial differential equations from a purely analytical point of view, it is more convenient to use the same notation as Monge and Ampère, which will be adopted in this chapter.

The function will therefore be designated z, and the independent variables x and y. The following equations will then be postulated:

$$p = \frac{\partial z}{\partial x}, \qquad q = \frac{\partial z}{\partial y},$$
$$r = \frac{\partial^2 z}{\partial x^2}, \qquad s = \frac{\partial^2 z}{\partial x \partial y}, \qquad t = \frac{\partial^2 z}{\partial y^2}.$$

From these expressions we would pass on to those which will be used in the following chapters, replacing z by u, y by t, p by $\frac{\partial u}{\partial x}$ and q by $\frac{\partial u}{\partial t}$ etc.

22. Any relationship between the function z, the variables x and y and the derivatives p, q, r, s, t etc. is a partial differential equation with two independent variables. The order of the equation is the same as the order of the highest derivatives.

Of the equations of order n, those which are *linear with respect to the derivatives of the highest order* have specific properties and must be differentiated from the others.

An equation is *perfectly linear* when all of its derivatives and the original function are involved in it only in linear form, such that the coefficients of z, p, q, r, s, t etc. are nothing more than functions of x and y.

23. Any function z of x and y which can be substituted into the equation without changing it is called an *integral function of the equation*.

Perfectly linear equations have one particular remarkable property: the sum of two integrals is another integral.

A partial differential equation may have an infinite number of integrals. In certain very specific cases, all the integrals, with the exception of a few so-called *singular integrals*, can be expressed as a single analytic function which may be explicit or implicit and is known as a *general integral*.

However, this is not normally the case; we must limit our search to the general properties of the integrals of a given equation, except when determining the analytic function of the integral which gives the solution for each specific problem.

Any integral of a partial differential equation with two independent variables can be viewed as the equation of a surface relative to three rectangular co-ordinate axes Ox, Oy and Oz. It can be very useful to consider this *integral surface*.

The partial differential equation is the expression of a geometric property common to all integral surfaces.

II. - Characteristic curves - the case where the equation is linear relative to the highest order derivatives

24. The integral surfaces of a given equation can be considered as being formed by the motion of a particular curve which changes shape as it moves and which Monge called the *characteristic curve*.

Characteristic curves play an important role, not just from the point of view of generating surfaces, but also in problems of motion.

In order to define these curves in detail, we shall first consider an equation which is linear with respect to the highest order derivatives. To organise our ideas a little we shall assume that this is a second order equation, although the generalisation poses no major difficulty.

Thus let

$$\mathrm{H}r + 2\mathrm{K}s + \mathrm{L}t + \mathrm{M} = 0$$

a second order equation, linear in r, s, t where H, K, L and M designate any functions of x, y, z, p and q.

Let $z_1 = f_1(x, y)$ be an integral of this equation: it can be shown as a surface. The characteristics of this surface will be taken to be the curves along which this surface may be intersected by another integral surface, $z_2 = f_2(x, y)$ such that the first order derivatives are equal along the line of intersection, while the same does not apply for the second order derivatives. The two surfaces z_1 and z_2 would therefore touch along their intersection.

We shall prove that these curves exist and that they satisfy the same differential equation.

25. By hypothesis, the following three equations are satisfied at the intersection of the two surfaces z_1 and z_2

$$z_1 - z_2 = 0 \ ,$$

$$p_1 - p_2 = 0 \ , \qquad q_1 - q_2 = 0 \ .$$

The last two equations can be differentiated by considering y as a function of x defined by the first equation, which is obviously the horizontal projection of the intersection of the two surfaces, such that $\frac{dy}{dx}$ is the angular coefficient of the tangent to this line. By differentiation we obtain:

(1) $$r_1 - r_2 + (s_1 - s_2)\frac{dy}{dx} = 0$$

(2) $$s_1 - s_2 + (t_1 - t_2)\frac{dy}{dx} = 0$$

In addition, by substituting z_1 and z_2 into the above equation and subtracting the results, since the derivatives p and q are equal at the intersection, we obtain the following equation:

(3) $$H(r_1 - r_2) + 2K(s_1 - s_2) + L(t_1 - t_2) = 0$$

Equations (1), (2) and (3) must be true at all points along the interface. They are also uniform relative to the difference between $r_1 - r_2$, $s_1 - s_2$ and $t_1 - t_2$ Therefore, if the second order derivatives are not equal along the intersection, the determinant is cancelled out, giving the following:

(4) $$H\left(\frac{dy}{dx}\right)^2 - 2K\left(\frac{dy}{dx}\right) + L = 0$$

This is an ordinary first order differential equation which determines the characteristics of the surface z, when this is known, since the coefficients H, K and L are then known functions of x and y.

For each system of values of x and y, in other words for each point on the surface z_1, the angular coefficient of the characteristic curve has two values corresponding to the roots of the previous equation. So two characteristics curves pass through each point on the surface.

The characteristic curves of the surface can be considered to form two distinct systems, each corresponding to the roots of equation (4). It is clear that the surface z_1 can be considered as being formed by one or other of these two systems of curves.

A surface such as z_1 can be formed by an infinite number of systems of curves. However, the characteristic curves have a specific property for all these systems. Indeed, if we consider another integral surface z_2 which connects up with the first along the line of intersection, this line becomes a characteristic common to the two surfaces z_1 and z_2. In other words C, the curve of intersection. When the characteristic curve of the same system changes shape and moves to generate surface z_1, there comes a moment when it merges with C; here it becomes possible to pass from surface z_1 to surface z_2 without changing generation mode.

The surface obtained will consist of a part of z_1 and a part of z_2 and will still have all the properties of the integral surfaces. However, a discontinuity will be observed in the second derivatives at all points along the curve C.

The characteristic curve generation mode is therefore to be preferred to all others because it offers the possibility of passing from one integral surface to another.

When the integral surface z_1 is not given, equation (4) is usually insufficient to give a few notions regarding the shape of the curve. However, if the partial differential equation is perfectly linear, the coefficients of equation (4) depend only on x and y, such that the horizontal projections of the characteristic curves are the same for all integral surfaces and can be determined by integrating an ordinary first order differential equation.

Whenever the proposed equation is not perfectly linear , the same property subsists inasmuch as the coefficients of the highest derivatives are independent of z, p and q.

26. There is a second equation which exists along a specific characteristic curve; it also corresponds to a new property which could be used to define this curve.

If we consider a definite curve plotted on an integral surface, we obtain the following by differentiating the first order derivatives along this curve,

$$\frac{dp}{dx} = r + s\frac{dy}{dx}$$

$$\frac{dq}{dx} = s + t\frac{dy}{dx}$$

If we multiply these equations by λ and μ and add them together, we obtain:

$$\lambda\frac{dp}{dx} + \mu\frac{dq}{dx} = \lambda r + s\left(\lambda\frac{dy}{dx} + \mu\right) + \mu\frac{dy}{dx}$$

We can try to solve λ and μ, such that the second term becomes identical, give or take a constant factor, to all terms which are, in the proposed equation, affected by second order derivatives. For this, the following must be true:

$$\lambda - \alpha H = 0, \ \lambda\frac{dy}{dx} + \mu - 2\alpha K = 0, \ \mu\frac{dy}{dx} - \alpha L = 0$$

These equations are uniform in λ, μ and α; they cannot support a system with solutions other than zero unless the determinant can be cancelled out, giving the following

$$H\left(\frac{dy}{dx}\right)^2 - 2K\frac{dy}{dx} + L = 0$$

which is the equation of the characteristics. Therefore, the only prerequisite for us to be able to determine λ and μ, in accordance with the conditions we have adopted, is that the curve must be considered as a characteristic.

Taking α = 1, we obtain the following

$$\lambda = H, \qquad \mu = 2K - H\frac{dy}{dx}$$

and subsequently along the characteristic curve,

$$H\frac{dp}{dx} + \left(2K - H\frac{dy}{dx}\right)\frac{dq}{dx} = Hr + 2Ks + Lt$$

However, from the equation proposed, we know that the second term is equal to – M. Therefore

(5) $$H\frac{dp}{dx}+\left(2K-H\frac{dy}{dx}\right)\frac{dq}{dx}+M=0$$

This is the second equation for the characteristic curves.

27. In addition to equations (4) and (5), which are specific to the characteristic curves, another equation is produced along any curve plotted on the surface. This equation is as follows:

(6) $$\frac{dz}{dx}=p+q\frac{dy}{dx}$$

Along a characteristic curve, the values of y, z, p and q can be considered to be functions of the variable x which can only be defined using four equations. We, however, have only the three equations (4), (5) and (6) which are insufficient for determining the four unknown functions; this means that along one of these curves it is generally possible to choose one of the four functions y, z, p or q; normally, much work needs to be carried out to determine the characteristic curves.

However, there are certain exceptions. Thus, when the coefficients of second order derivatives are functions of x and y only, the function y is determined, except for an arbitrary constant, and we can only assume one of the functions z, p or q. It may even happen that the shape of the characteristic curves is fully determined.

This is the case when one considers the equation

$$r+2as+a^2t=0,$$

the first equation of the characteristic curves has two equal roots and can be reduced to

$$\frac{dy}{dx}=a, \quad \text{hence } y=ax+b.$$

The two other equations are

$$\frac{dp}{dx}+a\frac{dq}{dx}=0, \qquad \frac{dz}{dx}=p+aq.$$

The first gives $p+aq=\alpha$, where α is an arbitrary constant and the second equation subsequently stipulates that

$$z = \alpha x + \beta.$$

The characteristics then become straight lines. Only the functions p and q cannot be determined, one of which can be selected at will.

In the case of the equation we have just examined, the two roots of equation (4) are equal. The partial differential equations, one of the properties of which are merged characteristics, can be clearly distinguished from the other equations and should be studied separately, which is not our task here since equations of this sort will not be encountered in problems concerning the propagation of motion.

I shall make no further study of cases where the characteristic curve equations can be used to determine the general integral of the proposed equation. Such research would be of little use in studying problems concerning the propagation of motion in bodies, which is our task here.

28. The theory concerning second order equations is so easy to apply to general cases, that we shall limit ourselves to setting out the results relating to a linear equation of order n as to the derivatives of the highest order.

The characteristics of an integral surface z_1 are the curves along which this surface can be intersected by another integral z_2, such that along this line of intersection the derivatives of the $n-1$ first orders of the functions z_1 and z_2 will be equal, although this will not be the case for derivatives of order n. The two surfaces z_1 and z_2 which intersect under these conditions have the intersection curve as a common characteristic.

The characteristics of an equation of order n are therefore the natural generatrices of its integral surfaces.

i.e.

$$A\frac{\partial^n z}{\partial x^n} + B\frac{\partial^n z}{\partial x^{n-1}\partial y} + C\frac{\partial^n z}{\partial x^{n-2}\partial y} + \cdots R\frac{\partial^n z}{\partial y^n} + M = 0$$

the proposed equation; the equation for the horizontal projection of the characteristic curves is

$$A\left(\frac{dy}{dx}\right)^n - B\left(\frac{dy}{dx}\right)^{n-1} + C\left(\frac{dy}{dx}\right)^n - \cdots \pm R = 0 \tag{7}$$

There are n systems of characteristics.

The second equation specific to these curves is obtained in the same way as the second order equation. i.e. $\omega_1, \omega_2, \ldots, \omega_n$ the n derivatives of order n - 1; this second equation is

$$(8)\qquad \begin{cases} A\dfrac{d\omega_1}{dx} - \left(A\dfrac{dy}{dx} - B\right)\dfrac{d\omega_2}{dx} \\ \qquad + \left[A\left(\dfrac{dy}{dx}\right)^2 - B\dfrac{dy}{dx} + C\right]\dfrac{d\omega_3}{dx} - \ldots + M = 0 \end{cases}$$

Lastly, the equations obtained by differentiating the function z and its derivatives to the power n - 2 along any curve, namely

$$(9)\qquad \begin{cases} \dfrac{dz}{dx} = p + q\dfrac{dy}{dx} \\ \dfrac{dp}{dx} = r + s\dfrac{dy}{dx}, \quad \dfrac{dq}{dx} = s + t\dfrac{dy}{dx}, \end{cases}$$

etc.

Since there are n - 1 derivatives of order n - 2, there are $\dfrac{n(n-1)}{2}$ equations (9). The number of equations which occur along the characteristic is equal to $\dfrac{n(n-1)}{2} + 2$.

However, these equations contain the unknown functions y, z and the derivatives of z up to the order n - 1; the number of unknown functions is therefore equal to $\dfrac{n(n+1)}{2} + 1$.

The number of functions is larger than that for the equations by a figure of

$$\frac{n(n+1)}{2} + 1 - \frac{n(n-1)}{2} - 2 = n - 1.$$

In general we may therefore arbitrarily obtain n - 1 of these functions along the curve; the others will then be determined. There are, of course, special cases where some of the functions are given in advance. If, for instance, the equation proposed is perfectly linear, the function y is determined by equation (7), except for any arbitrary constants.

For the first order, the number of functions is the same as the number of equations and the characteristic curves are fully determined.

III. - Characteristic curves of equations which are not linear with regard to the highest order derivatives

29. The definitions of characteristics given above must be modified when the proposed equation is not linear with regard to the highest order derivatives. In other words, this modification is no more than a generalisation, because the new definition provides the same results for linear equations as the previous definition. The linear equations were at first only considered to simplify the explanation.

In order to avoid over-complex notation and formulae, we shall limit ourselves to setting out the theory for a second order equation.

(1) $$F(x, y, z, p, q, r, s, t) = 0$$

Let z be an integral of this equation. $z + \delta z$ will be used to denote an infinitesimally close integral, where δz is the variation experienced by the function when, with x and y remaining constant, the transition is made from the first integral to the second. We should obtain

$$F(x, y, z+ \delta z, p + \delta p, q + \delta q, r + \delta r, s + \delta s, t + \delta t) = 0$$

which gives us

(2) $$\frac{\partial F}{\partial r}\delta r + \frac{\partial F}{\partial s}\delta s + \frac{\partial F}{\partial t}\delta t + \frac{\partial F}{\partial p}\delta p + \frac{\partial F}{\partial q}\delta q + \frac{\partial F}{\partial z}\delta z = 0$$

However

$$\delta r = \delta\frac{\partial^2 z}{\partial x^2} = \frac{\partial^2 \delta z}{\partial x^2}, \qquad \delta s = \delta\frac{\partial^2 z}{\partial x \partial y} = \frac{\partial^2 \delta z}{\partial x \partial y},$$

Equation (2) can therefore be considered as an equation with second order partial derivatives in which the unknown function is δz. The coefficients of δr, δs, ..., δz are, it is true, functions not only of x and y but also of z, p, q, r, s and t. However, since the integral z is assumed to be known, these coefficients can be expressed in terms of x and y and equation (2) then becomes perfectly linear. We can therefore immediately write the equation for its characteristics, i.e.

(3) $$\frac{\partial F}{\partial r}\left(\frac{dy}{dx}\right)^2 - \frac{\partial F}{\partial s}\frac{dy}{dx} + \frac{\partial F}{\partial t} = 0$$

Given an integral surface, these curves are completely defined by the above equation; they shall be considered as the characteristic curves of this surface.

The characteristic curves of the integral surfaces of a non-linear equation with respect to the derivatives of the highest order are the curves along which each of these surfaces can be intersected by an infinitesimally close integral surface, such that, since the first order derivatives are strictly equal along the intersection, the second order derivatives exhibit a discontinuity. Besides, this discontinuity is infinitesimally small; in other words, it converges to zero at the same time as the difference δz of the ordinates for infinitesimally close surfaces.

30. It is possible, as for the linear equation in r, s and t, to form a second equation specific to the characteristic curves. If we represent the four third order derivatives by u, w, ω and v, then along any curve plotted on the surface

$$(4) \qquad \begin{cases} \dfrac{dr}{dx} = u + w\,\dfrac{dy}{dx}, \\[2ex] \dfrac{ds}{dx} = w + \omega\,\dfrac{dy}{dx}, \\[2ex] \dfrac{dt}{dx} = \omega + v\,\dfrac{dy}{dx}, \end{cases}$$

Differentiating the initial equation with respect to x, we obtain

$$(5) \qquad u\,\frac{\partial F}{\partial r} + w\,\frac{\partial F}{\partial s} + \omega\,\frac{\partial F}{\partial t} + \mathrm{N} = 0$$

where N is the set of independent terms of the third order derivatives.

By multiplying the first two equations (4) by λ, μ and adding them together, an attempt can be made to arrange these coefficients in such a way that the second term of the equation will reproduce, within a constant factor, all the terms which, in (5), have third order derivatives. For this the following must be true

$$\lambda - \alpha\frac{\partial F}{\partial r} = 0, \qquad \lambda\frac{dy}{dx} + \mu - \alpha\frac{\partial F}{\partial r} = 0, \qquad \mu\frac{dy}{dx} - \alpha\frac{\partial F}{\partial t} = 0$$

i.e. three homogeneous equations involving λ, μ and α. For these equations to be compatible, the determinant must be cancelled out. Once this has been done, we are left with equation (3) for the characteristic curves.

If the curve considered is a characteristic curve, the problem can be solved; substituting $\alpha = 1$, we obtain

$$\lambda = \frac{\partial F}{\partial r}, \qquad \mu = \frac{\partial F}{\partial s} - \frac{\partial F}{\partial r}\frac{dy}{dx}.$$

The second equation for the characteristics is therefore

$$(6) \qquad \frac{\partial F}{\partial r}\frac{dr}{dx} + \left(\frac{\partial F}{\partial s} - \frac{\partial F}{\partial r}\frac{dy}{dx}\right)\frac{ds}{dx} + N = 0$$

If we had considered the last two equations (4) and had differentiated the proposed equation (1) with respect to y, we would have obtained another equation which can be written

$$(7) \qquad \left(\frac{\partial F}{\partial s} - \frac{\partial F}{\partial t}\frac{dx}{dy}\right)\frac{ds}{dx} + \frac{\partial F}{\partial t}\frac{dt}{dx} + N_1\frac{dy}{dx} = 0$$

However, if we add equations (6) and (7) and if equation (3) is taken into account, we obtain

$$\frac{\partial F}{\partial r}\frac{dr}{dx} + \frac{\partial F}{\partial s}\frac{ds}{dx} + \frac{\partial F}{\partial t}\frac{dt}{dx} + N + N_1\frac{dy}{dx} = 0$$

which is simply the proposed equation differentiated along the characteristic curve.

Equation (7) is therefore not new, since it is a consequence of theproposed equation and equation (6). Equation (7) should preferably be replaced by the proposed equation since the latter is laid out in integrated form with respect to the unknown functions.

31. The equations for the characteristic curves here contain seven unknown functions of x, i.e. y, z, p, q, r, s and t which must satisfy the equation put forward, equations (3) and (6) and the following equations

$$\frac{dz}{dx} = p + q\frac{dy}{dx}, \quad \frac{dp}{dx} = r + s\frac{dy}{dx}, \quad \frac{dq}{dx} = s + t\frac{dy}{dx}$$

There are therefore only six equations for determining seven unknown functions, in other words one of the functions still remains arbitrary. The indeterminacy remains as in the linear equations.

32. The theory of characteristic curves can be extended without difficulty to equations of order n. There are two equations specific to these curves and they are analagous to equations (3) and (6): to this we should add the proposed equation and the equation obtained by differentiating the derivatives of the first n - 1 orders along the curve. The number of equations obtained is thus equal to

$$3+\frac{n(n+1)}{2}.$$

However, these equations contain unknown functions of x, firstly y and z, and then the derivatives up to the order n; the number of unknown functions is therefore equal to

$$1+\frac{(n+1)(n+2)}{2}$$

The number of unknown functions is greater than the number of equations by a figure of

$$\frac{(n+1)(n+2)}{2}+1-\frac{n(n+1)}{2}-3=n-1.$$

The indeterminacy is the same as if theproposed equation were linear with regard to the highest order derivatives. We can arbitrarily obtain n - 1 unknown functions. Only in the first order is the determination of functions complete.

It can also be seen that the characteristics must be regarded as the natural generatrices of the integral surfaces. Indeed, when the characteristic curve changes shape and moves to produce a surface, it is possible, by passing from one characteristic to the next, to go from the surface in question to another infinitesimally close integral surface. This allows us to consider two surfaces which have all their derivatives equal along the line of intersection up to order n as a single integral surface.

IV. - Application to first order equations

33. In order to clarify the general theories developed above, we will apply them to those first order equations whose characteristics have been fully determined; since the number of equations is then equal to the number of unknown functions.

i.e. first of all the first order linear equation.

$$Pp+Qq=R$$

where P, Q and R are functions of x, y and z

The two equations specific to the characteristic curves are

$$P\frac{dy}{dx} - Q = 0\ , \qquad \frac{dz}{dx} = \frac{R}{p}\ ,$$

thereafter we obtain the following for the differential equations for these curves:

$$\frac{dx}{p} = \frac{dy}{\alpha} = \frac{dz}{R}\ ,$$

constituting the well known correlative system.

Returning to the definition given for these curves, it is evident that the characteristics for the first order linear equations are the curves along which the integral surfaces intersect.

They can also be defined as the curves belonging to the integral surfaces, along which the plane of the tangent may experience a discontinuity.

When the integrals of the correlative system can be obtained, they are in the following form

$$\alpha = \varphi(x, y, z), \qquad \beta = \psi(x, y, z)$$

where α and β are two arbitrary constants. It is sufficient to establish a relationship between α and β for the characteristic curve to move according to a given law and produce an integral surface. The general integral is therefore

$$\alpha = \Phi(\beta)$$

where Φ is any function.

34. Let us now consider a first order non-linear equation

(1) $$F(x, y, z, p, q) = 0$$

to simplify the expressions, let us assume

$$\frac{\partial F}{\partial x} = X\ , \quad \frac{\partial F}{\partial y} = Y\ , \quad \frac{\partial F}{\partial z} = Z \quad \frac{\partial F}{\partial p} = P\ , \quad \frac{\partial F}{\partial q} = Q\ .$$

The two equations specific to the characteristic curves are in this case

$$P\,\frac{dy}{dx} - Q = 0,$$

$$P\frac{dp}{dx}+\left(Q-P\frac{dy}{dx}\right)\frac{dq}{dx}+Zp+X=0$$

or

(2) $$\mathrm{P}\ \frac{dy}{dx}-\mathrm{Q}=0,$$

(3) $$\mathrm{P}\ \frac{dp}{dx}+\mathrm{Zp}+\mathrm{X}=0$$

to which we should add the suggested equation and the following equation

(4) $$\frac{dz}{dx}=p+q\ \frac{dy}{dx}\ .$$

For greater symmetry, the suggested equation can be replaced by the following

(5) $$\mathrm{Q}\ \frac{dq}{dx}+\mathrm{Zq}+\mathrm{Y}=0$$

which is obtained by combining equation (3) with the proposed equation differentiated along the characteristic curve. However, it must be pointed out that by doing this, the frame of the problem is extended; equations (2), (3), (4) and (5) would remain the same, the second term of equation (1) being replaced by an arbitrary constant.

Equations (2), (3), (4) and (5) define the four functions y, z, p and q of x; they can be expressed in the form

$$\frac{dx}{P}=\frac{dy}{Q}=\frac{dz}{Pp+Qq}=\frac{dp}{-X-Zp}=\frac{dq}{-Y-Zq}$$

and make up the correlative system.

We only need to obtain three integrals to determine the four functions y, z, p and q if the proposed equation is added; these functions will be expressed in terms of x and three arbitrary constants in the form

(6) $$\begin{cases} y=f_1(x,\alpha,\beta,\gamma) \\ z=f_2(x,\alpha,\beta,\gamma) \\ p=f_3(x,\alpha,\beta,\gamma) \\ q=f_4(x,\alpha,\beta,\gamma) \end{cases}$$

35. We shall not show here how these equations are used to deduce the *complete integral* of the proposed equation, which, as we know, contains two arbitrary constants, or how the general integral (i.e. the envelope of the surface generated by the complete integral) is formed when an arbitrary relationship is established between the two constants contained in this integral.

It will be sufficient to show how the characteristic curves move as the integral surfaces are generated.

We shall first point out that at a given point in space, there are an infinite number of characteristic curves — the first two equations (6) contain three arbitrary constants. These characteristic curves pass through a single point and also constitute a surface which shows there a conical apex . This surface will be shown below to be an integral.

First order non-linear equations can be clearly distinguished from linear equations, since in the case of linear equations there is only one characteristic which passes through a given point in space.

To study the generation of integral surfaces using characteristic curves, an attempt will be made to find the condition in which two infinitesimally close characteristic curves belong to a single integral surface.

Let α, β and γ be constants which correspond to the first characteristic; in the case of an infinitesimally close characteristic curve, they will have the values $\alpha + d\alpha$, $\beta + d\beta$, $\gamma + d\gamma$. The expression

$$dz = pdx + qdy$$

must be satisfied at the point of transition (x, y, z) from the first curve to the point $(x + dx, y + dy, z + dz)$ on the second curve. Therefore

$$dy = \frac{\partial y}{\partial x} dx + \frac{\partial y}{\partial \alpha} d\alpha + \frac{\partial y}{\partial \beta} d\beta + \frac{\partial y}{\partial \gamma} d\gamma \ ,$$

$$dz = \frac{\partial z}{\partial x} dx + \frac{\partial z}{\partial \alpha} d\alpha + \frac{\partial z}{\partial \beta} d\beta + \frac{\partial z}{\partial \gamma} d\gamma \ ,$$

and the condition becomes

$$\frac{\partial z}{\partial x} dx + \frac{\partial z}{\partial \alpha} d\alpha + \frac{\partial z}{\partial \beta} d\beta + \frac{\partial z}{\partial \gamma} d\gamma = pdx + q\left(\frac{\partial y}{\partial x} dx + \frac{\partial y}{\partial \alpha} d\alpha + \frac{\partial y}{\partial \beta} d\beta + \frac{\partial y}{\partial \gamma} d\gamma\right).$$

The coefficient of dx cancels out since it is equal to $\frac{\partial z}{\partial x} - p - q\,\frac{\partial y}{\partial x}$, the derivatives being taken from along the first characteristic curve. This leaves us with

$$(7) \qquad \left(\frac{\partial z}{\partial \alpha} - q\frac{\partial y}{\partial \alpha}\right)d\alpha + \left(\frac{\partial z}{\partial \beta} - q\frac{\partial y}{\partial \beta}\right)d\beta + \left(\frac{\partial z}{\partial \gamma} - q\frac{\partial y}{\partial \gamma}\right)d\gamma = 0 .$$

This is the condition we have been looking for. It must be independent of x, such that the variable cannot be part of a common factor. This can be verified as follows:

The values (6) of y, z, p and q make identical the proposed equation as well as the correlative system. The proposed equation, differentiated with respect to α gives

$$Y\frac{\partial y}{\partial \alpha} + Z\frac{\partial z}{\partial \alpha} + P\frac{\partial p}{\partial \alpha} + Q\frac{\partial q}{\partial \alpha} = 0 ;$$

moreover, the correlative system gives us

$$Y = -Zq - P\,\frac{\partial q}{\partial x}, \qquad Q = P\,\frac{\partial y}{\partial x} ;$$

thence

$$Z\left(\frac{\partial z}{\partial \alpha} - q\frac{\partial y}{\partial \alpha}\right) + P\left(\frac{\partial p}{\partial \alpha} - \frac{\partial q}{\partial x}\frac{\partial y}{\partial \alpha} + \frac{\partial q}{\partial \alpha}\frac{\partial y}{\partial x}\right) = 0 .$$

However, the equation $p = \frac{\partial z}{\partial x} - q\,\frac{\partial y}{\partial x}$ gives

$$\frac{\partial p}{\partial \alpha} = \frac{\partial^2 z}{\partial x \partial \alpha} - \frac{\partial q}{\partial \alpha}\frac{\partial y}{\partial x} - q\frac{\partial^2 y}{\partial x \partial \alpha} ,$$

and the previous equation becomes

$$Z\left(\frac{\partial z}{\partial \alpha} - q\,\frac{\partial y}{\partial \alpha}\right) + P\left(\frac{\partial^2 z}{\partial x\, \partial \alpha} - q\,\frac{\partial^2 y}{\partial x\, \partial \alpha} - \frac{\partial q}{\partial x}\,\frac{\partial y}{\partial \alpha}\right) = 0 ;$$

this is immediately integrated and gives

$$\frac{\partial z}{\partial \alpha} - q\frac{\partial y}{\partial \alpha} = \mathrm{A}\, e^{-\int \frac{Z}{P}dx},$$

since A is independent of x and, consequently, a function of α, β and γ only. Similar expressions would be found for the coefficients of $d\beta$ and $d\gamma$ in equation (7), which can then be written

$$e^{-\int \frac{Z}{P}dx}(\mathrm{A}d\alpha + \mathrm{B}d\beta + \mathrm{C}d\gamma) = 0.$$

The variable x is involved in only the common factor which will normally not be zero, and the required condition will be reduced to

(8) $$\mathrm{A}d\alpha + \mathrm{B}d\beta + \mathrm{C}d\gamma = 0.$$

It is not our task to consider the exceptions to the rule which could occur. We shall limit ourselves to the following comments:

If an arbitrary relationship is established between the constants α, β and γ, one of these constants could be eliminated; it would then be possible to integrate the previous equations and thence provide another finite relationship between the constants. α, β and γ can then be eliminated from the two relationships and from the first two equations (6). This would provide us with the equation for an integral surface.

We could, for instance, state that this surface contains a given curve. The first relationship between α, β and γ will be obtained by assuming that the curves defined by the first two equations (6) meet the curve in question.

It is easy to verify that equation (7) or its equivalent (8) provides the necessary and sufficient condition for two infinitesimally close characteristic curves to meet in such a way that at the point where two infinitesimally close characteristic curves meet they belong to the same integral surface. As a result, all the characteristic curves for a single surface have an envelope which forms its *edge of regression.*

It can also be seen that all the characteristic curves which pass through a single point generate one of the integral surfaces. In this case, the edge of regression is reduced to this point.

V. - Concerning one of the properties of second order linear equations

36. In the case of second order equations we shall limit our discussion to one rather curious property of perfectly linear equations; this property is fairly interesting from the point of view of its applications to pure physics.

If φ and ψ represent two functions of x, y, z, p and q, the equation

$$\frac{\partial\varphi}{\partial x} = \frac{\partial\psi}{\partial y} \tag{1}$$

where, for the purposes of differentiation, z, p and q are considered to be functions of x and y, is a second order linear equation in r, s and t.

Given an integral z of this equation, we have a function u of x and y such that

$$du = \psi dx + \varphi dy ,$$

since the equation proposed is none other than the integration condition of the previous equation. The function u is known, except for one arbitrary constant; it is continuous as long as φ and ψ remain finite, and it is evident that the functions z and u are completely linked to each other.

Let us consider two integrals of the proposed equation, z_1 and z_2; the representative surfaces intersect along a curve whose horizontal projection is defined by the equation $z_1 - z_2 = 0$. Could the surface whose ordinate would be represented on one side of the curve of intersection by z_1 and on the other side by z_2 be considered in certain respects as a single integral surface? The corresponding function, designated (z_1, z_2), would be continuous; however, its derivatives would experience a discontinuity along the curve of intersection.

Any integral z defines a continuous function u, therefore in order for (z_1, z_2) to be considered as an integral, the function u which it defines would have to be continuous along the intersection since the functions φ and ψ vary suddenly along the intersection but generally never tend towards infinity.

The integrals z_1 and z_2 have two corresponding functions u_1 and u_2; it is possible to obtain the arbitrary constants that these latter functions contain, such that u_1 and u_2 become equal at one point on the curve of intersection.

The same will apply to all the other points if, as one moves along the curve, du_1 and du_2 are equal. For this to be the case, the following, obviously, must apply

(2) $$(\psi_1 - \psi_2)\,dx + (\varphi_1 - \varphi_2)\,dy = 0\,.$$

But the equation $u_1 - u_2 = 0$ when differentiated along the curve gives

(3) $$(p_1 - p_2)dx + (q_1 - q_2)dy = 0\,.$$

The above equations must come into existence at the same time and must produce equal values for $\dfrac{dy}{dx}$.

37. Not all types of integrals can be associated in this manner to make new integrals. It would be useful to study the curves of intersection because they can be considered to be a generalisation of the characteristics. It is indeed easy to see that if p_2 and q_2 converge towards p_1 and q_1 respectively, the curves of intersection would tend to merge with the characteristic curves as defined above.

However, when equation (1) is perfectly linear or, more generally, when the coefficients of the second order derivatives are functions of only x, y and z, the curve of intersection is indeed a characteristic curve common to both surfaces.

i.e. let

$$\varphi = ap + bq + c,\ \phi = a'p + b'q + c',$$

and the functions a, b, c, a', b' and c' would depend only on x, y and z, and equation (2) would become

(4) $$[a'(p_1 - p_2) + b'(q_1 - q_2)]\,dx + [a\,(p_1 - p_2) + b\,(q_1 - q_2)]\,dy = 0.$$

Equations (3) and (4), which must come into existence at the same time, are homogeneous with regard to the values of $p_1 - p_2$ and $q_1 - q_2$. In order for these values not to equal zero at every point on the curve of intersection, the determinant must cancel out, giving the following condition

$$a\left(\frac{dy}{dx}\right)^2 - (b - a')\,\frac{dy}{dx} - b' = 0$$

it is easy to verify that this equation merges with that of the characteristic curves.

In the special case we have just examined, two integral surfaces which intersect along a common characteristic curve can be combined to make a single integral.

38. The results we have just obtained depend mainly on the particular form of the equation put forward. However, when this is a perfectly linear equation, we are dealing with a general property which depends on the nature of the equation and in no way on the specific form in which it is written.

This is what results from the following theorem, demonstrated by R. Liouville.

For a second order linear equation

$$Hr + 2Ks + Lt + Pp + Qq + Zz + R = 0,$$

there are an infinite number of factors μ, functions of x and y, such that when multiplied by μ (the first term of the equation), the equation becomes

$$\frac{\partial\varphi}{\partial x} - \frac{\partial\psi}{\partial y} = 0, \tag{1}$$

with functions φ and ψ in the form

$$\varphi = ap + bq + cz + h, \quad y = a'p + b'q + c'z + h'.$$

This is demonstrated by identifying the first term of the equation proposed, multiplied by μ, with the first term of (1). This produces seven equations which can be used to eliminate the eight functions a, b, c, h, a', b', c' and h' of x and y, producing the following equation

$$\frac{\partial^2(\mu H)}{\partial x^2} + 2\frac{\partial^2(\mu K)}{\partial x\partial y} + \frac{\partial^2(\mu L)}{\partial y^2} - \frac{\partial(\mu P)}{\partial x} - \frac{\partial(\mu Q)}{\partial y} + \mu Z = 0,$$

which has second order derivatives in μ and has the same characteristic curves as the equation proposed.

In this way, we have shown that any integral of the equation in μ has the required property.

Multiplying a partial differential equation by a factor which is a function of x and y obviously does not alter it in any way: we can therefore state that two integrals from a perfectly linear second order equation can be combined to make a single integral when they intersect along a common characteristic curve.

Chapter II: Producing Differential Equations

I. - Properties of perfect gases

39. The main subject of this paper being the motion of perfect gases, we should recapitulate on some of the properties of these bodies.

A relationship exists between the specific volume v, the external pressure p and the absolute temperature T of a particular body of the form

$$f(v, p, \mathrm{T}) = 0,$$

which is called the *characteristic equation*. The characteristic equation for a perfect gas is

$$pv = \mathrm{RT},$$

where R is a constant whose value depends on the nature of the gas. This relationship is nothing more than a translation of Mariotte's law and Gay-Lussac's law.

In addition, for perfect gases, the specific heat at constant pressure C and the specific heat at constant volume c are independent of temperature and pressure: the ratio $\frac{C}{c} = m$ must also therefore be considered as a constant. It is always greater than 1.

40. *Adiabatic expansion*. When a gas expands extremely slowly, so that the velocity of its particles is always negligible, the external work done is represented by $\int p\, dv$. This integral can only be solved if a second relationship between specific volume, temperature and pressure is added to the characteristic equation.

If, in particular, it is assumed that the walls of an envelope are impermeable to heat, the expansion is termed *adiabatic*, and the following relationship exists between pressure and specific volume

$$pv^m = \text{const.}$$

Let p_0 and v_0, be the pressure and the specific volume at a particular moment, taken as the initial moment,

$$pv^m = p_0 v_0{}^m;$$

the external work done since the initial moment is

$$\int_{v_0}^{v} p_0 v_0^m \frac{dv}{v^m} = \frac{1}{m-1} p_0 v_0^m \left(\frac{1}{v_0^{m-1}} - \frac{1}{v^{m-1}} \right).$$

41. *Internal energy.* When, in the above expression, the volume v increases indefinitely, the external work done converges to the limit

$$\frac{p_0 v_0}{m-1}.$$

This expression relates to the unit of weight of the gas, but it can be extended to any gaseous mass, provided that the total volume of the mass is used for v_0 rather than the specific volume.

So, if p is taken to be the pressure, and v the volume of a gaseous mass, the external work it develops during an indefinite expansion which takes place extremely slowly without either adding or removing heat, is expressed by $\frac{pv}{m-1}$. This is the amount of internal energy which would be used to produce the external work without modifying the composition of the gas, provided that the characteristic equation remains true and the specific heat remains at a constant value and this can be verified down to absolute zero, a temperature which the gas approaches during the indefinite adiabatic expansion.

In the remainder of this chapter, the value $\frac{pv}{m-1}$ will simply be termed the *internal energy*, thereby leaving out of consideration the amount of this energy which could be used only if the composition of the gas were changed.

42. *Total energy.* When expansion does not occur extremely slowly, the particles of gas can travel at fairly high velocities. The active force which corresponds to these velocities can in part be taken at the expense of the internal energy, still assuming that heat is neither added nor removed.

We shall suppose, in accordance with the equivalence principle, that in the transformation undergone by a space element of a gas, the drop in internal energy corresponds exactly to the active force acquired by the elements and to the external work produced.

To simplify our sentence structure, the term *total energy* will occasionally be used to designate the sum of the active force corresponding to the noticeable motion and the internal energy $\frac{pv}{m-1}$. We can then state the following principle:

During the basic transformation of a mass particle of the gas which is accomplished without heat being added or removed, the drop in total energy is equal to the external work produced.

II. - Motion of a cylindrical body in parallel strata

43. Let us consider a cylindrical body in an envelope of the same shape which is assumed to be strong enough to withstand all variations in the pressure of the body it contains without noticeable deformation. In the case of a solid, the envelope can be removed, provided that the variations in section are very slight.

Such a body is said to move *in parallel strata* when all the molecules located in the section perpendicular to the generatrices are, at a given moment, travelling at the same velocity and parallel to these generatrices. Motion of this type is the only one which will be studied here and we shall assume that it is always possible, although in certain cases the presence of the envelope may slow down the molecules which come into contact with it. We shall therefore not consider the friction of the body against the envelope or, if it is taken into account, we shall assume that it has the same effect on the motion of a stratum as if the friction were replaced by a force distributed in a uniform manner across the base of the stratum.

The different sections of the body, perpendicular to the generatrices, will be designated by their distance along the x-axis counted from an arbitrary origin and parallel to the generatrices, these abscissae being taken at a given moment, considered the initial moment.

If u is the distance travelled by section x at time t: the function u depends on two variables x and t and the problem consists in determining this function.

Let us consider the stratum between x and $x + dx$ on the x-axis, the two bases of the stratum have at time t travelled distances of u and $u + \frac{\partial u}{\partial x} dx$ respectively. The length of the stratum, originally equal to dx, has now become $dx + \frac{\partial u}{\partial x} dx$, i.e. it has increased in length by the amount $\frac{\partial u}{\partial x}$, relative to the length.

The partial derivative $\frac{\partial u}{\partial x}$ thus represents the *linear expansion* of the body; and the partial derivative $\frac{\partial u}{\partial t}$ obviously represents the velocity.

44. By applying the theorem of balance of momentum, we can easily obtain an equation which the function u satisfies.

Let us consider the stratum between the sections x and $x+dx$, the velocity of the section x, equal to $\frac{\partial u}{\partial t}$ at time t has become $\frac{\partial u}{\partial t}+\frac{\partial^2 u}{\partial t^2}dt$ at time $t+dt$, and thereafter increases at a rate of $\frac{\partial^2 u}{\partial t^2}dt$. This increase remains the same for all points on the stratum if the infinitesimally small second order values are ignored, and if the variation in the motion of the stratum is

$$\rho_0\,\omega\,\frac{\partial^2 u}{\partial t^2}dx\,dt,$$

where ω is the section of the body and ρ_0 the density corresponding to section ω at time $t=0$.

The product with respect to dt of the projection along the x-axis of the forces applied to the stratum must equal the above value. The external forces can be represented by ωR dx, this obviously incorporating friction from the envelope or viscosity. The value of R generally depends on the point along the x-axis, the velocity $\frac{\partial u}{\partial t}$, the expansion $\frac{\partial u}{\partial x}$ and the pressure p exerted in the stratum.

Since friction has been taken into account, the pressures exerted by the envelope are perpendicular to the x-axis: we therefore need only consider the pressures exerted at the bases. These are equal to ωp and to $-\omega\left(p+\frac{\partial p}{\partial x}dx\right)$. The equation provided by the theorem of balance of momentum is therefore as follows, once the common factors have been removed:

$$\rho_0\frac{\partial^2 u}{\partial t^2}=R-\frac{\partial p}{\partial x}. \tag{1}$$

However, this contains p, another unknown function of x and y, in addition to the function u; this new function depends on the nature of the body.

45. In certain cases, the function p is known in advance, and expressed in terms of x and $\frac{\partial u}{\partial x}$. Assuming that in general p is a known function of a, t, u, $\frac{\partial u}{\partial x}$ and $\frac{\partial u}{\partial t}$, the above equation is a second order partial differential equation in u; the equation is also linear with regard to the second order derivatives.

When the term R equals zero and the initial density ρ_0 is constant, and also provided that p depends only on $\frac{\partial u}{\partial x}$ and $\frac{\partial u}{\partial t}$, the partial differential equation is in the form

$$\mathrm{H}\frac{\partial^2 u}{\partial x^2} + 2\mathrm{K}\frac{\partial^2 u}{\partial x \partial t} + \mathrm{L}\frac{\partial^2 u}{\partial t^2} = 0,$$

where the differentials H, K, and L depend only on $\frac{\partial u}{\partial x}$ and on $\frac{\partial u}{\partial t}$. It will be shown below that it is always possible to find an integral containing an arbitrary function which is of particular importance.

When the body considered is an elastic solid, the pressure is proportional to the expansion and equal to $-\mathrm{E}\frac{\partial u}{\partial x}$, where E is Young's *modulus of elasticity.* Equation (1) from the previous chapter then becomes

$$\rho_0 \frac{\partial^2 u}{\partial t^2} = E\frac{\partial^2 u}{\partial x^2} + \frac{dE}{dx}\frac{\partial u}{\partial x},$$

When the solid is uniform, ρ_0 and E are constants, which provides us with the well known equation

$$\frac{\partial^2 u}{\partial t^2} = \frac{E}{\rho_0} + \frac{\partial^2 u}{\partial x^2},$$

which also applies to the motion of liquids.

46. Let us now return to perfect gases. We must find a second equation which, together with equation (1) will permit us to determine the two functions u and p of x and t. In connection with this, let us state that during the period of time dt, the reduction in the total energy of the stratum, or the increase in this energy is equal to the work of the external forces.

The volume of the stratum at time t is $\omega\, dx\left(1 + \frac{\partial u}{\partial x}\right)$ and its mass equals $\rho_0\, \omega\, dx$: the total energy is therefore

$$\frac{\rho_0\, \omega\, dx}{2}\left(\frac{\partial u}{\partial t}\right)^2 + \frac{p\omega\, dx\left(1 + \frac{\partial u}{\partial x}\right)}{m - 1}.$$

The variation in the total energy over time dt can be obtained by differentiating the above expression with respect to t. This leaves just the external work to calculate.

The work of the force ωR dx is obviously

$$\mathrm{w\,R}\,dx\,\frac{\partial u}{\partial t}\,dt,$$

since the distance travelled from the point of application is $\frac{\partial u}{\partial t}\,dt$. The work performed by normal pressure from the envelope is zero. The work of the pressure ωp which is carried out on stratum x is $\omega p\frac{\partial u}{\partial t}dt$, that of the pressure

$$\omega\left(p+\frac{\partial p}{\partial x}\,dx\right)$$

which is exerted on stratum $x+dx$ is equal to the previous value but with the sign changed and increased by its derivative with respect to x, in other words it equals

$$-\omega p\frac{\partial u}{\partial t}dt-\omega\frac{\partial p}{\partial x}\frac{\partial u}{\partial t}dxdt-\omega p\frac{\partial^2 u}{\partial x\partial t}dxdt.$$

This gives us the following equation when the common factor ω $dx\,dt$ is eliminated

$$\rho_0\,\frac{\partial u}{\partial t}\,\frac{\partial^2 u}{\partial t^2}+\frac{\dfrac{\partial p}{\partial t}\left(1+\dfrac{\partial u}{\partial x}\right)}{m-1}+\frac{p\dfrac{\partial^2 u}{\partial x\,\partial t}}{m-1}-\mathrm{R}\,\frac{\partial u}{\partial t}+\frac{\partial p}{\partial x}\,\frac{\partial u}{\partial t}+p\,\frac{\partial^2 u}{\partial x\,\partial t}=0 \tag{2}$$

which, in view of equation (1) can be reduced to

$$\frac{\partial p}{\partial t}\left(1+\frac{\partial u}{\partial x}\right)+mp\,\frac{\partial^2 u}{\partial x\,\partial t}=0 \tag{3}$$

Equation (2) can thus be split into two parts: equation (1) and equation (3), thereby resulting in an interpretation.

If equation (1) is multiplied by $\omega \frac{\partial u}{\partial t} dx\, dt$, it becomes

$$\rho_0 \omega \frac{\partial u}{\partial t} \frac{\partial^2 u}{\partial t^2} dxdt = \omega R \frac{\partial u}{\partial t} dxdt - \omega \frac{\partial p}{\partial x} \frac{\partial u}{\partial t} dxdt.$$

The first term is the variation in the active force of the stratum; the first expression of the second term is the work of the external force ωR dx over time dt; finally, the last expression is the work of the difference between the pressures on the two bases corresponding to a distance travelled by the stratum of $\frac{\partial u}{\partial t} dt$.. This equation is obviously obtained by assimilating the stratum with an invariable solid.

Equation (3), however, shows that the change in internal energy is equal to the external work done by expansion of the stratum, as can easily be seen, as if the expansion had taken place without the molecules reaching any real velocities.

47. Equation (3) can be expressed in the following form

$$\frac{\frac{\partial p}{\partial t}}{p} = -m \frac{\frac{\partial^2 u}{\partial x \partial t}}{1 + \frac{\partial u}{\partial x}};$$

and is integrated immediately to give

$$p = f(x)\left(1 + \frac{\partial u}{\partial x}\right)^{-m} \tag{4}$$

where $f(x)$ is an arbitrary function which must be determined by the initial conditions.

We could have immediately attempted to solve equation (4) by stating that in each stratum the pressure varies as in adiabatic expansion, while the molecules reach no real velocity. However, such a hypothesis, which is normally accepted without argument, is far from evident. It is no longer accurate when a discontinuity occurs in the law of variation in velocity or of expansion of strata. In this specific case, the validity of the previous analysis becomes dubious, as we shall show soon.

48. Eliminating p from equations (1) and (4) gives

$$\rho_0 \frac{\partial^2 u}{\partial t^2} = \mathrm{R} + m f(x) \left(1 + \frac{\partial u}{\partial x}\right)^{-m-1} \frac{\partial^2 u}{\partial x^2} - f'(x) \left(1 + \frac{\partial u}{\partial x}\right)^{-m}.$$

In many cases, the external forces can be disregarded, as too can the effects produced by friction and viscosity, such that R is cancelled out. The independent term for the second order derivatives may then be removed from the equation.

This would make the equation as follows

$$\rho_0 \frac{\partial^2 u}{\partial t^2} = - \frac{\partial}{\partial x} \left[f(x) \left(1 + \frac{\partial u}{\partial x}\right)^{-m} \right],$$

let us assume

$$u_1 = x + u, \qquad x' = \varphi(x)$$

which would give us

$$\rho_0 \frac{\partial^2 u_1}{\partial t^2} = - \varphi'(x) \frac{\partial}{\partial x'} \left\{ f(x) [\varphi'(x)]^{-m} \left(\frac{\partial u_1}{\partial x'} \right)^{-m} \right\}$$

If we determine the function φ such that the product $f(x)\,[\varphi'(x)]^{-m}$ remains constant, we will have

$$f(x)\,[\varphi'(x)]^{-m} = \mathrm{A}, \qquad \varphi'(x) = \left[\frac{f(x)}{A} \right]^{\frac{1}{m}},$$

and the equation will become

$$\rho_0 \left[\frac{f(x)}{A} \right]^{-\frac{1}{m}} \frac{\partial^2 u_1}{\partial t^2} = - \frac{\partial}{\partial x'} \left[\mathrm{A} \left(\frac{\partial u_1}{\partial x'} \right)^{-m} \right].$$

The factor $\rho_0 \left[\frac{f(x)}{\mathrm{A}} \right]^{-\frac{1}{m}}$ can be expressed in terms of x'. If it is designated by $\phi(x')$ and if we assume $u' = u_1 - x'$, the equation finally becomes

$$\phi(x') \frac{\partial^2 u'}{\partial t^2} = m \mathrm{A} \left(1 + \frac{\partial u'}{\partial x'}\right)^{-m-1} \frac{\partial^2 u'}{\partial x'^2}.$$

The expression independent of the second order derivatives has disappeared.

The above transformation amounts to changing the initial state which all the others are referred to. The pressure and the density corresponding to this new initial state have still to be calculated. The length of a stratum, equal to dx in the first initial state, has become dx' in the second. So, if ρ'_0 designates the new initial density and F (x') the new initial pressure,

$$\rho'_0\, dx' = \rho_0\, dx, \qquad F(x')\,(dx')^m = f(x)\, dx^m.$$

But $dx' = \varphi'(x)\, dx$; we can see from this that F (x) is equal to the constant A. Furthermore

$$\rho'_0 = \frac{\rho_0}{\varphi'(x)} = \rho_0 \left[\frac{f(x)}{A}\right]^{-\frac{1}{m}} = \phi(x').$$

So, in the new initial state, the pressure is constant within the gas; the same does not apply for the density, since ρ'_0 is normally a function of x'. The absolute temperature T is inversely proportional to the density ρ'_0 since the pressure remains constant; this temperature varies from one point to the next and, although in the new initial state the pressure remains constant, the gas would be in equilibrium only if its own conductivity was always zero.

To ensure that the density and temperature are constant in this state, the product $\rho_0\, [f(x)]^{-\frac{1}{m}}$ would also have to be constant.

49. From the above, we can see that if we disregard the external forces, friction and viscosity, the motion of a perfect gas enclosed in a cylindrical tube always satisfies a partial differential equation of the form

$$\rho_0 \frac{\partial^2 u}{\partial t^2} = m p_0 \left(1 + \frac{\partial u}{\partial x}\right)^{-m-1} \frac{\partial^2 u}{\partial x^2}, \tag{5}$$

where p_0 is an arbitrary constant and ρ_0 a function of x defined as follows. Let us consider a particular state of the gas, the expansion of the various strata is modified in such a way that the pressure in each of them becomes constant and equal to p_0. During this modification, which is assumed to have been made without any discontinuity, the pressure of a stratum is always related to its volume according to the adiabatic law. The function ρ_0 is then the function which, after modification, shows the density of a stratum in terms of its value on the x-axis. Each unit is designated by the point on the axis corresponding to the state just defined; the function u is the distance travelled by the stratum x at time

t, such that the x-axis value of this stratum has, at this point, become equal to $x + u$.

Equation (5) would no longer represent the motion of the gas if a discontinuity were introduced at a particular moment into the velocity or the expansion.

50. Until now, we have only spoken of perfect gases, but the above results could be applied to general cases and extended to cover all bodies which have insignificant specific conductivity.

Let us therefore consider any body:

$$f(v, p, \mathrm{T}) = 0$$

is its characteristic equation. If we take $d\mathrm{Q}$ to represent the amount of heat absorbed per unit weight of the bodies in a basic transformation where the volume increases by dv and the pressure by dp, without the bodies attaining any real velocity,

$$d\mathrm{Q} = h\ dv + k\ dp,$$

where h and k are functions of v,p and T which can be written as functions of v and p only, thanks to the characteristic equation. If $d\mathrm{Q}$ equals zero, the transformation is adiabatic, and we obtain a differential equation which can be integrated to provide the pressure p and the specific volume v as a relationship in the form

$$\varphi\ (p, v) = \alpha,$$

where α is an arbitrary constant whose value depends on the initial state, $\alpha = f\ (p_0, v_0)$.

The above equation can be solved with respect to p and arranged in the form

$$p = \mathrm{F}'\ (v, \alpha),$$

where F' is the derivative, with respect to v, of a particular function of v and α. In any adiabatic transformation, the value of α remains constant.

The external work performed in a basic transformation is equal to $p\ dv$. In the case of a finite adiabatic transformation, in which the volume changes from v_0 to v, the external work done equals

$$\int_{v_0}^{v} F'(v, \alpha)\, dv = F(v, \alpha) - F(v_0, \alpha),$$

since α remains constant during the transformation.

When the volume v increases indefinitely, the value of $F(v, \alpha)$ converges to a finite limit; otherwise, the unit of weight for a body would, as it expanded, produce an infinite amount of work, without the addition of heat. Let $\Phi(\alpha)$ be this limit, it is clear that the expression

$$\Phi(\alpha) - F(v, \alpha)$$

represents the external work that a body with specific volume v and pressure $p = F'(v, \alpha)$ is able to produce during an indefinite expansion.

By analogy with what was done for the perfect gases, this value will be called the *internal energy*. Thus, the internal energy of a body of mass ϖ and volume V is

$$\varpi\left[\Phi(\alpha) - F\left(\frac{V}{\varpi}, \alpha\right)\right].$$

Lastly, we shall use the term *total energy* of a mass particle to designate the sum of its active force and its internal energy and, by virtue of the principle of equivalence, we shall assume that during the transformation of a mass particle which is carried out without heat being added or removed, the reduction in total energy is equal to the external work performed.

51. Let us now consider a cylindrical body which moves in parallel strata; applying the theorem of balance of momentum will, as before, give the equation

$$\rho_0 \frac{\partial^2 u}{\partial t^2} = R - \frac{\partial p}{\partial x}.$$

Furthermore, since the specific conductivity of the body is assumed to be insignificant, as is the conductivity of the envelope, the increase in the total energy of a stratum between sections x and $x + dx$ can be expressed as being equal to the work done by the external forces.

The volume of the stratum is $\left(1 + \frac{\partial u}{\partial x}\right)$ its mass is $\rho_0\, \omega\, dx$, and its weight is $\rho_0\, g\, \omega\, dx$ where g represents the density. The total energy is therefore

$$\frac{\rho_0\, \omega\, dx}{2}\left(\frac{\partial u}{\partial t}\right)^2 + \rho_0\, g\omega\, dx\left[\Phi(\alpha) - \mathrm{F}\left(\frac{1+\dfrac{\partial u}{\partial x}}{\rho_0 g}, \alpha\right)\right],$$

where α is also related to the pressure p by the equation

$$p = \mathrm{F}'\left(\frac{1+\dfrac{\partial u}{\partial x}}{\rho_0 g}, \alpha\right).$$

Differentiating the above expression with respect to t, and making it equal to the work of the external forces during time dt (this work was calculated when we dealt with the perfect gases), we obtain an equation which, when compared with the equation for balance of momentum, can be reduced after elimination of the common factors to

$$\left[p - \mathrm{F}'\left(\frac{1+\dfrac{\partial u}{\partial x}}{\rho_0\, g}, \alpha\right)\right]\frac{\partial^2 u}{\partial x\, \partial t} +$$

$$\rho_0\, g\left[\Phi'(\alpha) - \frac{\partial}{\partial \alpha}\mathrm{F}\left(\frac{1+\dfrac{\partial u}{dx}}{\rho_0\, g}, \alpha\right)\right]\frac{d\alpha}{dt} = 0.$$

The first term of this equation is equal to zero by virtue of the relationship existing between p and α; the same must apply to the second term. The value between brackets cannot be cancelled out, otherwise the internal energy would be independent of α and, hence, independent of the pressure. Therefore $d\alpha/dt = 0$ and the value of α remains unchanged during the transformation, such that the relationship between the specific volume and the pressure is the same as if the transformation had taken place without the molecules attaining any real velocity.

This theorem, first demonstrated purely for the perfect gases, is now also extended to cover other bodies; we are, however, well aware that it would no longer be true if the stratum in question could receive heat from an outside source or if the velocity of the stratum could undergo a finite variation in an infinitesimally small space of time.

52. From the above, we can conclude that when a body with a sufficiently low specific conductivity to be considered insignificant is enclosed in a cylindrical envelope impermeable to heat and undergoes motion in strata, the pressure p of a particular stratum is related to the expansion $\frac{\partial u}{\partial x}$, as long as there are no discontinuities in the velocity, by a relationship of the form

$$p = F'\left(\frac{1+\frac{\partial u}{\partial x}}{\rho_0\, g}, \alpha\right),$$

in which the value of α is independent of the time; however, α varies with x as well as with ρ_0 if the pressure and density are not constant throughout the body at the initial moment. The pressure p must therefore in general be viewed as being dependent on x and on $\frac{\partial u}{\partial x}$, in such a way that we can postulate the following relationship

$$p = \psi(x, \frac{\partial u}{\partial x})$$

The partial differential equation which governs this motion is therefore

$$\rho_0 \frac{\partial^2 u}{\partial t^2} = -\frac{d\psi}{dx} - R,$$

where $\frac{d\psi}{dx}$ is a total derivative.

When the body is in a uniform state and the density and pressure are the same at all points at the initial moment, p depends only on $\frac{\partial u}{\partial x}$, and we have the following

$$p = -\varphi\left(1 + \frac{\partial u}{\partial x}\right);$$

and the partial differential equation governing the motion then becomes, assuming R is equal to zero,

(6) $$\rho_0 \frac{\partial^2 u}{\partial t^2} = \varphi'\left(1 + \frac{\partial u}{\partial x}\right)\frac{\partial^2 u}{\partial x^2}.$$

This is the generalised version of equation (5).

53. The envelope has until now been considered as being impermeable to heat. Assuming, on the contrary, that this envelope acts during motion as a perfect source of heat, maintaining the temperature of the body constant, the transformation of each stratum would be *isothermal* instead of adiabatic.

In the case of a perfect gas, the relationship between the volume and the pressure becomes

$$pv = p_0 v_0,$$

such that we only need to postulate that $m = 1$ in the general equation (5) to obtain the equation for the motion. Thus we obtain

$$\rho_0 \frac{\partial^2 u}{\partial t^2} = p_0 \left(1 + \frac{\partial u}{\partial x}\right)^{-2} \frac{\partial^2 u}{\partial x^2}.$$

This equation is closer to the true state of affairs than equation (5) when the gas is enclosed in a very small bore pipe. The case where $m = 1$ is also interesting from the point of view of the discontinuities which can be produced in the velocities.

III. -Motion of a cone-shaped body

54. In general, the motion of a body is governed by a second order partial differential equation with two independent variables when there is an axis $0x$ such that the molecules contained in a plane perpendicular to $0x$ are travelling, at any particular moment, at the same velocity as each other, and parallel to this straight line. We shall limit ourselves here to giving a few examples of this.

Let us first consider a solid elastic body whose section ω perpendicular to $0x$ varies with x, the pressure exerted on a section x is ωp, that exerted on the section $x + dx$ is

$$\omega p + \frac{\partial}{\partial x}(\omega p)dx.$$

Since the mass of the stratum contained between the two sections is equal to $\rho_0\, \omega\, dx$, applying the theorem of balance of momentum gives us the following

$$\rho_0 \omega \frac{\partial^2 u}{\partial t^2} = -\frac{\partial}{\partial x}(\omega p).$$

Representing Young's modulus of elasticity by E, we obtain

$$p = -\mathrm{E}\,\frac{\partial u}{\partial x},$$

and thereafter,

$$\rho_0 \omega \frac{\partial^2 u}{\partial t^2} = \omega E \frac{\partial^2 u}{\partial x^2} + \frac{d(\omega E)}{dx}\frac{\partial u}{\partial x}.$$

The terms assigned to the second order derivatives are the same as for a cylindrical body.

55. For our second example, we shall take a fluid, gas or liquid, enclosed in an envelope in the shape of a circular cone. The axis of the cone will be taken to be the x-axis and the apex will be the origin of the coordinates. β is the angle at the apex of the cone. We shall attempt to obtain the partial differential equation governing the motion of the fluid, assuming all the points of a stratum perpendicular to the axis are travelling at the same velocity as each other at the same moment and are moving parallel to this straight line. The specific conductivity of the fluid will be assumed to be zero, as will that of the envelope; furthermore, to simplify our calculations, we shall disregard friction, viscosity and external forces which would only introduce terms into the equation which were independent of the second order derivatives.

Having postulated this, the stratum whose end sections had the ordinates x and $x + dx$ at the initial moment would then have the volume

$$\pi x^2 \operatorname{tang}^2 \beta \, dx,$$

and its mass would equal $\pi \rho_0 x^2 \operatorname{tang}^2 \beta \, dx$.

At time t, the ordinate on the x-axis has become $x + u$, and the volume is

$$\pi (x + u)^2 \operatorname{tang}^2 \beta \left(1 + \frac{\partial u}{\partial x}\right) dx.$$

The expansion, relative to the volume is thus

$$\frac{u^2 + 2ux}{x^2}\left(1 + \frac{\partial u}{\partial x}\right).$$

Before applying the theorem of balance of momentum, we will first point out that the increase in velocity of all points of the stratum during time dt is, give or take the infinitesimally small second order values, equal to $\frac{\partial^2 u}{\partial t^2} dt$.

Furthermore, if we project the pressure exerted on both the bases and the circumference towards the x-axis, it becomes clear that the sum of the projections is equal to $-\pi (x-u)^2 \operatorname{tang}^2 \beta \frac{\partial p}{\partial x} dx$. This gives us the following equation

$$\rho_0 x^2 \frac{\partial^2 u}{\partial t^2} = -(x+u)^2 \frac{\partial p}{\partial x} .$$

It now only remains to calculate the function p.

56. Starting from the principle that the variation in the total energy is equal to the external work performed, we shall show, as we did for motion in a cylinder, that the pressure is related to the specific volume v, as if the expansion took place without the molecules attaining any real velocity, such that we have the following equation

$$p = \mathrm{F}'(v, \alpha),$$

where α is a function of x independent of t.

The specific volume v has the following value at the moment in question,

$$\frac{(x+u)^2}{\rho_0 g x^2}\left(1+\frac{\partial u}{\partial x}\right).$$

Therefore

$$p = \mathrm{F}'\left[\frac{(x+u)^2}{\rho_0 g x^2}\left(1+\frac{\partial u}{\partial x}\right), \alpha\right],$$

and when this value is substituted into the above equation, it transforms it into a second order partial differential equation which is linear with respect to the second order derivatives. The terms which depend on the second order derivatives are

$$\rho_0 x^2 \frac{\partial^2 u}{\partial t^2} + \frac{(x+u)^4}{\rho_0 g x^2} F'' \left[\frac{(x+u)^2}{\rho_0 g x^2} \left(1 + \frac{\partial u}{\partial x}\right), \alpha \right] \frac{\partial^2 u}{\partial x^2}.$$

If the fluid is homogeneous at the initial moment, and if the pressure and density are constant, α and ρ_0 are independent of x; this gives

$$p = -\varphi\left[\left(\frac{x+u}{x}\right)^2 \left(1 + \frac{\partial u}{\partial x}\right)\right],$$

and the terms which depend on the second order derivatives are

$$\rho_0 x^2 \frac{\partial^2 u}{\partial t^2} - \frac{(x+u)^4}{x^2} \varphi' \left[\left(\frac{x+u}{x}\right)^2 \left(1 + \frac{\partial u}{\partial x}\right)\right] \frac{\partial^2 u}{\partial x^2}.$$

57. In the specific case where the fluid in question is a perfect gas, the relationship between the pressure and the specific volume is simply

$$pv^{m} = p_0\, v^{m\,;0},$$

such that the following value is obtained for the pressure,

$$p = p_0 \left[\frac{(x+u)^2}{x^2} \left(1 + \frac{\partial u}{\partial x}\right)\right]^{-m}$$

or, assuming that $x + u = z$, where z is the point on the x-axis where the stratum is located at time t,

$$p = p_0\, z^{-2m}\, x^{2m} \left(\frac{\partial z}{\partial x}\right)^{-m}.$$

The equation provided by the theorem of balance of momentum then becomes

$$\rho_0 x^2 \frac{\partial^2 z}{\partial t^2} = -z^2 \frac{\partial p}{\partial x},$$

and the following is obtained when the value of p is substituted

$$\rho_0 x^2 \frac{\partial^2 z}{\partial t^2} = - z^2 p_0 \frac{\partial}{\partial x}\left[z^{-2m} x^{2m}\left(\frac{\partial z}{\partial x}\right)^{-m}\right]. \tag{7}$$

This is the partial differential equation we are looking for. The terms which contain the second order derivatives are

$$\rho_0 x^2 \frac{\partial^2 z}{\partial t^2} - m p_0 z^{-2m+2} x^{2m}\left(\frac{\partial z}{\partial x}\right)^{-m-1} \frac{\partial^2 z}{\partial x^2}.$$

58. When a fluid consisting of homogeneous spherical and concentric layers moves in such a way that the different points in a single layer are all travelling at the same velocity perpendicular to the layer at a given moment, it is evident that if we isolate, in theory, a portion of the fluid contained in a circular cone of infinite spread and with its apex at the centre of the spherical layers, the motion of the fluid contained in this cone is the same as that studied in the previous headings.

Conversely, the equations obtained by the cone are applied without any modification to the motion in concentric spherical layers; x then designates the initial radius of a layer and z is the radius of this layer at time t.

In particular, equation (7) governs the motion of a perfect gas which travels in spherical concentric layers.

Chapter III: Theory of the Propagation of Motion

I. - General remarks

59. The theory which will be explored in this chapter only applies to motions in parallel strata. It was shown in the previous chapter that in the case of non-conductive fluids and elastic solids, motions of this sort are governed by second order partial differential equations with two independent variables and linear with respect to the second order derivatives, at least while there are no discontinuities in the velocity of the molecules.

Each of the integrals of the partial differential equation represents a possible motion for the body in question. In other words, if the body is of infinite length in both directions and if the motion represented by the integral is present at a given moment, the said motion will continue indefinitely.

When we arrive at the general integral of the partial differential equation, we can ascertain all the possible motions from this. However, we still have to determine in each particular problem which of all these integrals best answers the question. This poses problems which are often the same as those encountered when attempting to determine the general integral.

Thus, the equation for the motion of a homogeneous, cylindrical and elastic cylinder is

$$\rho_0 \frac{\partial^2 u}{dt^2} = E \frac{\partial^2 u}{\partial x^2} ,$$

when there are no external forces. The general integral of this equation

$$u = \varphi\left(x + \sqrt{\frac{E}{\rho_0}}\, t\right) + \phi\left(x - \sqrt{\frac{E}{\rho_0}}\, t\right)$$

has been known since d'Alembert's time, and yet it is only in the last few years that we have been able to fully solve these specific problems regarding the motion of elastic cylindrical solids. We may, for example, cite the case where a solid of this type is fixed at one end and at the other it carries an additional weight assimilated with an invariable solid supposed to be moving at a certain velocity, while the rest of the rod is still in a state of rest. This problem was raised by Navier who provided a solution using trigonometric series which was manifestly inadequate because it indicated none of the circumstances of the motion.

Phillips, in a paper published in the *Journal de Liouville* (1864 edition), was the first to show that motion could be represented using d'Alembert's general integral, determining the arbitrary functions as necessary. Colonel Sébert and I, in a paper included in the *Comptes rendus de l'Académie des Sciences* (1882 edition), recently gave formulae for calculating the integrals corresponding to the various moments of the motion using the recursive method. Even more recently, the question was dealt with in great detail by de Saint-Venant, Flamant and Boussinesq in the *Comptes rendus de l'Académie des Sciences* and in the paper of the French translation of Clebsch's *Theory of Elasticity*.

It was only through long hard work that this apparently most simple problem could be fairly satisfactorily solved.

60. The problem of determining integrals which suit a given problem is normally posed in the following way.

Let us consider a limited portion of a body whose initial state is known; conditions are imposed at the two ends, expressed by equations; our task is to determine the integral which represents the motion.

This is one of the most inappropriate ways of stating the question and appears to be the cause of many difficulties encountered in this sort of research.

The initial state is in fact normally defined by giving the position and the velocity of all the points, in other words at time $t = 0$, the following two relations are given

$$u = \mathrm{f}_1(x), \qquad \frac{du}{dt} = f_2(x). \tag{1}$$

In addition, let x_1 and x_2 be the co-ordinates along the x-axis of the two end strata: the conditions imposed at the ends are generally of the form

$$\begin{cases} \mathrm{F}_1\left(x,\ t,\ u,\ \dfrac{\partial u}{\partial x},\ \dfrac{\partial u}{\partial t},\ \dfrac{\partial^2 u}{\partial x^2},\ \dfrac{\partial^2 u}{\partial x\,\partial t},\ \dfrac{\partial^2 u}{\partial t^2}\right) = 0, \\ \mathrm{F}_2\left(x,\ t,\ u,\ \dfrac{\partial u}{\partial x},\ \dfrac{\partial u}{\partial t},\ \dfrac{\partial^2 u}{\partial x^2},\ \dfrac{\partial^2 u}{\partial x\,\partial t},\ \dfrac{\partial^2 u}{\partial t^2}\right) = 0, \end{cases} \tag{2}$$

the first has to be satisfied for $x = x_1$, and the second for $x = x_2$, irrespective of the value of t.

If a specific integral exists for the partial differential equation which suits the problem, it should therefore, at $t = 0$, satisfy the two equations in (1), when $x = x_1$, satisfy the first equation in (2) and, when $x = x_2$, also satisfy the second of these equations.

It is, however, well known that the general integral of a second order partial differential equation with two independent variables which can be expressed by means of arbitrary functions, contains only two such functions. The indeterminacy in the general integral is no greater when it cannot be expressed using arbitrary functions; it is therefore extremely difficult to see how, in view of the indeterminacy of the general integral, it would be possible to obtain a particular integral which satisfied the four conditions (1) and (2).

In fact, it is generally impossible to find a single analytical expression which satisfies these four conditions; normally, this can only be achieved by combining an infinite number of integrals which can be substituted for each other according to a particular law.

Thus, the equations in (1) are sufficient for determining a particular integral of the partial differential equation, but the expression obtained will usually not satisfy the equations in (2).

61. It would be worthwhile clarifying the above notions by means of a simple example.

The motion of a cylindrical uniform rod is governed by the equation

$$\frac{\partial^2 u}{dt^2} = a^2 \frac{\partial^2 u}{\partial x^2} ,$$

where we have taken $a^2 = \dfrac{E}{\rho_0}$ to simplify the equations. Let the end strata cross the x-axis at zero and λ, and let

$$u = 0, \qquad \frac{\partial u}{\partial t} = 0 .$$

be the equations which define the initial state. It is obvious that the corresponding particular integral is

$$u = 0.$$

However, this is only the answer if the ends are constantly in a state of rest.

When the ends are not in a state of rest, a combination of integrals must be used to obtain the answer required. If, for instance, the conditions imposed at the ends are

$$u = 0 \text{ when } x = 0, \qquad u = \frac{at^2}{2} \text{ when } x = \lambda,$$

then irrespective of the value of t, the motion is represented by an infinite number of integrals which succeed each other according to a law which can easily be deduced. At all times, the rod is divided into two parts which have two corresponding separate integrals. At the point of intersection, the values of u are equal, as are those of $\dfrac{\partial u}{\partial x}$ and $\dfrac{\partial u}{\partial t}$; however, the values of the second derivatives are not the same. The two integrals retain the same analytical expression for a period of time equal to $\dfrac{\lambda}{a}$, at the end of which one of the integrals is changed.

When t ranges between zero and $\dfrac{\lambda}{a}$, the two integrals are

$$u = 0, \qquad u = \frac{\alpha}{2a^2}(x + at - \lambda)^2 \ .$$

When t ranges between $\frac{\lambda}{a}$ and $\frac{2\lambda}{a}$, these become

$$u = \frac{2\alpha x}{a^2}(at - \lambda) \ , \qquad u = \frac{\alpha}{2a^2}(x + at - \lambda)^2 \ .$$

When t ranges between $\frac{2\lambda}{a}$ and $\frac{3\lambda}{a}$,

$$u = \frac{2\alpha x}{a^2}(at - \lambda) \ ,$$

$$u = \frac{\alpha}{2a^2}(x + at - \lambda)^2 + \frac{2\alpha x}{a^2}(at - \lambda) + \frac{2\alpha\lambda}{a^2}(2\lambda - x - at)$$

and so on.

These formulae determine the value of u which corresponds to all the positive values of t and to all values of x between zero and λ. From certain points of view, the function u defined in this way is continuous and, like its two first integrals, could be considered as a single integral since it satisfies the partial differential equation at all points, except for certain values of x and t where the values of the second derivatives are ambiguous.

However, this way of looking at things has no real advantage, since it is impossible to express the function u using a single finite analytical expression.

When we represent the function u by a surface, taking x and t as the horizontal co-ordinates, it becomes evident that the surface obtained is made up of an infinite number of portions of hyperbolic paraloboids which connect up with one another, but which do not have the same indicatrix at the points of connection. It would be unusual to view this assembly as a single surface.

62. In a certain number of cases, analogous to the case which has just been examined, it was indeed possible to represent the entire solution using a single formula by means of trigonometric series. The use of these series is self-justificatory when the body in question is subjected to a periodic motion, as in the case of acoustic vibrations. These series then proceed along the sines and cosines of the multiples of a single arc and highlight the different possible values which the period of the motion can assume.

However, attempts have frequently been made to introduce these series into problems where they are nowhere near so useful. When the motion ceases to be periodic, the numbers by which the arc is multiplied can no longer be integers. They are usually an infinite number of roots from a particular transcendental equation, which also varies depending on the nature of the problem.

Series of this sort were first used by Fourier in his remarkable Treatise on heat, and the problems in which he used the series could not be solved without them until now. However, there are various disadvantages associated with the use of these series for the motion of bodies:

1) We have only a partial knowledge of the conditions in which these series can represent the functions whose development they are supposed to express.

2) It may not always be possible to differentiate these series: so, in certain cases, we cannot use them to calculate the velocities and expansions, which are, however, almost always the most important values to ascertain.

3) The trigonometric series provide no indication of the mechanism behind these phenomena, in other words the propagation and reflection of motion.

4) Lastly, it proves impossible to search for them when the partial differential equation is no longer linear.

63. The solutions expressed using the trigonometric series are obviously imperfect compared with those where the values of u, of $\frac{\partial u}{\partial x}$ and $\frac{\partial u}{\partial t}$ are expressed in finite form using analytical functions. It was Phillips who, in his important paper published in the *Journal de Liouville* (1864), first showed the superiority of these latter solutions. He was soon followed along this track by de Saint-Venant in his paper on the longitudinal shock of elastic prisms (*Journal de Liouville*, 1867). However, these eminent geometricians did not provide a general method for determining the solutions; what is more, they limited their discussion to a consideration of the motions governed by an equation with a known general integral, i.e. the equation $\frac{\partial^2 u}{\partial t^2} = a^2 \frac{\partial^2 u}{\partial x^2}$. Moreover, this general integral is expressed in terms of arbitrary functions which only require to be suitably detrmined as in the two examples dealt with above.

However, as we shall see, there is a general method which may be applied, even to equations with an unknown general integral. What is more, this alternative method brings an altogether new light to bear on the phenomena of the propagation of motion.

64. If we refer to the specific example dealt with above, we can see that the elastic prism was initially in a state of rest, in other words its motion was represented by the specific integral $u = 0$. At end λ, an accelerated uniform motion was applied, the velocity being zero in the beginning; a new integral

$$u = \frac{\alpha}{2a^2}(x + at - \lambda)^2$$

immediately came into existence at the end of the body, such that the body was divided into two parts: the part closest to the origin, still in a state of rest, and the other whose motion is represented by the above equation. The section of separation has $x = \lambda - at$ as its x-axis value; the section is equal to λ when $t = 0$ and it travels towards the origin at a velocity of a. The new integral therefore continuously stretches out at the expense of the first integral; the motion it represents propagates at a velocity of a. It must be pointed out that the shape of the new integral obviously depends only on the motion applied at the end and on the state of the body at the moment it came into existence.

If the body was indefinite on the side of the negative *x*s, the phenomenon would always remain the same, in other words integral (3) would continue to advance at a velocity of a.

However, the body is limited and, at the moment when integral (3) reaches the end $x = 0$, the nature of the phenomenon is altered by the fact that this point is fixed. Indeed, at the point $x = 0$, a new integral develops whose form is determined by the condition imposed at the end $x = 0$ and by the integral in (3). This new integral is

$$u = \frac{2\alpha x}{a^2}(at - \lambda) , \tag{4}$$

and the rod is again divided in two; the part closest to the origin moves according to the equation in (4), the other part moves as represented by the integral in (3). The x-axis co-ordinate for the point of separation is $at - \lambda$; zero at time $t = \frac{\lambda}{a}$, where the integral in (3) reached the end $x = 0$, this x-axis increases gradually and moves towards the right at a velocity of a. The integral in (4) therefore stretches at the expense of the integral in (3); it propagates with a velocity of a and finishes by completely eliminating the integral in (3) at time $t = \frac{2\lambda}{a}$; the motion of the entire rod is then represented by (4) and this would continue indefinitely if the end $x = \lambda$ was moved by the motion defined by this equation.

However, as this is not the case, a new integral comes into existence at the point λ; this is

$$(5) \qquad u = \frac{\alpha}{2a^2}(x + at - \lambda)^2 + \frac{2\alpha x}{a^2}(at - \lambda) + \frac{2\alpha\lambda}{a^2}(2\lambda - x - at),$$

and the shape of this results only from the integral in (4) and from the condition imposed on the motion of point λ.

The integral in (5) in turn extends at the expense of (4); in other words, it propagates into the latter at a velocity of a, until it reaches the end $x = 0$ where a new integral will come into existence.

65. We have just seen how integrals can come into existence at the ends of the bodies. When the motion of the prism is represented by a particular integral and the condition imposed at the end is incompatible with the motion defined by this integral, another integral will come into existence at the end and will develop at the expense of the first, thereby propagating into the body at a velocity which, as we shall see later, is always equal to a in the case of elastic prisms. The new integral can only, of course, depend on the primitive integral and the condition imposed at the end.

However, new integrals can still come into existence at other points on the body; and before we bring the above results into general application, we must study this new phenomenon.

Let us consider, as before, a homogeneous and elastic cylinder with extremities on the x-axis at zero and λ; it is in a state of rest at the initial moment; let us suppose that the two ends are being subjected to uniformly accelerated motion, such that

$$\text{when } u = \frac{1}{2}\,\beta t^2\ x = 0, \qquad u = \frac{1}{2}\,\alpha t^2 \quad \text{when } x = \lambda.$$

irrespective of the value of t.

Under these conditions, the motion of the cylinder shall at all times be represented by three integrals which will also vary at intervals of $\frac{\lambda}{2a}$.

Between $t = 0$ and $t = \frac{\lambda}{2a}$, the three integrals are

$$u = \frac{\beta}{2a^2}(at - x)^2, \qquad u = 0, \quad u = \frac{\alpha}{2a^2}(at + x - \lambda)^2;$$

between $t = \dfrac{\lambda}{2a}$ and $t = \dfrac{2\lambda}{2a}$, they are

$$u = \frac{\beta}{2a^2}(at - x)^2,$$

$$u = \frac{\alpha}{2a^2}(at + x - \lambda)^2 + \frac{\beta}{2a^2}(at - x)^2,$$

$$u = \frac{\alpha}{2a^2}(at + x - \lambda)^2;$$

the two end ones have remained unchanged, the middle one only was changed. We can easily see that between $t = \dfrac{2\lambda}{2a}$ and $t = \dfrac{3\lambda}{2a}$ there are another three integrals; the middle one is the same as for t in the range $\dfrac{\lambda}{2a}$ to $\dfrac{2\lambda}{2a}$, but the other two have been changed, *et sequens*.

In this case, the rod is initially at rest, while the two ends are subjected to uniformly varied motions with an initial zero velocity.

Thus, at each of these ends an integral comes into existence which extends continuously at the expense of the primitive integral $u = 0$. Each of these integrals propagates at a velocity a, one from the zero point to the point λ, the other in the opposite direction; consequently, at time $t = \dfrac{\lambda}{2a}$, the two integrals

$$u = \frac{\beta}{2a^2}(at - x)^2, u = \frac{\alpha}{2a^2}(at + x - \lambda)^2$$

have reached the middle of the rod, where they meet. The integral $u = 0$ has completely disappeared.

It is clear that a particular phenomenon must occur at this point, and indeed, a new integral comes into existence here; this integral is written

$$u = \frac{\alpha}{2a^2}(at + x - \lambda)^2 + \frac{\beta}{2a^2}(at - x)^2,$$

and we can see that, in this particular case, it is equal to the sum of the first two expressions; however, this is not generally the case. In any case, this new integral obviously comes into existence because the first two cannot exist alongside one another. Its expression can only depend on the integrals which have met in the middle of the rod.

The new integral, first limited at the middle point, extends in both directions at the expense of the first two expressions; the velocity at which the limiting points move is still equal to a, such that at time $t = \frac{\lambda}{a}$ it will have reached both ends and will continue only in the rod.

However, as it is not compatible with the conditions imposed at the two ends, new integrals will come into existence at these two points and will extend at the expense of the previous one until it is completely eliminated; thereafter, another integral will come into being at the middle of the rod, *et sequens*.

66. The above examples, although they only deal with the equation for the motion of an elastic rod in the shape of a prism, which is the simplest of all, is adequate for helping us to understand how integrals come into existence and develop. Two distinct cases must be considered.

Let us first suppose that the whole body is subjected to a motion represented by the integral

$$u = f(x, t) \tag{6}$$

from the corresponding partial differential equation, if one of the ends, $x = \lambda$ for instance, is subjected, in response to the specific conditions imposed on it, to the motion

$$u = f(\lambda, t)$$

or, more usually, if, when the condition imposed at the end in question is

$$F\left(x, t, u, \frac{\partial u}{\partial x}, \frac{\partial u}{\partial t}, \frac{\partial^2 u}{\partial x^2}, \ldots\right) = 0, \tag{7}$$

the expression in terms of u in (6) also perfectly satisfies the previous equation when making $x = \lambda$, and no particular phenomenon will occur at this point.

However, when this is not the case, a new integral comes into existence at this point, which is, necessarily, fully determined. The only conditions imposed on it are: 1) to satisfy the condition in (7), when $x = \lambda$; 2) to be able to propagate into the integral in (6).

It is very important that we give a very precise definition here of exactly what we mean by the propagation of one integral into another.

Let us consider a body whose motion is represented at a given moment by two separate integrals, and let ξ be the point of separation of these two integrals on the x-axis at a given moment t. If at the next moment $t + dt$, the motion is still represented by the same two integrals we shall take them to be mutually *compatible*. The point ξ will move to the right, for example, by a distance $d\xi$, and the derivative $\frac{d\xi}{dt}$ will be taken to represent the velocity at which the first integral propagates into the second.

In the previous number we saw how the two integrals given were not always mutually compatible since, in the example cited, there came into existence at the common point of the two integrals

$$u = \frac{\beta}{2a^2}(at - x)^2, u = \frac{\alpha}{3a^2}(at + x - \lambda)^2,$$

a new integral which developed and extended at the expense of the first ones. The new integral obtained was then compatible with the other two; however, this compatibility obviously depends on a specific condition. This condition will be determined below.

Let us come back to the case where an integral comes into existence at one end, the conditions which determine it in full are as follows:

1) to satisfy the condition in (7) when $x = \lambda$;

2) to be compatible with (6).

Conversely, when the new integral comes into existence at a point in common with two other integrals, it is fully determined by the condition of being compatible with the first two.

II. Integral compatibility condition

67. We showed in the previous chapter how certain partial differential equations representing the motion of bodies in strata may cease to be exact when the velocity of a particular stratum experiences a finite variation in an infinitesimally small period of time. In order to find the integral compatibility condition, we shall assume that at the common point the values of u, $\frac{\partial u}{\partial x}$ and

$\dfrac{\partial u}{\partial t}$ are equal, except in certain cases where discontinuities arise, which we shall examine later.

Let, as a general rule,

$$(1) \qquad L \frac{\partial^2 u}{\partial t^2} + 2K \frac{\partial^2 u}{\partial x \, \partial t} + H \frac{\partial^2 u}{\partial x^2} + M = 0$$

be the partial differential equation which governs the motion of the body; u_1 and u_2 are two integrals which, at time t, represent the motion of two portions of the body with a common point ξ.

If the integrals are compatible, the common point ξ is given in terms of t by the equation

$$(2) \qquad u_1 - u_2 = 0\,;$$

furthermore, at the common point, by virtue of the continuity

$$\frac{\partial u_1}{\partial x} - \frac{\partial u_2}{\partial x} = 0, \qquad \frac{\partial u_1}{\partial t} - \frac{\partial u_2}{\partial t} = 0.$$

These equations must always occur at the common point, because the continuity cannot disappear; we can therefore differentiate them by considering ξ to be a function of t defined by the equation in (2); this gives

$$(3) \qquad \left(\frac{\partial^2 u_1}{\partial x \, \partial t} - \frac{\partial^2 u_2}{\partial x \, \partial t} \right) + \left(\frac{\partial^2 u_1}{\partial x^2} - \frac{\partial^2 u_2}{\partial x^2} \right) \frac{d\xi}{dt} = 0,$$

$$(4) \qquad \left(\frac{\partial^2 u_1}{\partial t^2} - \frac{\partial^2 u_2}{\partial t^2} \right) + \left(\frac{\partial^2 u_1}{\partial x \, \partial t} - \frac{\partial^2 u_2}{\partial x \, \partial t} \right) \frac{d\xi}{dt} = 0.$$

Furthermore, the two integrals u_1 and u_2 satisfy the equation in (1). If u_1 and then u_2 are substituted into this equation in the place of u, and x is made equal to ξ, corresponding to the common point at time t, the values of L, K, H and M which are only dependent on x, t, u, $\dfrac{\partial u}{\partial x}$ and $\dfrac{\partial u}{\partial t}$ will become equal, such that when the equations obtained are subtracted, we are left with

$$(5)\quad \begin{cases} L\left(\dfrac{\partial^2 u_1}{\partial t^2} - \dfrac{\partial^2 u_2}{\partial t^2}\right) + 2K\left(\dfrac{\partial^2 u_1}{\partial x\,\partial t} - \dfrac{\partial^2 u_2}{\partial x\,dt}\right) + \\ \qquad\qquad H\left(\dfrac{\partial^2 u_1}{\partial x^2} - \dfrac{\partial^2 u_2}{\partial x^2}\right) = 0\ . \end{cases}$$

Equations (3), (4) and (5) are simultaneous. They are uniform with respect to the differences

$$\frac{\partial^2 u_1}{dt^2} - \frac{\partial^2 u_2}{\partial t^2}\,, \qquad \frac{\partial^2 u_1}{\partial x\,\partial t} - \frac{\partial^2 u_2}{\partial x\,\partial t}\,, \qquad \frac{\partial^2 u_1}{\partial x^2} - \frac{\partial^2 u_2}{\partial x^2}\ ;$$

therefore, if these three differences are not equal to zero, the determinant must cancel out.

Let us first put away the case where three second derivatives of the functions would always be equal at the common point, this gives us the equation

$$(6)\qquad L\left(\frac{d\xi}{dt}\right)^2 - 2K\,\frac{d\xi}{dt} + H = 0,$$

which is the required compatibility condition.

Given one of the integrals u_1 or u_2, we shall take L, K and H to equal the values corresponding to the given integral and which are, consequently, known functions of x and t. If we then replace x with ξ, the equation in (6) becomes a first order differential equation which can be integrated to give ξ as a function of t. The other integral will have to be such that the equation $u_1 - u_2 = 0$ is satisfied when x is replaced by the function of t which is obtained in this way.

68. We still have to examine what happens when the second derivatives of the two functions are always equal at the common point. In such a case, we deduce from the first second derivatives

$$\left(\frac{\partial^3 u_1}{\partial t^3} - \frac{\partial^3 u_2}{\partial t^3}\right) + \left(\frac{\partial^3 u_1}{\partial x\,\partial t^2} - \frac{\partial^3 u_2}{\partial x\,\partial t^2}\right)\frac{d\xi}{dt} = 0,$$

$$\left(\frac{\partial^3 u_1}{\partial x\,\partial t^2} - \frac{\partial^3 u_2}{\partial x\,\partial t^2}\right) + \left(\frac{\partial^3 u_1}{\partial x^2\,\partial t} - \frac{\partial^3 u_2}{\partial x^2\,\partial t}\right)\frac{d\xi}{dt} = 0.$$

If we then differentiate the partial differential equation with respect to t, and substitute in u_1 and then u_2, and make x equal to ξ and subtract, we obtain

$$L\left(\frac{\partial^3 u_1}{\partial t^3} - \frac{\partial^3 u_2}{\partial t^3}\right) + 2K\left(\frac{\partial^3 u_1}{\partial x\, \partial t^2} - \frac{\partial^3 u_2}{\partial x\, dt^2}\right) + L\left(\frac{\partial^3 u_1}{\partial x^2\, \partial t} - \frac{\partial^3 u_2}{\partial x^2\, \partial t}\right) = 0 ,$$

and, since the determinant is not equal to zero, this leads us to conclude that the three third derivatives considered are equal. It will then be fairly easy to prove that

$$\frac{\partial^3 u_1}{\partial x^3} = \frac{\partial^3 u_2}{\partial x^3} ,$$

and there is nothing to prevent us from continuing with this reasoning step by step for all the orders.

According to this, if condition (6) in § 67 is not satisfied, in order for the integrals u_1 and u_2 to be compatible, their derivatives of all orders would always have to be equal at the common point. The functions u_1 and u_2 are not really two separate functions in the neighbourhood of the common point. If they are developable in this neighbourhood, they must, according to Taylor's formula, be absolutely equal.

69. Thus, the condition for the two distinct integrals to be compatible is that expressed by the equation in (6). We can interpret this in geometrical terms in a quite remarkable manner.

As we did in § 61, we shall represent the value u of an integral from the partial differential equation by the ordinate of a surface where x and t represent the horizontal axes, and where the three co-ordinate axes are designated Ox, Ot and Ou. This surface can be considered to represent the motion of the body; if we cut it in planes parallel to xOu, we obtain the state of the body at the various moments. If we cut it in planes parallel to tOu, we obtain the motion at a given point.

The compatibility condition in (6), for a given integral, then represents the differential equation of the horizontal projection of a curve plotted on the integral surface. If ξ is replaced by x, this equation can be written

$$H\left(\frac{dt}{dx}\right)^2 - 2K\frac{dt}{dx} + L = 0;$$

it is the same as the differential equation for the horizontal projections of the characteristic curves.

Thus, the characteristic curves of the surface which represent a motion are the curves along which the surfaces representing motions compatible with the first motion join up with the first surface.

Any integral surface which joins up with the first surface along one of its characteristic curves will also have the curve of connection as one of its characteristic curves. So, two surfaces representing compatible motions will have a common characteristic curve. We saw in Chapter I that two integral surfaces which have a common characteristic curve can in certain respects be considered to form a single integral surface. By the same reasoning, we can consider that all the compatible integrals which represent the motion of a body form a single integral.

III - Velocity of propagation of the integrals

70. If, as earlier, we take ξ to designate the point along the x-axis of the common point with two compatible integrals, the velocity at which one of the integrals propagates into the other is equal to $\frac{d\xi}{dt}$ and, when one of these integrals is known, the compatibility condition can be used to calculate this velocity without integrating.

This is a quadratic equation in $\frac{d\xi}{dt}$; consequently, at each point on the body and at all times there are two velocities of propagation. When K is equal to zero, the two velocities of propagation are equal and opposite, in other words the propagation may occur in both directions with the same velocity. This is what happens for all the bodies which we will examine later.

The analytical expression of the velocity of propagation depends solely on the coefficients of the second order derivatives; from this we should not, however, conclude that those terms which do not depend on the second order derivatives have no effect on the velocity of propagation. This velocity actually depends on the integral under consideration, and the form of the integrals changes with the independent terms of the second order derivatives.

When the coefficients of the second order derivatives are solely functions of x and t, the velocity of propagation is independent of the particular integral being studied; it depends only on the point on the body and on the moment in question.

When the coefficients of the second order derivatives contain only x, the two values of the velocity of propagation are determined for each point on the body.

Finally, when these coefficients are constant, the two values of the velocity of propagation are also constant.

71. When there are no external forces acting on the molecules in the body, the body can obviously remain in a state of rest, such that $u = 0$ is an integral from the partial differential equation which governs its motion. In the case of this particular integral, the values of H, K and L are known and, subsequently, we already know the velocities of propagation for the integrals compatible with $u = 0$ without integrating and without having to solve these integrals.

The integrals compatible with $u = 0$ represent all the motions which can propagate in the body which is assumed to be in a state of rest, without, however, introducing discontinuities, and their velocity of propagation is the real *velocity of propagation of the motion in the body.*

The velocity of propagation of the motion in a body is therefore independent of the nature of the motion which is being propagated, provided that there are no discontinuities. This velocity can be ascertained by a very simple calculation as soon as the partial differential equation governing the motion of the body is known.

Besides, it is clear that this velocity depends only on the terms which involve the second order derivatives. The friction and viscosity cannot modify it in any way since these types of resistance come into play in a simple manner in the independent terms of the second order derivatives.

72. In general, when the motion of a body is represented by a particular integral u, the values of H, K and L which correspond to it are known in terms of x and t; we therefore know in advance the velocity of propagation of the integrals compatible with u and which represent all the motions which can propagate into the first without introducing discontinuities. The velocity of propagation in this integral u is the real generalisation of the velocity of the propagation of the motion, as in the previous case; it is independent of the nature of the motion being propagated; it just needs to be compatible with the first.

73. We can interpret the velocity of propagation of one particular motion into another in geometrical terms in a quite remarkable manner.

Since, as above, the original integral is represented by a surface, if the characteristic curves of this surface are projected along the plane xOt, the differential equation of the projections obtained in this way is

$$L\left(\frac{d\xi}{dt}\right)^2 - 2K\frac{d\xi}{dt} + H = 0;$$

therefore if we consider x as the ordinate of a point on one of the curves corresponding to abscissa t, $\frac{d\xi}{dt}$ represents the angular coefficient of the tangent to the curve; this leads to the following theorem:

Given an integral represented by a surface, the velocity of propagation of the motion in this integral is represented, in the case of known values of x and t, by the angular coefficient of the tangent to the horizontal projection of the characteristic curve which passes through the corresponding point on the surface.

Two characteristic curves pass through each point on the integral surface, each curve has its own velocity of propagation.

When the projection of the characteristic curve is a straight line, the velocity of propagation is constant; in the opposite case, it should be considered as a function of abscissa ξ of the point common to both integrals.

IV - Determining motion in a body

74. The problem with determining the motion of a body which takes place in parallel strata can now be postulated with complete precision. We shall see that, in fact, it is not a single problem which has to be solved, but a series of problems which must be dealt with in a stepwise manner and in a specific order.

To simplify matters, we shall assume that the partial differential equation which governs the motion does not contain a term assigned to $\frac{\partial^2 u}{\partial x \, \partial t}$; this is, as we have already pointed out, the case for all natural bodies which we will have to consider.

We shall first take the case where the body is in a state of rest at the initial moment and where the condition imposed at one of the ends, λ for instance, is compatible with the state of rest. If the condition imposed at the other end, $x = 0$, was also compatible with the state of rest, no motion would be generated in the body.

However, when this is not the case, a motion must come into existence at the point $x = 0$; this motion is given by an integral which is compatible with the state of rest and which also satisfies the condition imposed at the end.

Let us represent the integrals by surfaces; the state of rest, defined by $u = 0$, is represented by the horizontal plane xOt which, *a priori*, has continuous characteristic curves. The integral surface which represents the motion which will come into existence therefore connects up with the first along the characteristic curve which passes through the origin of the co-ordinates. However, two characteristic curves from the surface $u = 0$ also pass through this point; since the motion should be propagated in the direction of the positive x's, we must take the characteristic curve which has a positive angular coefficient.

There is, however, an infinite number of integral surfaces which connect with $u = 0$ along the characteristic curve thus defined and their expressions include that indeterminacy which the presence of an arbitrary function would introduce, since the condition imposed at the end is random and subject to the sole restriction of not introducing any discontinuities and, in all cases, there is only one integral surface which fulfils the problem because only one type of motion is possible. If, for instance, we take the motion of the end to be $x = 0$, the analytical expression of this motion can be chosen at will, provided, however, that there are no sudden variations in the velocity and that the velocity is not zero at the origin; in other words, if $f(t)$ represents this motion, the second derivative $f''(t)$ is a continuous function, but also completely arbitrary.

If we can obtain the general expression of the integrals which connect with $u = 0$ along the characteristic curve passing through the origin, we then only need to determine the arbitrary function such that when we put $x = 0$ into the integral, it complies with the condition applied at the end.

We can even imagine cases where, with all of these integrals unknown, it remains possible to determine the integral which suits the particular problem we are dealing with; nothing could be more easy than verifying that it satisfies the conditions imposed. This is the main advantage of this method.

75. Since the integral u_1, which comes into existence at the end, is supposed to be known, we also know the velocity at which it propagates into the first integral $u = 0$; in other words, $\frac{d\xi}{dt}$ is a known function of ξ, which is the same as saying that we can consider point ξ—common to the two integrals—as if it were a function of t. The motion of the body is therefore determined until the moment when the new integral reaches the end λ of the body.

The condition imposed at the end λ, which was compatible with the state of rest, will not usually be compatible with the new integral u_1. A specific phenomenon therefore occurs at this point, called *reflection of motion* , which consists in an integral u_2 coming into existence at point λ, as defined by the condition of being compatible with u_1 and of satisfying the condition imposed at the end λ.

If we represent the motion, depending on the conditions adopted, by a surface and if t_1 is used to designate the time at which the integral u_1 reaches the point λ, we can see that one point on the integral u_1 has the horizontal co-ordinates t_1 and λ. For the integral u_2 to be compatible with u_1, the former must have a characteristic curve which is common to the latter and which passes through the point (t_1, λ). However, the integral u_1 must propagate towards the negative xs, and so the angular coefficient of this characteristic must be negative, which defines it completely. It must be pointed out that of the two characteristic curves of the surface u_1, which passes through the point (t_1, λ), one of them is actually the curve along which u_1 connects with the horizontal plane $u = 0$, which represents the original state of rest of the body. The two surfaces u_1 and u_2 connect up along the characteristic curve of the other system.

The integral u_2 is defined by the condition of connecting up with the integral u_1 along the characteristic curve which we have just defined, and of satisfying the condition imposed at the end λ. Since only one motion comes into existence as a result of reflection, the integral in question has therefore been determined in full.

Let us suppose that this known integral u_2 also has a known velocity of propagation into u_1. We therefore know the motion of the body until the moment t_2 when u_2 reaches the end $x = 0$, where an integral u_3 will come into existence, defined by the condition of connecting up with u_2 along the characteristic curve of this positive angular coefficient which passes through the point with the horizontal co-ordinates $x = 0$, and $t = t_2$, and hence of satisfying the condition imposed at the end $x = 0$.

Continuing in the same way, we can easily see how the integrals u_4, u_5, ... , which come into existence alternately at the two ends, are defined; each of them can be obtained only after the previous one has been determined. Any one of the integrals, u_n for instance, has a common characteristic with the previous integral u_{n-1} and with the subsequent one u_{n+1}; these two characteristic curves do not belong to the same system. The point of intersection of these two curves crosses the x-axis at either zero or λ, depending on whether n is even or odd.

76. It is easy to generalise the above method. Let us again consider a body where x and λ are its end points on the x-axis, and which is subject at the initial moment to a motion represented by an integral u_0; let us also suppose that the conditions imposed at the ends are incompatible with u_0 but such that they will not introduce any discontinuities in the first derivatives $\frac{\partial u}{\partial x}$ and $\frac{\partial u}{\partial t}$. A new integral comes into existence at each end; the integral u_1, which starts from $x = 0$, meets u_0 along the characteristic curve with a positive angular coefficient which passes through the point $x = 0$, $t = 0$ on the surface representing this integral. The integral u_2, which begins from the point λ, connects up with u_0 along the characteristic curve with a negative angular coefficient which passes through the point $x = \lambda$, $t = 0$, on this same surface.

These two integrals propagate in the opposite direction to the first at a known velocity and meet at a certain point x_1 at time t_1; the integral u_0 has then completely disappeared.

Since the two integrals u_1 and u_2 are generally mutually incompatible, a new integral must come into existence at the point x_1, i.e. u_3, the new integral; the latter must propagate into integral u_1 at a negative velocity and into integral u_2 with a positive velocity.

Let us consider the surface which represents the motion and let us draw the characteristic curve with a negative angular coefficient of integral u_1 through the point $x = x_1$, $t = t_1$, the characteristic curve with a positive angular coefficient of integral u_2, then integral u_3 should connect up with these two surfaces along one of these characteristic curves. Since only one motion is possible, the integral u_3 must be fully determined by these conditions alone.

From time t_1, the integral u_3 extends in both directions at the expense of the integrals u_1 and u_2. At a particular time t_2, one of the latter, u_1 for instance, disappears and the integral u_3 reaches the end $x = 0$, where another integral u_4 comes into existence, *et sequens*.

77. The method which has just been set out gives a precise indication of the general procedure which should be adopted to solve motion problems and shows the difficulties which can be encountered.

So that these difficulties can be overcome in all cases, we must be able, for the second order partial derivatives or at least for those which represent a motion in parallel strata, to solve the following problem: how to determine all the integral surfaces which connect with a given integral surface along a given characteristic curve.

This problem, by virtue of its general nature, is substantially the same as that of how to integrate partial differential equations which represent motion, a problem which seems far from resolved.

It must, however, be pointed out that we can find many specific solutions, each of which represents a particular case of the propagation of motion into another and thereby constitutes a milestone in the study of natural phenomena.

In Chapter IV we shall show the fruitfulness of this method which can be used to establish the entire theory of the propagation of motion in fluids by means of integrals which have been known for many years, but which have never until now been used to study the properties of motion.

78. We have until now assumed that we knew the integral representing the motion of the body at the initial moment. However, we sometimes define the initial state by giving the position of all the units, together with their velocities by two equations of the form

$$u = \mathrm{F}(x), \quad \frac{du}{dt} = \mathrm{F}_1(x),$$

which occur at $t = 0$.

In this case, we must first determine the corresponding integral, in other words the integral which is reduced to $\mathrm{F}(x)$ when $t = 0$, and whose derivative with respect to t then becomes equal to $\mathrm{F}_1(x)$.

This problem is not very difficult when the general integral is known; however, when this is not the case, it is impossible to provide a general solution: we shall limit ourselves to indicating its geometric translation.

Since the motion is represented by a surface, we shall give its intersection as the plane xOu; we shall also state that from $u = \mathrm{F}(x)$ we can deduce that

$$\frac{\partial u}{\partial x} = \mathrm{F}'(x),$$

we therefore know the plane of the tangent all along the curve. This data will be used to ascertain the integral surface. We are certain that there is only one such surface, provided that only one type of motion is possible.

If we accept that the required integral u can be developed according to Taylor's formula, it is easy to calculate the terms of the development; however, it is pointless covering this subject in any greater depth since, in reality, there are very few cases where the integral corresponding to the initial state is not known in advance.

V. - Application - the propagation of motion in solids

79. It was shown in Chapter II, § 53 that the motion of an elastic solid, whose section ω is a function of *x*, was governed by the equation

$$\rho_0 \, \omega \, \frac{\partial^2 u}{dt^2} = \omega E \, \frac{\partial^2 u}{\partial x^2} + \frac{d\,(E\omega)}{dx} \, \frac{\partial u}{\partial x} \; .$$

The compatibility equation, which is combined with the equation for the characteristic curves is now

$$\rho_0 \, \omega \left(\frac{d\xi}{dt} \right)^2 - \omega \, \mathrm{E} = 0.$$

From this we can deduce that

$$(1) \qquad \frac{d\xi}{dt} = \pm \sqrt{\frac{E}{\rho_0}} \, .$$

The velocity of propagation of the motion is therefore independent of ω and, consequently, remains constant irrespective of the shape of the body. Its square is equal to the ratio of Young's modulus to density.

The values of E and ρ_0 are in this case functions in *x* only, hence the velocity of propagation has a constant value at each point of the body.

When the ratio $\dfrac{E}{\rho_0}$ is constant, the velocity of propagation is constant.

Integrating the equation in (1) gives us the horizontal projections of the characteristic curves of the two systems which, since the equation is linear, are independent of the integrals. When representing motion by a surface, we can therefore plot in advance the horizontal projections of the curves along which the various integral surfaces connect .

When the ratio of $\dfrac{E}{\rho_0}$ is a constant, the characteristics are projected horizontally along straight lines.

80. Let us now assume that the solid is cylindrical and the ratio $\dfrac{E}{\rho_0}$ is constant; if $\dfrac{E}{\rho}$ is made equal to a^2, then the equation can be reduced to

$$\frac{\partial^2 u}{dt^2} = a^2 \frac{\partial^2 u}{\partial x^2} .$$

The general integral is known and is written as follows

$$u = \varphi\,(x + at) + \psi\,(x - at)\,,$$

where φ and ψ are arbitrary functions. The compatibility condition is then capable of simple interpretation as far as the arbitrary functions are concerned.

Two compatible integrals u_1 and u_2 have a single common characteristic curve; consequently, along the intersection, one of the values $x + at$ or $x - at$ remains constant. Let us assume, for instance, that $x - at = k$ represents the equation of the line of intersection, if the functions u_1 and u_2 are equal along this line, then

$$\varphi_1\,(x + at) - \psi_1\,(k) = \varphi_2\,(x + at) - \psi_2\,(k)$$

or

$$\varphi_1\,(x + at) - \varphi_2\,(x + at) = \psi_2\,(k) - \psi_1\,(k)\,.$$

Since the second term is independent of the point in question on the intersection, the same should apply for the first, which can only be the case if the difference $\varphi_1 - \varphi_2$ is equal to a constant that we can also assume has been introduced into one of the functions ψ_1 or ψ_2.

Thus, when two integrals connect along a characteristic curve $x - at = k$, the two corresponding functions of φ are equal. It is obvious that the required condition is also sufficient.

We should also see that if two integrals connect along a characteristic curve from the second system $x + at = k'$, the two corresponding functions of ψ are equal.

When an integral is compatible with two others, it loans one of them its function φ, and the other its function ψ, and is thus ascertained, except for an arbitrary constant which is determined by the condition that the values of u are the same at the common points.

81. An elastic cylinder is subjected at time t_1 to a motion represented by the integral

$$u_1 = \varphi_1 (x + at) + \psi_1 (x - at),$$

we can easily see how to determine the new integral which comes into existence at one end. Let us assume, for instance, that this end is at a point λ on the x-axis and is turned towards the positive values, if we take the new integral to be

$$u_2 = \varphi_2 (x + at) + \phi_2 (x - at),$$

it must be compatible with the first and the corresponding surfaces will connect along a characteristic curve from the second system with the following equation

$$x + at = \lambda + at_1 .$$

We therefore have $\psi_2 = \psi_1$, and we still need to determine the function φ_2. Then let

$$F\left(x, t, u, \frac{\partial u}{\partial x}, \frac{\partial u}{\partial t}, \frac{\partial^2 u}{\partial x^2}, \frac{\partial^2 u}{\partial x\, \partial y}, \frac{\partial^2 u}{\partial t^2}\right) = 0$$

be the condition imposed on the end λ. We shall replace u in the previous equation with u_2, giving ψ_2 the known value of ψ_1, and we shall then make $x = \lambda$. This gives us a second order differential equation in which the unknown function will be $\varphi\ (\lambda + at)$. Making $\lambda + at = y$, and integrating the equation, we obtain the function φ of y, in which y must then be replaced by $x + at$.

The function φ will contain two arbitrary constants which we shall determine by making the values of u and of $\dfrac{\partial u}{dt}$ when $x = \lambda$ and when $t = t_1$ equal for the two integrals u_1 and u_2.

The determination of the integral u_2 therefore depends, as we have seen, on the integration of a second order differential equation which is always easy to form.

Futhermore, in Chapter V, we shall show that the above method continues to be applicable for the motion of elastic cylinders when discontinuities are introduced into the derivatives of the function u.

VI. - Motion of a heavy vertical rod fixed at its top end and carrying a heavy mass at its opposite end

82. As a specific example, and in order to fully understand this method, we shall deal with the following problem:

A heavy vertical rod is fixed at its top end and reaches an equilibrium under the action of its own weight. A body of weight Q is fixed to its bottom end, similar to an invariable solid, and the system is abandoned without velocity. We are asked to determine the motion of the system.

This problem was dealt with by Poncelet (*Introduction à la Mécanique industrielle*, 3e édition, p. 419), but he disregarded the mass of the rod. Under these conditions, the motion of the weight is a simple vibratory motion with isochronous oscillations.

It is true that, in a note on page 460, the author provides a general solution to the problem by taking into account the inertia of the rod; however, this solution is given by a trigonometric series in which the arcs are multiplied by an infinite number of roots of the transcendental equation

$$yL \; tang \; yL = K,$$

where L and K are constants, and which provides no information on the way in which the motion develops and is propagated.

83. Since the rod is subjected to an external force, which is that of gravity, it is easy to see that the partial differential equation which governs its motion is, when the x-axis is forced downwards,

$$\frac{\partial^2 u}{dt^2} = a^2 \frac{\partial^2 u}{\partial x^2} + g,$$

where g is taken to be the acceleration of gravity.

The equation has one particular integral which is independent of the time, i.e.

$$u_0 = -\frac{g}{2a^2} x^2 + Ax + B,$$

where A and B are constants.

We shall determine the constants in such a way that the integral represents the state of equilibrium of the rod when it is subject only to its own weight.

Let us suppose that the co-ordinates have their origin at the point of suspension, let λ be the length of the rod; u_0 must be zero when $x = 0$ and $\frac{\partial u_0}{\partial x}$ must also be zero when for $x = \lambda$; therefore $B = 0$, $A = \frac{g\lambda}{a^2}$. This gives us

$$u_0 = -\frac{g}{2a^2}\,x^2 + \frac{g\lambda}{a^2}\,x\,.$$

and the value of u_0 is the value corresponding to the state of the rod before the additional mass was left unsupported.

Let us now postulate that $u = u' + u_0$, the equation which determines u' is

$$\frac{\partial^2 u'}{dt^2} = a^2 \frac{\partial^2 u'}{\partial x^2}\;;$$

it is reduced to the ordinary form; the distance travelled is the distance travelled by a point on the rod, starting from the position of equilibrium. It is therefore pointless to retain the accents in the above equation; however, once the integrals have been obtained, u_0 will have to be added to each of them to ascertain the real state of the rod.

84. We still have to find the integrals for the equation

$$\frac{\partial^2 u}{dt^2} = a^2 \frac{\partial^2 u}{dx^2}\,,$$

which are in keeping with the problem. The rod is initially in a state of rest and, by virtue of the transformation which has just taken place, it can be represented by the integral $u = 0$.

The condition imposed at the fixed end is obviously $u = 0$ when $x = 0$, irrespective of the value of t.

To obtain the condition imposed at the fixed end, we need just express the condition that the weight is in equilibrium under the action of its power of inertia - $\frac{Q}{g}\frac{\partial^2 u}{dt^2}$, by its weight Q and the tension of the rod, which is forced upwards and equals $E\omega\frac{\partial u}{\partial x}$. This gives us the following for $x = \lambda$, irrespective of the value of t

$$(1) \qquad \frac{Q}{g}\frac{\partial^2 u}{\partial t^2} = \mathrm{Q} - \mathrm{E}\omega\frac{\partial u}{\partial x} \qquad \text{or} \qquad \frac{\partial^2 u}{\partial t^2} + \alpha\frac{\partial u}{\partial x} - g = 0,$$

postulating that $\frac{\mathrm{E}\omega g}{\mathrm{Q}} = \alpha$.

When an integral comes into existence at a fixed point, it connects with the integral which represents the motion of the body at that moment along a characteristic curve $x - at = K$; consequently, it takes the function φ from the first integral, and we need only determine the function ψ.

Similarly, when an integral comes into existence at the end λ, we need only determine the corresponding function φ.

85. At the beginning of this phenomenon, an integral comes into existence at the end λ; its function ψ is zero since it is the same as for $u = 0$; therefore, if we designate the corresponding function φ as φ_1,

$$a^2\varphi_1''(\lambda + at) + \alpha\,\varphi'_1(\lambda + at) - g = 0,$$

we obtain the following by integrating

$$\varphi_1(\lambda + at) = \mathrm{C}e^{-\frac{\alpha}{a^2}(\lambda + at)} + \frac{g}{\alpha}(\lambda + at) + \mathrm{C}'.$$

The arbitrary constants C and C' must be determined by the condition that, when $t = 0$, the value of u is zero, as is the derivative $\frac{\partial u}{\partial t}$, thereby reducing the previous expression to

$$\varphi_1(\lambda + at) = \frac{gat}{\alpha} - \frac{ga^2}{\alpha^2}\left(1 - e^{-\frac{\alpha t}{a}}\right).$$

Let us now make $\lambda + at = z$, and replace z with $x + at$; this gives us the integral u_1 we are looking for,

$$u_1 = \frac{g}{\alpha}(x + at - \lambda) - \frac{ga^2}{\alpha^2}\left(1 - e^{\frac{\alpha(\lambda - x - at)}{a^2}}\right).$$

The value $\varphi_1(\lambda + at)$ represents the motion of the additional mass, as long as the integral u_1 continues in the neighbourhood of the point λ, in other words for a period of time equal to $\frac{2\lambda}{a}$. It is easy to see that the velocity, which is nil at the start of the motion, increases continuously.

The integral u_1 reaches the origin after a time of $\frac{\lambda}{a}$. In order to determine the new integral u_2, we need just observe that it takes its function φ from u_1; this gives us

$$\varphi_2(x + at) = \varphi_1(x + at) = u_1 .$$

Since u_2 cancels out when $x = 0$, we also know that

$$\varphi_2(at) + \psi_2(-at) = 0,$$

$$\psi_2(-at) = -\varphi_1(at) = -\frac{g}{\alpha}(at - \lambda) + \frac{a^2 g}{\alpha^2}\left(1 - e^{\frac{\alpha(\lambda - at)}{a^2}}\right).$$

If we replace $-at$ in this equation by z and then replace z with $x - at$, which is the same as replacing $-at$ with $x - at$, we find that

$$\psi_2(x - at) = -\frac{g}{\alpha}(at - x - \lambda) + \frac{ga^2}{\alpha^2}\left(1 - e^{\frac{\alpha(x - at + \lambda)}{a^2}}\right).$$

The integral u_2 is thus determined in full.

86. This integral u_2 reaches the point λ at the moment $t = \frac{2\lambda}{a}$. The integral u_3 which comes into existence at the point λ at this moment has the same function ϕ as the integral u_2; thus

$$\psi_3 = \psi_2.$$

We then need to determine the function $\varphi_3(x + at)$. However, according to (1), we know that

$$\frac{\partial^2 u_3}{\partial t^2} + \alpha \frac{\partial u_3}{\partial x} - g = 0,$$

when $x = \lambda$, this gives us

$$a^2\varphi''_3(\lambda + at) + \alpha\,\varphi'_3(\lambda + at) = g - a^2\,\phi''_3(\lambda - at) - \alpha\,\phi'_3(\lambda - at),$$

or, by replacing $\psi_3(\lambda - at)$ with its known value

$$a^2\,\varphi''_3(\lambda + at) + \alpha\,\varphi'_3(\lambda + at) = 2\,g e^{\frac{\alpha}{a^1}(2\lambda + at)}.$$

This equation can be integrated without any problem and gives

$$\varphi_3(\lambda + at) = -\frac{2g}{\alpha}(\lambda + at)\,e^{-\frac{\alpha t}{a}} + C e^{-\frac{\alpha t}{a}} + C'.$$

The two constants can be determined by assuming that the values of u and of $\frac{\partial u}{dt}$ at the moment $t = \frac{2\lambda}{a}$, provided by the integrals u_1 and u_3, are equal when $x = \lambda$.

Since the expression we arrive at for $\varphi_3(\lambda + at)$ is fairly complex, we shall not reproduce it here. After calculating it, we need just replace t by $\frac{x + at - \lambda}{a}$ to obtain $\varphi_3(x + at)$, and, from there the integral u_3.

We can then easily obtain the integral u_4, and our next task would be to find the integral u_5, this would finally provide us with a linear equation which it would be fairly easy to integrate in finite terms. The calculations can be taken as far as one likes, but they become increasingly tedious.

VII. -The velocity of propagation of motion in fluids

87. Let us first assume that the fluid in question is a perfect gas. The partial differential equation governing the motion of this body when it is enclosed in a cylindrical envelope was established in § 48. The set of second order terms is

$$\rho_0 \frac{\partial^2 u}{\partial t^2} - m f(x)\left(1 + \frac{\partial u}{\partial x}\right)^{-m-1} \frac{\partial^2 u}{\partial x^2};$$

where $f(x)$ represents the pressure corresponding to the point on the x-axis for the original state of the body.

The formula which gives the velocity of propagation for the integrals here becomes

$$\left(\frac{d\xi}{dt}\right)^2 = \frac{m f(x)}{\rho_0}\left(1 + \frac{\partial u}{\partial x}\right)^{-m-1}$$

or, by postulating that $f(x) = p_0$,

$$\left(\frac{d\xi}{dt}\right)^2 = \frac{m p_0}{\rho_0}\left(1 + \frac{\partial u}{\partial x}\right)^{-m-1}.$$

When there are no external forces, the gas can remain in a state of rest; $u = 0$ is therefore an integral for which $\frac{\partial u}{\partial x}$ is nil. The velocity of propagation of a motion in a gas at rest is therefore given by the formula

$$\left(\frac{d\xi}{dt}\right)^2 = \frac{m p_0}{\rho_0},$$

when there are no discontinuities.

This expression has been known since the time of Laplace, but it has never been thoroughly proven.

It is normally proven in the following manner: one first assumes that the gas is homogeneous and of uniform temperature when in a state of rest, we shall then ignore the friction and the viscosity; this, obviously, reduces the partial differential equation to

$$\rho_0 \frac{\partial^2 u}{dt^2} = m\, p_0 \left(1 + \frac{\partial u}{\partial x}\right)^{-m-1} \frac{\partial^2 u}{\partial x^2}. \tag{1}$$

Consideration is then confirned to the case where the expansion $\frac{\partial u}{\partial x}$ would be negligible as compared to unity, and the above equation reduced to

$$\rho_0 \frac{\partial^2 u}{dt^2} = m\, p_0 \frac{\partial^2 u}{\partial x^2}, \tag{2}$$

with no further justification for this simplification. The equation thus obtained is the same as that for the solids and the known value of the velocity of sound may then be deduced easily.

However, this method is not in the least thorough. While we may eliminate insignificant values in finite relationships and retain the rest, we must try not to do the same in differential equations, unless we can assign a limit to the error we make.

Furthermore, any simplification of a differential equation amounts to changing the hypotheses previously adopted; the original equation postulated that each stratum extended according to the adiabatic law in such a way that the product pv^m remained constant. The pressure was therefore inversely proportional to the power m of the volume.

However, if in the relationship

$$p = p_0 \left(1 + \frac{\partial u}{\partial x}\right)^{-m}$$

we develop the binomial, stopping at the first two terms, we are left with

$$p = p_0 - m p_0 \frac{\partial u}{\partial x}.$$

The equation (2) supposes that this is the relationship which exists between the volume and the expansion of a stratum. Thus, when we adopt the equation (2) as the equation for the motion of gases, we implicitly accept that the pressure varies in proportion to the expansion, which clearly contradicts the law of adiabatic expansion.

Our above demonstration of the formula for the velocity of propagation is, on the contrary, extremely thorough. It also shows that the formula is in no way altered by the viscosity or even by friction on the envelope, at least supposing that these latter still generally support motion in parallel strata.

88. When, with a gas which is indefinite in all directions, the motion takes place in spherical layers, the partial differential equation is, as we saw in § 54,

The limits of this paper do not permit a discussion of the formulae obtained. We simply wished to show how the method of determining integrals leads quite naturally to the complete solution of a problem which had not been solved until now.

$$\rho_0 x^2 \frac{\partial^2 z}{\partial t^2} = -z^2 p_0 \frac{\partial}{\partial x}\left[z^{-2m} x^{2m}\left(\frac{\partial z}{\partial x}\right)^{-m}\right];$$

where x is the original radius of the layer in question, and z is its radius at time t. The theory of the propagation of motion can be applied to this equation as it was to the equation which represented motion in the cylinder; since the set of terms assigned to the second order derivatives is

$$\rho_0 x^2 \frac{\partial^2 z}{\partial t^2} - m p_0 z^{-2m+2} x^{2m}\left(\frac{\partial z}{\partial x}\right)^{-m-1} \frac{\partial^2 z}{\partial x^2},$$

the velocity of propagation of the integrals becomes

$$\left(\frac{d\xi}{dt}\right)^2 = \frac{m p_0}{\rho_0} x^{2m-2} z^{-2m+2}\left(\frac{\partial z}{\partial x}\right)^{-m-1}.$$

When the gas is in a state of rest, the integral is $z = x$, hence we can deduce that $\frac{\partial z}{\partial x} = 1$. The velocity of propagation of the motion in a gas which is at rest and indefinite in all directions is therefore provided by the formula

$$\left(\frac{d\xi}{dt}\right)^2 = \frac{m p_0}{\rho_0},$$

as if the gas were enclosed in a cylindrical pipe.

The formula remains unchanged when the viscosity is taken into account, since this added resistance can only alter the independent terms of the second order derivatives in the partial differential equation.

Furthermore, it is only applicable as long as there are no discontinuities in the velocities.

It as long been demonstrated that the propagation of motion in an indefinite gas took place at the same velocity as in a cylinder; however, for this demonstration, the equations for the motion were always reduced to linear from, which amounted to attributing properties to the gases which were very different from those they actually have.

89. The above results relate to the case where the motion propagates in a gas at rest; however, it is easy to generalise these results and establish a general theorem.

To this end, we shall first alter the definition of the velocity at which one motion propagates into another. When this velocity is represented by $\frac{d\xi}{dt}$, this means that during the period of time dt, the point common to the two integrals passes from the stratum which initially crosses the x-axis at ξ, to the stratum which initially crosses the initial x-axis at $\xi + d\xi$. However, at the moment in question, these strata cross the x-axis at $\xi + u$ and at $\xi + d\xi + u + \frac{\partial u}{\partial x} d\xi$, where u is the distance travelled by the stratum ξ. The distance actually travelled in the fluid by the point common to the two integrals is therefore not $d\xi$, but $d\xi \left(1 + \frac{\partial u}{\partial x}\right)$.

We can see from this that $\frac{d\xi}{dt}$ represents the velocity of propagation represented at the initial state, whereas if we represent it as at the actual state of the body, it has the value

$$\frac{d\xi}{dt}\left(1 + \frac{\partial u}{\partial x}\right).$$

If we adopt this last definition, then the theorem in question becomes:

When a motion propagates into another, the velocity of propagation is always equal, in the case of gases, to $\sqrt{\frac{mp}{\rho}}$, *where* p *is the pressure, and* ρ *is the density at the point at which the two motions separate.*

This theorem of course presumes that there are no discontinuities in the velocities at the point common to the two integrals.

To prove this, we shall first consider the case where the gas is enclosed in a cylindrical envelope. We then have

$$\left(\frac{d\xi}{dt}\right)^2 \left(1 + \frac{\partial u}{\partial x}\right)^2 = \frac{mp_0}{\rho_0}\left(1 + \frac{\partial u}{\partial x}\right)^{-m+1}.$$

However,

$$p = p_0\left(1 + \frac{\partial u}{\partial x}\right)^{-m}, \quad \text{hence} \quad p_0 = p\left(1 + \frac{\partial u}{\partial x}\right)^{m},$$

$$\rho = \rho_0\left(1 + \frac{\partial u}{\partial x}\right)^{-1}, \quad \text{hence} \quad \rho_0 = \rho\left(1 + \frac{\partial u}{\partial x}\right),$$

and, if we substitute the values of p_0 and ρ_0 into the above equation, we have

$$\left(\frac{d\xi}{dt}\right)^2 \left(1 + \frac{\partial u}{\partial x}\right)^2 = \frac{m\,p}{\rho},$$

which proves the theorem for this particular case.

When the motion takes place in spherical layers, we have

$$\left(\frac{d\xi}{dt}\right)^2 \left(\frac{\partial z}{\partial x}\right)^2 = \frac{mp_0}{\rho_0}\, x^{2m-2} z^{-2m+2} \left(\frac{\partial z}{\partial x}\right)^{-m+1}$$

In this case,

$$p = p_0 \left[\frac{z^2}{x^2}\left(\frac{\partial u}{\partial x}\right)\right]^{-m}, \text{ hence } \quad p_0 = p\left[\frac{z^2}{x^2}\,\frac{\partial u}{\partial x}\right]^{m},$$

$$\rho = \rho_0 \left[\frac{z^2}{x^2}\,\frac{\partial u}{\partial x}\right]^{-1}, \quad \text{hence} \quad \rho_0 = \rho\left(\frac{z^2}{x^2}\,\frac{\partial u}{\partial x}\right),$$

and, if we substitute the values of p_0 and ρ_0 into the above equation, we have

$$\left(\frac{d\xi}{dt}\right)^2 \left(\frac{\partial z}{\partial x}\right)^2 = \frac{m\,p}{\rho},$$

as before.

90. In the case of any homogeneous fluid enclosed in a cylindrical tube, the partial differential equation which governs the motion is (§ 51)

$$\rho_0\,\frac{\partial^2 u}{\partial t^2} = \varphi'\left(1 + \frac{\partial u}{\partial x}\right)\frac{\partial^2 u}{\partial x^2},$$

hence we have

$$\left(\frac{\partial \xi}{dt}\right)^2 = \frac{1}{\rho_0}\,\varphi'\left(1 + \frac{\partial u}{\partial x}\right).$$

The velocity at which the motion propagates into a fluid at rest can be obtained by making $\dfrac{\partial u}{\partial x} = 0$ in this equation.

When the same fluid is subjected to motion in spherical waves, the terms assigned to the second order derivatives in the partial differential equation are (§ 55)

$$\rho_0 x^2 \frac{\partial^2 u}{\partial t^2} - \frac{(x+u)^4}{x^2} \varphi' \left[\frac{(x+u)^2}{x^2} \left(1 + \frac{\partial u}{\partial x}\right) \right] \frac{\partial^2 u}{\partial x^2},$$

since the function φ' is the same as in the previous case if, at the initial moment, the specific volume and the pressure are the same. We then have

$$\left(\frac{d\xi}{dt}\right)^2 = \frac{1}{\rho_0} \frac{(x+u)^4}{x^4} \varphi' \left[\frac{(x+u)^2}{x^2} \left(1 + \frac{\partial u}{\partial x}\right) \right],$$

and the velocity of propagation into the fluid at rest can be obtained by putting $\frac{\partial u}{\partial x} = 0$, $u = 0$ into this expression. It then becomes obvious that the value of the velocity of propagation $\frac{d\xi}{dt}$ assumes the same value as in the case of the cylinder.

When we consider the velocity $\frac{d\xi}{dt}\left(1 + \frac{\partial u}{\partial x}\right)$ which is represented at the actual state of the body, instead of the velocity of propagation of the motion $\frac{d\xi}{dt}$ which is represented at the initial state, we can establish a general theorem for any fluid similar to that proven for the perfect gases.

When one motion propagates into another, the velocity of propagation is always, in the case of a non-conductive fluid, represented by a single function of the pressure p and the density ρ at the point at which the two motions separate. This function can only depend on the initial state.

To simplify our demonstration, we shall presume that the fluid is contained in a cylindrical envelope.

Since the conductivity of the fluid is insignificant, as is that of the envelope, the pressure is related to the specific volume, provided there are no discontinuities, by a relationship of the form

$$p = F'\left(\frac{1 + \frac{\partial u}{\partial x}}{\rho_0 g}, \alpha\right)$$

(§ 51), where the constant α depends only on the initial state. From this we can deduce that

$$\frac{\partial p}{\partial x} = F''\left(\frac{1 + \frac{\partial u}{\partial x}}{\rho_0 g}, \alpha\right)\frac{\partial^2 u}{\partial x^2}\,\frac{1}{\rho g_0}.$$

If we take the value of $\dfrac{1 + \frac{\partial u}{\partial x}}{\rho_0 g}$ from the first equation in terms of p and α, and place it into the second, we will obviously find that

$$\frac{dp}{dx} = -\frac{1}{\rho_0}\,\psi(p, \alpha)\,\frac{\partial^2 u}{dx^2}\,,$$

such that the partial differential equation for the motion can be written

$$\rho_0\,\frac{\partial^2 u}{\partial t^2} = -\frac{1}{\rho_0}\,\psi(p, \alpha)\,\frac{\partial^2 u}{\partial x^2}\,,$$

This will give us the following for the velocity of propagation,

$$\left(\frac{d\xi}{dt}\right)^2\left(1 + \frac{\partial u}{\partial x}\right)^2 = \frac{1}{\rho_0^2}\,\psi(p, \alpha)\left(1 + \frac{\partial u}{\partial x}\right)^2.$$

Furthermore

$$\rho_0 = \rho\left(1 + \frac{\partial u}{\partial x}\right).$$

Therefore

$$\left(\frac{d\xi}{dt}\right)^2\left(1 + \frac{\partial u}{\partial x}\right)^2 = \frac{1}{\rho^2}\,\psi(p, \alpha),$$

which will prove the theorem.

(To be continued.)

On the Propagation of Motion in Bodies *and in Perfect Gases in Particular* - II[1]

H. Hugoniot, Translated from the *Journal de l'École Polytechnique,* Vol. 58, pp. 1-125 (1889)

Chapter IV: Motion of a Fluid in a Cylindrical Pipe

I. - Outline of the problem

90. The fluids whose motion will be studied in this section are presumed to be enclosed in a cylindrical envelope which is impermeable to heat. External forces are considered to be negligible, as are friction against the envelope and viscosity. Lastly, we shall assume that the specific conductivity of the fluid is practically nil.

The only case considered will be that where, at the initial moment, the pressure and temperature are constant inside the fluid. It has been proven (Chapter II) that, under these conditions, the motion was governed by an equation of the form

$$\frac{\partial^2 u}{\partial t^2} = \varphi\left(\frac{\partial u}{\partial x}\right)\frac{\partial^2 u}{\partial x^2}.$$

The motions we propose studying are those which are generated when, with the initial fluid in a state of rest, one end of the column is subjected to a particular condition, for instance applying a specific motion at the end stratum, such that the velocity is zero at the origin and thereafter varies in a continuous manner.

The corresponding integrals are those which are compatible with the integral $u = 0$, which corresponds to the column in a state of rest. If we represent the motion of the body by a surface represented at three rectangular axes Ox, Ot, Ou, the integral $u = 0$ is the horizontal plane xOt.

The differential equation for the characteristic curves of this surface is

---

[1] See Part One, LVIIth Booklet, p. 3.

$$\left(\frac{dx}{dt}\right)^2 \frac{\varphi(0)}{\rho_0}$$

where φ (0) designates what the function φ later becomes when we make $\frac{\partial u}{\partial x}$ = 0. If we make $\frac{\varphi(0)}{\rho_0} = a^2$, the equation for the characteristic curves becomes

$$x - at = \text{K}, \qquad x + at = \text{K}',$$

and the integral surfaces we are trying to find are those which connect with the horizontal plane along a characteristic curve from one of these systems or the other.

If the connection takes place along a characteristic curve from the first system, the velocity of propagation of the motion is a; it equals $-a$ when this connection takes place along a characteristic curve from the second system.

In other words λ_1 and λ_2 are the points at which the ends of the column cross the x-axis; let us assume that $\lambda_1 < \lambda_2$, if the moved end is λ_1, the velocity of propagation must be positive, and the connection takes place along the straight line $x - at = \lambda_1$. If, on the contrary, we assume that the moved end is λ_2, the connection takes place along the straight line $x + at = \lambda_2$. The phenomena must obviously be the same in both cases and what is more, it cannot depend on the value of x at the moved end. We shall therefore obtain all the integral surfaces by considering either one of the two ends, λ_1 for instance, and attributing a known value to λ_1.

In the following, we shall assume that $\lambda_1 = 0$, and that λ_2 is positive.

The integral surfaces which have to be found therefore connect with the horizontal plane along the straight line

$$x = at,$$

which they both have as a characteristic.

91. If there are no discontinuities at the point common to the two integrals, the motion which comes into existence at the origin reaches the end λ_2 after a period of time equal to $\frac{\lambda_2}{a}$. At this point, a reflection phenomenon occurs whose nature depends on the condition imposed at the end λ_2 and from which a new integral originates. This phenomenon will not be studied here; a complete analysis would require the use of integrals which we have not yet been able to discover, at least while the function $\varphi\left(\frac{\partial u}{\partial x}\right)$ remains arbitrary.

We shall therefore limit ourselves to studying motion in a fluid in a state of rest. The integrals representing the motion which is being propagated are suitable until the opposite end of the column is reached. When the column is of infinite length, these integrals are fully representative of the motion of the fluid, unless discontinuities arise in the motion; as we shall see, this can indeed happen, even when the velocity of the moved stratum is nil at the origin and then varies in a continuous manner.

92. We shall deal especially with perfect gases , using our equation

$$\varphi\left(\frac{\partial u}{\partial x}\right) = mp_0\left(1 + \frac{\partial u}{\partial x}\right)^{-m-1},$$

and we shall show the extent to which the normal theory, which assumes that the motion of these gases is represented with sufficient accuracy by the equation

$$\frac{\partial^2 u}{\partial t^2} = \frac{mp_0}{\rho_0}\frac{\partial^2 u}{\partial x^2}$$

distorts the phenomena which actually occur.

The equation

$$\frac{\partial^2 u}{\partial t^2} = \frac{mp_0}{\rho_0}\left(1 + \frac{\partial u}{\partial x}\right)^{-m-1}\frac{\partial^2 u}{\partial x^2},$$

which governs the motion of perfect gases was integrated for certain values of m using arbitrary functions and, for the other values, using definite integrals (see the paper by Darboux, *Comptes rendus de l'Académie des Sciences*, session of 10 July 1882). Unfortunately the equations were only integrated after the linear equation had been obtained using Legendre's transformation, which, as we know, consists in making

$$u_1 = u - x\frac{\partial u}{\partial x} - t\frac{\partial u}{\partial t}, \qquad x_1 = \frac{\partial u}{\partial x}, \qquad t_1 = \frac{\partial u}{\partial t},$$

such that inverse transformations would have to be carried out on the integrals, indeed a troublesome procedure.

We shall in no case use these integrals in what follows, although they may be used later in determining mutually compatible motions and in analysing certain cases of reflection. This is a point which I shall return to in a subsequent paper.

II. - Search for integrals compatible with the body's state of rest

93. Let us consider the equation for the motion of a fluid

$$\frac{\partial^2 u}{\partial t^2} = \varphi\left(\frac{\partial u}{\partial x}\right)\frac{\partial^2 u}{\partial x^2} \tag{1}$$

and let us represent the integrals of this equation by surfaces; we know that, for each of these surfaces, the horizontal projections of the characteristic curves will satisfy the differential equation

$$\left(\frac{dx}{dt}\right)^2 = \varphi\left(\frac{\partial u}{\partial x}\right).$$

However, we can also find another equation which exists along each of the characteristic curves of the surface.

By differentiating the values of $\frac{\partial u}{\partial x}$ and $\frac{\partial u}{\partial t}$ along one of these curves, we must consider x as a function of t defined by the above equation, giving us

$$\frac{d}{dt}\left(\frac{\partial u}{\partial x}\right) = \frac{\partial^2 u}{\partial x\,\partial t} + \frac{\partial^2 u}{\partial x^2}\frac{dx}{dt},$$

$$\frac{d}{dt}\left(\frac{\partial u}{\partial t}\right) = \frac{\partial^2 u}{\partial t^2} + \frac{\partial^2 u}{\partial x\,\partial t}\frac{dx}{dt}.$$

If we multiply the first equation by $-\frac{dx}{dt}$ and add it to the second, when $\frac{dx}{dt}$ is replaced by its value and when proposed equation is taken into account, we obtain

$$\frac{d}{dt}\left(\frac{\partial u}{\partial t}\right) - \frac{dx}{dt}\frac{d}{dt}\left(\frac{\partial u}{\partial x}\right) = 0. \tag{2}$$

Making

$$\varphi\left(\frac{\partial u}{\partial x}\right) = \left[\psi'\left(\frac{\partial u}{\partial x}\right)\right]^2,$$

we have

$$\frac{dx}{dt} = \pm \psi'\left(\frac{\partial u}{\partial x}\right),$$

each of the signs corresponds to one of the systems of characteristic curves.

Equation (2), in which $\frac{dx}{dt}$ is replaced with its value, becomes

(3) $$\frac{d}{dt}\left(\frac{\partial u}{\partial t}\right) \pm \psi'\left(\frac{\partial u}{\partial x}\right)\frac{d}{dt}\left(\frac{\partial u}{\partial x}\right) = 0.$$

The equation can be integrated immediately. Since the first term is a function in t only along the characteristic curve, the integral must be equal to a constant; this gives us the following two equations, each of which corresponds to a system of characteristics,

(4) $$\begin{aligned} \frac{\partial u}{\partial t} + \psi\left(\frac{\partial u}{\partial x}\right) &= \alpha, \\ \frac{\partial u}{\partial t} - \psi\left(\frac{\partial u}{\partial x}\right) &= \beta. \end{aligned}$$

In the first of these equations, the value of α remains the same for all the points on a characteristic from the first system, but varies when we pass from one characteristic curve to the next. Similarly, the value of β remains the same along a characteristic curve from the second system.

It is easy to see that equation (3) is the second equation for the characteristic curves (see Chapter I); in this specific case, it can be integrated immediately; however, this does not greatly simplify our search for the general integral of the equation for the motion; it is just a means of making the equation linear by means of a transformation which it is pointless dwelling on here.

94. Having established this, we can easily prove the following theorem:

When the tangent plane has the same direction along all of the points of any known curve plotted on a surface from equation (1), this surface is developable.

The partial derivatives $\frac{\partial u}{\partial x}$ and $\frac{\partial u}{\partial t}$ represent the coefficients of the equation for the tangent plane of the integral surface. Therefore, if, for all the points on a curve plotted on the surface, the tangent plane has the same direction, $\frac{\partial u}{\partial x}$ and $\frac{\partial u}{\partial t}$ have the same value at all these points.

If the curve in question is not a characteristic, it is met by an infinite number of characteristic curves from the two systems. Since the values of $\frac{\partial u}{\partial x}$ and $\frac{\partial u}{\partial t}$ remain the same, α and β are constant all along the curve. However, since α remains constant along a characteristic curve from the first system and β along a characteristic curve from the second, α and β remain constant on the surface; hence, according to the equations (4), the same applies for $\frac{\partial u}{\partial x}$ and $\frac{\partial u}{\partial t}$. The tangent plane is therefore the same at all points on the surface and this surface must be reduced to a single plane. It is also obvious that the planes are integral surfaces.

Let us now assume that the curve is a characteristic curve, from the first system for instance; the second of the equations (4) shows that β is constant along the curve. However, a characteristic curve from the second system passes through each of the points on the first curve and β remains unchanged along each of these curves, which can be considered as the generatrices of the surface. Hence, the value of β is the same at all points on the surface and the surface satisfies the partial differential equation

$$(5) \qquad \frac{\partial u}{\partial t} - \psi\left(\frac{\partial u}{\partial x}\right) = \beta,$$

in which β is a constant and represents, as we know, one particular class of developable surfaces.

We would arrive at the same result by assuming that the curve in question is a characteristic curve from the second system; however, in this case the equation for the surfaces would be

$$(6) \qquad \frac{\partial u}{\partial t} + \psi\left(\frac{\partial u}{\partial x}\right) = \alpha,$$

where α is an arbitrary constant. Furthermore, the two equations (5) and (6) represent identical surfaces, since we pass from one to the other by changing the direction of the axis Ot.

It is evident that all the integrals from equation (5) satisfy the proposed equation. What is more, it can be verified without problem. From (5) we conclude that

$$\frac{\partial^2 u}{\partial t^2} = \psi'\left(\frac{\partial u}{\partial x}\right)\frac{\partial^2 u}{\partial x\,\partial t},$$

$$\frac{\partial^2 u}{\partial x\,\partial t} = \psi'\left(\frac{\partial u}{\partial x}\right)\frac{\partial^2 u}{\partial x^2}.$$

If we multiply the second equation by ψ' and add it to the first, noting in passing that $\psi'^2 = \varphi$, we obtain the equation put forward in (1).

95. The integrals compatible with the body in a state of rest are represented by surfaces which connect with the horizontal plane $u = 0$ along a straight line.

All along this straight line, the surfaces have the horizontal plane as their tangent plane; thus according to the theorem in the previous number, they are developable surfaces; they are therefore among the surfaces represented by the equations (5) and (6).

The equations for these surfaces must also be satisfied when we put $\frac{\partial u}{\partial t} = 0$ and $\frac{\partial u}{\partial x} = 0$ into them; in the case of equation (5), we must take

$$\beta = -\psi(0)$$

and, in the case of equation (6),

$$\alpha = \psi(0).$$

The following are the first order equations satisfied by the integrals which are compatible with the state of rest

(7) $$\frac{\partial u}{\partial t} = \psi\left(\frac{\partial u}{\partial x}\right) - \psi(0),$$

(8) $$\frac{\partial u}{\partial t} = \psi(0) - \psi\left(\frac{\partial u}{\partial x}\right).$$

The characteristics for these two systems of equations must be straight lines, as for any developable surfaces and, by forming the correlative system, we can see that the angular coefficient $\frac{dx}{dt}$ of the horizontal projection of these straight lines is equal to $-\psi'\left(\frac{\partial u}{\partial x}\right)$ for equation (7) and to $\psi'\left(\frac{\partial u}{\partial x}\right)$ for equation (8).

The angular coefficient of the characteristic common to the horizontal plane $u = 0$ and to the surface in question is therefore $-\psi'(0)$ for equation (7), or $\psi'(0)$ for equation (8). There is nothing to stop us assuming that $\phi'(0)$ is positive, since the sign of the function ψ' is not known in advance.

Thus, if we consider only those motions which have a positive velocity of propagation, then equation (8) provides all the integrals representing these motions.

The integrals in (8) are represented in geometric terms by developable surfaces which belong to the same *class*; it is easy to show that they all have the following property:

If we extend lines parallel to the generatrices of one of the surfaces through a point in space, the cone obtained is always the same, irrespective of the surface in question. This cone is also one of the integral surfaces.

The generatrices of a developable surface are tangents to a three-dimensional curve, known as the edge of regression, which separates the two nappes from the surface. The previous property can therefore be written as follows:

The edges of regression of developable surfaces (8) all have the same spherical indicatrix.

The results just obtained are summarised in the following theorem:

All motions which can propagate in a homogeneous and non-conductive fluid at rest and enclosed in a cylindrical envelope are represented in geometry by developable surfaces whose edges of regression have the same spherical indicatrix.

The first order partial differential equation which is satisfied by all these surfaces can easily be deduced without difficulty from the equation of the same order which governs the motion[1].

---

[1] The specific integral representing the classes of developable surfaces has been known for many years; however it was not widely used until now. In a paper published in Volume VII of the *Journal de l'Ecole Polytechnique*, Poisson considers a gas in a tube and assumes that, contrary to reality, the fluid obeys Mariotte's law; he then attempts to use this particular integral to show that the velocity of propagation in the fluid at rest is independent of the nature of the motion being propagated. Poisson makes use of Euler's formulae, such that his integral is in a different form than that which will be established in § 110; furthermore, he did not deduce any consequences as regards the properties of motion. The theorem which he arrives at can be proven in more general terms from the characteristic curves, as shown in Section III.

96. Any linear function of x and t is obviously an integral of equation (1). Let

$$u = \mathrm{A}x + \mathrm{B}t + \mathrm{C} \tag{9}$$

be one of these integrals: it expresses the fact that the expansion is constant and equal to A and the velocity is also constant and represented by B. The fluid has therefore remained homogeneous, but it is subjected to a translational motion parallel to the generatrices of the envelope.

The surface represented by this integral is a plane and its characteristics are obviously straight lines. We can conclude from this, as above, that the integrals compatible with the previous equation are also developable surfaces.

To obtain the first order equations representing these surfaces, the equations must be satisfied when $\frac{\partial u}{\partial x} = \mathrm{A}$, $\frac{\partial u}{\partial t} = \mathrm{B}$. The constants β and α for the equations (5) and (6) are therefore determinable by the equations

$$\beta = \mathrm{B} - \psi(\mathrm{A}), \qquad \alpha = \mathrm{B} + \psi(\mathrm{A}).$$

The considerations given mentioned in the previous section still apply. The theory of the propagation of motion in an integral, such as (9), is the same as in the state of rest. Only the class of developable surface depends on the coefficients of the linear integral or, more exactly, on the direction normal to the plane which represents it.

97. Let us return to motions which propagate into a fluid at rest with a positive velocity; we still have to determine how these motions are determined given the condition applied at the end.

We must first integrate equation (8) from § 95,

$$\frac{\partial u}{\partial t} = \psi(0) - \psi\left(\frac{\partial u}{\partial x}\right).$$

Since the integral surfaces are developable, they can be represented, as we know, in the form of envelopes from a mobile plane by the two equations

$$u = \gamma x + [\psi(0) - \psi(\gamma)]\, t + \mathrm{F}(\gamma),$$

$$0 = x - \psi'(\gamma)\, t + \mathrm{F}'(\gamma),$$

in which F is an arbitrary function, each form of which will give one of the required integrals when γ is eliminated. Assuming that the elimination has been carried out, we shall state that u cancels out at $x = 0$, $t = 0$; this determines the constant which can be considered as contained in F (γ).

However, we have to determine the function F such that it satisfies a condition imposed at the end $x = 0$. If, for instance, we gave the motion of the end $x = 0$ as a relationship $u = f(t)$ in advance, the equation obtained would have to be satisfied once γ had been eliminated by replacing x with zero and u with $f(t)$.

The calculation poses certain difficulties which disappear when we take either the velocity $\frac{\partial u}{\partial t}$ or the expansion $\frac{\partial u}{\partial x}$ as the unknown function of x and t instead of trying to determine u, the distance travelled, in terms of these two variables.

Moreover, it is evident that the motion is known in full when either the velocity or the expansion has been calculated in terms of x and t, since u, the distance travelled, is in each case deduced simply by simple quadrature.

98. Let us first take the velocity as our unknown function and postulate

$$\frac{\partial u}{\partial t} = v \quad ,$$

equation (8) from § 95 becomes

$$v = \psi(0) - \psi\left(\frac{\partial u}{\partial x}\right), \tag{10}$$

hence

$$\frac{\partial v}{\partial x} = \psi'\left(\frac{\partial u}{\partial x}\right)\frac{\partial^2 u}{\partial x^2} \; .$$

Furthermore, equation (1) can be written as follows, stating that $\varphi = \psi'^2$,

$$\frac{\partial v}{\partial t} = \left[\psi'\left(\frac{\partial u}{\partial x}\right)\right]^2 \frac{\partial^2 u}{\partial x^2} \; .$$

Eliminating $\frac{\partial^2 u}{\partial x^2}$ from the last two equations gives

$$\frac{\partial v}{\partial t} + \psi'\left(\frac{\partial u}{\partial x}\right)\frac{\partial v}{\partial x} = 0.$$

Furthermore, $\frac{\partial u}{\partial x}$ can be deduced in terms of v from equation (10); substituting this into the above equation will produce the following

(11) $$\frac{\partial v}{\partial t} + \theta(v)\frac{\partial v}{\partial x} = 0,$$

which is a first order equation with respect to $\frac{\partial v}{\partial t}$ and $\frac{\partial v}{\partial x}$.

If we take the expansion as the unknown function, the results will be even simpler. Indeed, if we assume that

$$\frac{\partial u}{\partial x} = z,$$

equation (8) becomes

$$\frac{\partial u}{\partial t} = \psi(0) - \psi(z),$$

from which we can deduce that

$$\frac{\partial^2 u}{\partial t^2} = -\psi'(z)\frac{\partial z}{\partial t},$$

The equation put forward can be written as follows

$$\frac{\partial^2 u}{\partial t^2} = [\psi'(z)]^2\frac{\partial z}{\partial x},$$

and by eliminating $\frac{\partial^2 u}{\partial t^2}$ we obtain the equation satisfied by the expansion, in other words

(12) $$\frac{\partial z}{\partial t} + \psi'(z)\frac{\partial z}{\partial x} = 0.$$

Like equation (11), this is a first order linear equation.

99. It is easy to integrate equations (11) and (12). Let us consider the first, for instance; the correlative system is

$$\frac{dt}{1} = \frac{\theta(v)}{dx} = \frac{dv}{0} \quad .$$

As we know, this defines the characteristics of the surface which in geometry represents the function v. In this case, these characteristics are obviously horizontal straight lines whose angular coefficient is just a function of the ordinate v.

The representative surface is thus a sort of conoid, fully determined, when it is made to pass through a known curve.

Integrating the correlative system gives us

$$v = \alpha, \qquad t - \frac{x}{\theta(\alpha)} = \beta,$$

where α and β are two constants which need only be sorted into an arbitrary relationship to obtain the general integral, which is thus in the form

$$(13) \qquad v = F\left[t - \frac{x}{\theta(v)}\right],$$

where F is an arbitrary function.

The surfaces represented by equation (12) are conoids of the same sort; their general integral is in the form

$$(14) \qquad z = F_1\left[t - \frac{x}{\psi'(z)}\right],$$

where F_1 is a new arbitrary function.

Integrals (13) and (14) can be extremely easily used to determine the arbitrary functions F and F_1 through the conditions applied at the end $x = 0$.

If we know the velocity of the end stratum at each moment, v can be expressed in terms of t when $x = 0$ using an equation of the form $v = f(t)$. However, if we put $x = 0$ into equation (13), we are left with $v = F(t)$. Therefore, in order to obtain the particular integral which corresponds to this case, we need just make $F = f$.

Similarly, we can see that if we know that the expansion of the end stratum for each moment obeys a relationship of the form $z = f_1(t)$, we have to put $F_1 = f_1$ to obtain the corresponding integral (14).

It is worth pointing out that if we know the velocity of the end stratum $x = 0$ in terms of the time, it follows that we also know its expansion, and *vice versa*, because the expansions are related to each other by the relationship in (9) which can be written

$$v = \psi(0) - \psi(z),$$

which is independent of x and t.

100. Thus, each integral can be represented by three surfaces, one for the distance travelled, a second for the velocity and a third for the expansion in terms of x and t. These three surfaces, defined by first order partial differential equations have rectilinear characteristics; the last two surfaces are always horizontal.

If we consider the three surfaces which are assigned to one and the same motion, the characteristics corresponding to the same values of x and t will be projected horizontally along the same straight line. Indeed, the angular coefficients of the horizontal projections of the characteristics for the three surfaces are

$$\frac{dx}{dt} = \psi'\left(\frac{\partial u}{\partial x}\right), \qquad \frac{dx}{dt} = \theta(v), \qquad \frac{dx}{dt} = \psi'(z).$$

However, $\theta(v)$ and $\psi'(z)$ are in fact the same as $\psi'\left(\frac{\partial u}{\partial x}\right)$ which proves the proposition. Moreover, this results from the fact that the values of $\frac{\partial u}{\partial x} = z$ and $\frac{\partial u}{\partial t} = v$ remain constant in the developable surface along each characteristic.

III. - A General study of motion

101. The locus of the points on the expansion surface where the ordinate is a constant describes a straight line. Thus, an expansion z originating from the end stratum $x = 0$ propagates into the body *unchanged* at a constant velocity $\phi'(z)$ which depends on the value of this expansion. The corresponding velocity v propagates in the same manner, *unchanged*. The propagation phenomenon is therefore analogous to the phenomenon which takes place in cylindrical elastic solids; only here the velocity of propagation of an expansion depends on its value and is no longer an absolute constant.

We can consider the phenomenon in yet another manner. There is nothing to prevent us considering the expansion of the end stratum $x = 0$ as remaining constant during each infinitesimally small interval dt and undergoing an infinitesimally small variation dz at the start of each interval. The first expansion dz comes into contact with a fluid in a state of rest and propagates unchanged at a velocity equal to $\phi'(0)$, since the expansion of the fluid it comes into contact with is zero. At the same time, the expansion communicates an infinitesimally small velocity dv to the fluid such that

$$dv = \psi(0) - \psi(dz) .$$

If this expansion dz is not followed by a further expansion, it would leave the fluid in a homogeneous state and subject to a uniform translational motion.

However, after a period of time dt, a new expansion dz_1 comes into existence at the end $x = 0$; the fluid into which it propagates still has the same expansion dz; consequently, its velocity of propagation is constant and equal to $\psi'(dz)$. Furthermore, it communicates a new infinitesimally small velocity to the fluid which is added to the previous velocity.

We can continue to consider matters in the same way: it is obvious that the expansion dz, which comes into existence at time t in the end stratum must be considered as propagating in a homogeneous fluid with an expansion z and which is subjected to a uniform translational motion, with a velocity v, which can in no way alter the nature of the propagation. The new expansion dz is thus propagated at a velocity $\phi'(z)$ and applies a new infinitesimally small velocity to the fluid which is added to the velocity

$$v = \psi(0) - \psi(z)$$

which it already had.

We are thus led to consider a succession of elementary waves, each of which propagates at a constant velocity. However, in this case, the velocity of propagation varies from one elementary wave to the next and this circumstance gives rise to a new phenomenon which can be encountered with elastic solids.

102. Let us assume, indeed, that the velocity of propagation of one elementary wave is greater than the previous one, such that

$$\psi'(z + dz) > \psi'(z),$$

the second wave will eventually encounter the first one. It is even simpler to calculate the distance at which the two waves will meet. Let X represent this distance, τ the time the first wave takes to cover it; this gives us the following two equations

$$\tau\psi'(z) = X, \qquad (\tau - dt)\,\psi'(z + dz) = X\,;$$

hence we can deduce that

$$\tau = \frac{\psi'(z)}{\psi''(z)} \frac{1}{\frac{dz}{dt}},$$

(1) $$X = \frac{[\psi'(z)]^2}{\psi''(z)} \frac{1}{\frac{dz}{dt}}.$$

If T also designates the moment at which the two waves meet, counted from the time the motion originated, and t is the time at which the wave z came into existence,

$$\tau = T - t,$$

and the equation

$$\tau\psi'(z) = X$$

becomes

(2) $$X = (T - t)\,\psi'(z)\,.$$

In equations (1) and (2), z is a known function of t which represents the expansion of the end stratum $x = 0$; by eliminating t from the two equations (1) and (2), we shall obtain a relationship between X and T, which is the geometric representation of a curve, the locus of the points at which two successive waves join together.

Obviously, this curve is the envelope of the projections of the characteristics for the three surfaces which represent the motion in question. In the case of the surface of the distances travelled, the characteristics intersect two by two in space, with the locus of their intersection describing the edge of regression. The curve represented by equations (1) and (2) is thus the projection of the edge of regression of the surface of the distance travelled.

The intersection of the projecting cylinder with the velocity surface and the surface of the distances travelled gives, in each of these cases, a curve which is obviously the locus of the feet of the shortest distances from two infinitesimally close generatrices. Each of these curves is the *line of striction* of the surface.

The edge of regression of the surface of the distances travelled and the lines of striction for the velocity and expansion surfaces are therefore on a single cylinder with its generatrices perpendicular to the plane xOt, which we could easily have predicted in advance.

103. Once we obtained the equation for the curve, horizontal projection of the edge of regression, we still have to look for the portion of this curve which is to be considered in a study of the motion. It generally contains a parasite branch from the generatrices of the developable surface which meet the plane tOu in the part of this plane which corresponds to the negative values of t and which have no relation to the motion of bodies. We can obtain the point of separation of these two parts of the curve by making $t = 0$ in equations (1) and (2); we must then consider all the points which correspond to positive values of t.

We should then attempt to find the first point on the curve which is affected by the motion. This is generally one of the points where the curve has a tangent parallel to Ox, in other words where the ratio $\frac{dX}{dT}$ is infinite. If, as will normally happen, the functions ψ' and $\frac{dz}{dt}$ remain finite, and this is also the case with their derivatives throughout the motion, we obtain the following equation for determining these points

$$\frac{\frac{d^2z}{dt^2}}{\left(\frac{dz}{dt}\right)^2} + \frac{\psi'''(z)}{\psi''(z)} - \frac{2\,\psi''(z)}{\psi'(z)} = 0. \tag{3}$$

We must attempt to find, for the values of t deduced from this equation, the direction of the concavity of the curve and to take the point which has the lowest value of T.

Finally, we will still have to compare the point obtained in this way to that corresponding to $t = 0$. If the value of T which corresponds to this point was the smallest, it is this point which will be the first affected by the motion.

104. The method just mentioned can therefore be used to determine the coordinates X_1 and T_1 from the point of the projection from the edge of regression for the surface of the distances travelled or from the lines of striction for the expansion and velocity surfaces which is the first affected by the motion. Let t_1, the corresponding value of t, be a value which can also equal zero. We then know that the two successive elementary waves which met first left the end $x = 0$ at times t_1 and $t_1 + dt$.

From the moment T_1, a new phenomenon comes into existence which will only be cursorily considered in this paper and to which I shall return later. We shall, nonetheless, show hereafter that from this moment discontinuities are introduced into the velocities and expansions.

105. If the surfaces representing the motion are intersected by planes perpendicular to Ot, for the corresponding time t, we know the distance travelled, the velocity or the expansion for all points on the body.

Let us only consider the curves obtained by continuing the velocity and expansion surface through these planes; obviously, if the velocity of propagation was the same for all the elementary waves, all the curves would be identical because they would correspond to the same surface; we would pass from one curve to another, transforming it parallel to itself. This is the case with solid elastic cylinders.

However, this is no longer the case when a fluid is involved. Since the elementary waves do not have the same velocities, the corresponding points on the two expansion curves for instance have the same ordinates; however, they usually cross the x-axis at different points.

The two curves, obtained by intersecting this surface through planes perpendicular to Ot at distances of t and t' from the origin, correspond to each other point by point, with corresponding points having the same ordinate z. The distance between the points at which they cross the x-axis can be obtained by multiplying the difference $t' - t$ by the velocity of propagation which corresponds to the ordinate z, in other words $\psi'(z)$; it is also equal to $(t' - t)\,\psi'(z)$.

The expansion curve thus moves inside the body, but alters as it does so according to a law which depends on the nature of the function $\psi'(z)$. The velocity curve undergoes similar changes.

106. If two elementary waves, starting from the origin at times t and $t + dt$, meet at time T and at a distance of X, it is evident that the expansion curve corresponding to time T has two infinitesimally close points located on a vertical line drawn from the point X; it may therefore have a vertical tangent for the value X of x , and the same applies for the velocity curve.

When we consider the remainder of the expansion curves and we limit our study to the part located between the origin and the point reached by the motion, it is clear that the first of these curves will not usually have a vertical tangent and if T is taken to designate the time when two elementary waves meet for the first time, the first expansion curve to have a vertical tangent will be the one corresponding to the time T_1. The tangent in question will always be a stationary tangent, unless the two elementary waves which meet at time T_1 are those which left the origin at the beginning of the motion. In this case, the vertical tangent is the tangent which corresponds to the point at which the curve crosses the x-axis.

Everything we have just said could be repeated for the velocity curve.

107. The formulae established above can obviously be applied until time T_1 when, for the first time, two elementary waves come into contact. From this moment, it is conceivable that a specific disruption will come about, resulting from the fact that all the elementary waves which left the origin at different times will successively catch up with each other. This new phenomenon comes into being, according to the above, when the motion passes through the edge of regression of the surface representing the distance travelled, in other words when the second nappe of the developable surface comes into play.

We stated above that this should bring about discontinuities in the velocities and the expansions. This can be proven very simply by choosing a suitable condition applied at the end $x = 0$.

To do this, we shall determine this condition such that all elementary waves leaving the end reach a point with abcissa X at the same time.

We saw previously (§ 102) that, when z is a function of t which represents the expansion of the end stratum $x = 0$,

$$\frac{[\psi'(z)]^2}{\psi''(z)} \frac{1}{\frac{dz}{dt}} = X.$$

If X is a constant, the above equation can be integrated; it can indeed be written

$$\frac{\psi''(z)\frac{dt}{dz}}{[\psi'(z)]^2} = \frac{1}{X}.$$

If we integrate this and make z equal to zero when $t = 0$, we obtain

$$\psi'(z) = \frac{X\,\psi'(0)}{X - t\,\psi'(0)}. \tag{4}$$

From this equation we can obtain z as a function of t and we need only represent the law of succession governing the expansions of the end stratum $x =$

0 by this function for all the elementary waves to come together at a distance X (which we can arbitrarily assign) and at time $T = \frac{X}{\psi'(0)}$.

Under these conditions, it is clear that the characteristic curves of the three surfaces representing the motion project in a horizontal plane along the straight lines which pass through the point with the co-ordinates X and T. The developable surface for the distance travelled can be reduced here to a cone with its apex at the point (X, T) and the velocity and expansion surfaces are two conoids with vertical axes.

108. The expansion z of the end stratum, which gives rise to the motion we have just studied is a known function of t; we can therefore deduce from this the velocity, v, of this stratum using the formula

$$x = \psi(0) - \psi(z) ;$$

the velocity can therefore be expressed in terms of time; and, to produce these conditions, we need just close off the pipe at the end $x = 0$ using a piston to which one applies a motion described by the above equation. However, we should point out that according to equation (4), the function $\psi'(z)$ becomes infinite when $t = \frac{X}{\psi'(0)}$ and, normally, the values of z, $\psi'(z)$ and $\psi(z)$ become infinite at the same time. The velocity given to the piston would therefore have to become infinite at time $\frac{X}{\psi'(0)}$.

However, we shall see that this is unnecessary. Indeed, when, owing to the condition applied at the end $x = 0$, a motion comes into existence in the fluid which is assumed to be initially at rest, we can try to find the new integrals for the equation which will be compatible with the equation for the first motion; these will represent those motions which can propagate into the first one.

For the sake of compatibility, however, the surface representing the new motion must have a characteristic in common with the first, in other words the one passing through the point $x = 0$, and which has a positive angular coefficient.

Since the first surface is developable, this characteristic is a straight line; and, according to the theorem postulated in § 94, the new surface is also developable since it connects with the first along this straight line and thus has the same tangent plane on all points along this straight line. Furthermore, this new surface belongs to the same class as the first. Indeed, as the velocity of propagation is positive, it must be one of the surfaces represented by the equation

$$\frac{\partial u}{\partial t} + \psi\left(\frac{\partial x}{\partial u}\right) = \alpha \,.$$

The constant α is the same as for the first surface, since they connect up; therefore $\alpha = \phi'(0)$.

In particular, the plane tangent to the developable surface, along the characteristic in question, itself represents an integral which is compatible with the first one.

When, given a fluid which is initially at rest, a velocity is applied at the end stratum defined by a particular analytical function of time $f(t)$, we can change , at a given moment t_1, the nature of the function, provided that no discontinuities are introduced into the velocity. Until time t_1, the motion is represented by the developable surface which corresponds to the function $f(t)$; however, from the moment t_1 onwards, a new integral comes into existence at the end $x = 0$; this is represented in geometrical terms by another developable surface which connects up with the first.

We can assume, for example, that from the moment t_1 onwards, the velocity of the end stratum remains constant; the new integral surface is then the plane at a tangent to the developable surface along its last characteristic.

In the case of velocity and expansion surfaces, this tangent plane has corresponding horizontal planes which pass through the last characteristic of the surfaces.

109. In this specific case where the velocity of the end stratum is the velocity corresponding to the expansion z defined by equation (4), there is no reason not to assume that the velocity v remains constant from t_1 onwards, before $\dfrac{X}{\psi'(0)}$ Under these conditions, all the elementary waves which start at the origin and continue until time t_1 will arrive at a point at a distance X from the origin at the same time.

If we consider the expansion curve at various moments during the phenomenon ranging between t_1 and $T = \dfrac{X}{\psi'(0)}$ this curve is partly rectilinear and partly curvilinear; the curvilinear part alters with time and disappears suddenly at time T to be replaced by a vertical straight line.

The velocity of the end stratum no longer tends towards infinity, making the experiment physically possible. We can see that at all moments before T, the expansion and velocity vary continually with x; however, at time T, the situation is quite different: the fluid is divided in two by the stratum X; one is still at rest, while the expansion and velocity are constant at all points in the other part, and may take any value we choose.

In this example we can see that a continuous motion applied to the end stratum and travelling at a velocity increasing from zero, while not undergoing any sudden changes, can however introduce a discontinuity into the motion which comes into existence very suddenly and can remain indefinitely, as we shall show in Chapter V.

We would be inviting serious errors if we tried to apply the developable integrals outside the points where the motion meets the edges of regression of the corresponding surfaces. The partial differential equation governing the motion may no longer even be suitable, since this equation was only produced for the case where there are no discontinuities in the velocity of the strata.

IV. - Motion of perfect gases

110. The general theory stated above will now be applied to perfect gases. Their partial differential equation is as follows

$$\frac{\partial^2 u}{\partial t^2} = \frac{m p_0}{\rho_0}\left(1 + \frac{\partial u}{\partial x}\right)^{-m-1} \frac{\partial^2 u}{\partial x^2}.$$

The value of the velocity at which the motion propagates into the gas at rest is as follows

$$\sqrt{\frac{m p_0}{\rho_0}} = a.$$

The function given above by $\psi'\left(\frac{\partial u}{\partial x}\right)$ here becomes

$$\psi'\left(\frac{\partial u}{\partial x}\right) = \sqrt{\frac{m p_0}{\rho_0}}\left(1 + \frac{\partial u}{\partial x}\right)^{-\frac{m+1}{2}} = a\left(1 + \frac{\partial u}{\partial x}\right)^{-\frac{m+1}{2}},$$

from which we can deduce that

$$\psi\left(\frac{\partial u}{\partial x}\right) = -\frac{2a}{m-1}\left(1 + \frac{\partial u}{\partial x}\right)^{-\frac{m-1}{2}}.$$

The partial differential equation for the developable surfaces which represent the integrals compatible with the state of rest and which propagate in a positive direction then become

$$\frac{\partial u}{\partial t} - \frac{2a}{m-1}\left(1 + \frac{\partial u}{\partial x}\right)^{-\frac{m-1}{2}} + \frac{2a}{m-1} = 0\,,$$

which, when we take the velocity to equal v and the expansion z, can be written

$$v = \frac{2a}{m-1}(1+z)^{-\frac{m-1}{2}} - \frac{2a}{m-1}$$

or, solving this with respect to z,

$$z = \left(1 + \frac{m-1}{2a}v\right)^{-\frac{2}{m-1}} - 1.$$

Since the ratio m of the specific heats is greater than one, z and v always have opposite signs. The same would apply for the rest if the value of m were less than one.

The above formulae no longer hold true when $m = 1$; in such a case, the gas would observe Mariotte's law as it expanded; this is what would happen, for instance, if the envelope acted on the gas as a perfect heat source in order to keep it at a constant temperature.

In this case,

$$a = \psi'(0) = \sqrt{\frac{p_0}{\rho_0}}\,,$$

$$\psi'\left(\frac{\partial u}{\partial x}\right) = a\left(1 + \frac{\partial u}{\partial x}\right)^{-1},$$

$$\psi\left(\frac{\partial u}{\partial x}\right) = a \log\left(1 + \frac{\partial u}{\partial x}\right).$$

The partial differential equation for the developable surfaces then becomes

$$\frac{\partial u}{\partial t} + a \log\left(1 + \frac{\partial u}{\partial x}\right) = 0,$$

which can then be written

$$v + a \log(1 + z) = 0$$

or

$$z = e^{-\frac{v}{a}} - 1.$$

111. When m is not equal to one, the integrals compatible with the state of rest of the gas are given by the two equations

$$u = \gamma\ x - \frac{2\ at}{m - 1}\left[1 - (1 + \gamma)^{-\frac{m-1}{2}}\right] + F(\gamma),$$

$$0 = x - at(1 + \gamma)^{-\frac{m+1}{2}} + F'(\gamma),$$

where F is an arbitrary function and γ is a parameter which can be cancelled out in each case to directly obtain the integral.

When $m = 1$, the integrals are given by the two equations

$$u = \gamma x - at \log(1 + \gamma) + F(\gamma),$$

$$0 = \gamma - \frac{at}{1 + \gamma} + F'(\gamma).$$

Since these integrals are not particularly convenient, in order to determine the arbitrary function according to the condition imposed at the end, we must, as in the general case, consider the functions v and z, representing velocity and expansion respectively, rather than the function u in x and t.

112. Let us first consider the velocity v, we know that it satisfies the linear equation

$$\frac{\partial v}{\partial t} + \theta(v)\frac{\partial v}{\partial x} = 0,$$

where $\theta\ (v')$ represents the function $\psi'\left(\frac{\partial u}{\partial x}\right)$ or $\psi'\ (z)$ in which z has been replaced by its value as a function of v. Hence

$$\psi'\ (z) = a\,(1 + z)^{-\frac{m+1}{2}}, \quad z = \left(1 + \frac{m-1}{2a}\,v\right)^{-\frac{2}{m-1}} - 1 \ ;$$

and hence,

$$\theta\,(v) = a\left(1 + \frac{m-1}{2a}\,v\right)^{\frac{m+1}{m-1}}$$

the partial differential equation in v then becomes

$$\frac{\partial v}{\partial t} + a\left(1 + \frac{m-1}{2a}\,v\right)^{\frac{m+1}{m-1}} \frac{\partial v}{\partial x} = 0.$$

It has the following general integral

$$v = \mathrm{F}\left[t - \frac{x}{a\left(1 + \frac{m-1}{2a}\,v\right)^{\frac{m+1}{m-1}}}\right],$$

where F is an arbitrary function.

In the specific case where $m = 1$, we have

$$\psi'\ (z) = \frac{a}{1+z}, \qquad 1 + z = e^{-\frac{v}{a}} \ ;$$

hence

$$\theta\,(v) = ae^{\frac{v}{a}}\ .$$

The partial differential equation in v is then

$$\frac{\partial v}{\partial t} + ae^{\frac{v}{a}} \frac{\partial v}{\partial x} = 0$$

and has the following general integral

$$v = \mathrm{F}\left(t - \frac{x}{a} e^{-\frac{v}{a}}\right).$$

113. We must still consider the function z which, as we know, satisfies the equation

$$\frac{\partial z}{\partial t} + \psi'(z) \frac{\partial z}{\partial x} = 0.$$

In the case of perfect gases, this equation becomes

$$\frac{\partial z}{\partial t} + a(1 + z)^{-\frac{m+1}{2}} \frac{\partial z}{\partial x} = 0,$$

when m is not equal to one, and

$$\frac{\partial z}{\partial t} + \frac{a}{1 + z} \frac{\partial z}{\partial x}$$

when $m = 1$. The general integrals are

$$z = \mathrm{F}_1\left[t - \frac{x}{a}(1 + z)^{\frac{m+1}{2}}\right]$$

and, when $m = 1$,

$$z = \mathrm{F}_1\left[t - \frac{x}{a}(1 + z)\right].$$

114. The important role played by the edge of regression of the expansion surface and the line of striction of the velocity or expansion surfaces was demonstrated earlier. These three curves have the same horizontal projection which, as has been proven, is represented by the equations

$$\mathrm{X} = \frac{[\psi'(z)]^2}{\psi''(z) \dfrac{dz}{dt}},$$

$$T = t + \frac{\psi'(z)}{\psi''(z)\,\dfrac{dz}{dt}},$$

where X and T are the current co-ordinates and z is a function of t representing the expansion of the end stratum $x = 0$ at various times during the motion.

In the case of perfect gases, we have

$$\psi'(z) = a\,(1+z)^{-\frac{m+1}{2}},$$

$$\psi''(z) = -\frac{(m+1)\,a}{2}\,(1+z)^{-\frac{m+3}{2}};$$

and hence,

$$X = -\frac{2\mathrm{a}\,(1+z)^{-\frac{m-1}{2}}}{(m+1)\,\dfrac{dz}{dt}},$$

$$T = t - \frac{2\,(1+z)}{(m+1)\,\dfrac{dz}{dt}}.$$

The value of X is negative if $\dfrac{dz}{dt}$ is positive and *vice-versa*.

Therefore if $\dfrac{dz}{dt}$ remains positive, in other words, if the expansion of the end stratum continually increases, the values of x will always be negative, and these points will obviously never be reached by the motion; there is therefore no reason to concern ourselves with this.

If, for a particular value of z, $\dfrac{dz}{dt}$ cancels out, the edge of regression has one of its points at infinity. The corresponding projection of the characteristic is a tangent to the curve at this point at infinity; in other words, the asymptote. This asymptote has an angular coefficient of $\psi'(z) = a\,(1+z)^{-\frac{m+1}{2}}$; it passes through the point where the co-ordinates are $x = 0$, $t = t_1$, where t_1 is the value of t which cancels out $\frac{dz}{dt}$.

When the gas expands according to Mariotte's law, $m = 1$; the above formulae become

$$X = -\frac{a}{\frac{dz}{dt}}, \qquad T = t - \frac{1+z}{\frac{dz}{dt}}.$$

When a given motion is given to the end $x = 0$, it is the velocity of the end stratum which can be ascertained directly and not its expansion. There is therefore no point replacing z with v in the formulae which provide the co-ordinates X and T for the different points on the curve.

However, if m is not equal to one, we have

$$1 + z = \left(1 + \frac{m-1}{2a}v\right)^{-\frac{2}{m-1}},$$

and hence

$$\frac{dz}{dt} = -\frac{1}{a}\left(1 + \frac{m-1}{2a}v\right)^{-\frac{m+1}{m-1}}\frac{dv}{dt},$$

and, by substituting this into the expressions for X and T,

$$X = \frac{2a^2\left(1 + \frac{m-1}{2a}v\right)^{\frac{2m}{m-1}}}{(m+1)\frac{dv}{dt}},$$

$$T = t + \frac{2a\left(1 + \frac{m-1}{2a}v\right)}{(m+1)\frac{dv}{dt}}.$$

When $m = 1$,

$$1 + z = e^{-\frac{v}{a}}, \qquad \frac{dz}{dt} = -\frac{1}{a} e^{-\frac{v}{a}} \frac{dv}{dt}$$

and the formulae become

$$\mathrm{X} = \frac{a^2 e^{\frac{v}{a}}}{\frac{dv}{dt}}, \qquad \mathrm{T} = t + \frac{\mathrm{a}}{\frac{dv}{dt}}.$$

115. The relationship

$$v = \frac{2a}{m-1}(1+z)^{-\frac{m-1}{2}} - \frac{2a}{m-1}$$

gives rise to a curious comment. The expansion z obviously cannot, when it is negative, be greater than one as an absolute value; if it did, the branch would be shortened beyond its initial length. In this equation, we must therefore assume z ranges between -1 and $+\infty$.

If z is increased continuously from -1 to $+\infty$, we can see that, when $z = -1$, the velocity v is positive and infinitely large; z decreases in terms of its absolute value, and the same applies to the velocity which cancels out when $z = 0$.

The expansion z becomes positive, the velocity becomes negative and the two values increase at the same time as the absolute value. However, when z increases indefinitely, the value of v converges to a finite limit equal to $-\frac{2a}{m-1}$. Thus, the velocity of the gas can only range between the two limits $+\infty$ and $-\frac{2a}{m-1}$.

The above considerations are applicable at the end branch. Let us assume that at the point $x = 0$ the gas is limited by a piston whose velocity, initially zero, is gradually increased in such a way as to expand the gas; when this velocity reaches the limit $\frac{2a}{m-1}$, the expansion of the end unit will have become infinite, and, if the velocity of the piston increases beyond $\frac{2a}{m-1}$, the end unit will cease to follow the piston as it moves.

Hence, the value of $\frac{2a}{m-1}$ represents the top speed at which the gas flows. In the case of simple gases, such as oxygen, hydrogen and nitrogen, and also air, m

is practically equal to 1.4. The velocity, the top flow rate for one of these gases, for which the pressure and the density are known, can be obtained by multiplying the speed of sound corresponding to this pressure and density by five.

It is evident that since the expansion increases indefinitely, the pressure and absolute temperature converge towards zero. The limit of the speed $\frac{2a}{m-1}$, which has just been defined, could not therefore be carried out in practice, since the properties admitted for perfect gases, properties which were used as the basis for producing the partial differential equation, cease to be applicable when the temperature reaches absolute zero.

This velocity $\frac{2a}{m-1}$ is, however, an important factor to consider; it is definitely greater than the flow velocity which could be achieved in practice and, consequently, gives one of its limits.

In the case of air, assumed to be at normal temperature and pressure, since the speed of sound, a, is almost equal to 330 m per second, the flow velocity would be 1650 m per second.

In the case where the gas would obey Mariotte's law, the situation would be quite different. We would actually have

$$v = -a \log(1 + z)$$

such that, as z increases indefinitely, the value of v would converge to $-\infty$.

116. In perfect gases, the pressure p is related to the expansion z by the formula

$$p = p_0 (1 + z)^{-m}.$$

The following relationship therefore exists between the velocity and the pressure

$$v = \frac{m-1}{2a}\left(\frac{p}{p_0}\right)^{\frac{m-1}{2m}} - \frac{2a}{m-1}.$$

The velocity v is positive as long as p is greater than p_0, and negative in the opposite case. When p increases indefinitely, the same applies for v; conversely, when p converges towards zero, v tends towards the limit $-\frac{2a}{m-1}$.

It can be considered that the pressure opposite the end stratum, initially equal to p_0, reduces gradually and converges towards zero. In these conditions, this stratum will assume a negative velocity, initially zero, but increasing continually in terms of its absolute value. This velocity will converge towards the limit $-\frac{2a}{m-1}$.

If the gas observed Mariotte's law, the relationship between the velocity and the pressure would be

$$v = -a \log\left(\frac{p_0}{p}\right).$$

The velocity would increase indefinitely in terms of its absolute value if the pressure converged towards zero.

V. - An examination of a few cases of motion in perfect gases

117. The general formulae will now be applied to some specific cases.

We shall first suppose that the column of gas is closed, at the end $x = 0$, by a piston which is being actuated by a motion with uniform acceleration, the velocity being initially equal to zero. Then

$$v = \alpha t$$

when $x = 0$. The expansion z of the end stratum is given by the formula

$$z = \left[1 + \frac{(m-1)\alpha}{2a} t\right]^{-\frac{2}{m-1}} - 1$$

and its pressure is

$$p = p_0 \left[1 + \frac{(m-1)\alpha}{2a} t\right]^{\frac{2m}{m-1}}.$$

When the velocity is positive, the expansion of the end stratum is negative and transforms into a contraction; the pressure is greater than p_0 and increases indefinitely with time.

Conversely, when the velocity is negative, since the coefficient α is also negative, the expansion z is positive and increases with time; the pressure is less than p_0 and decreases as t increases. If we make $\alpha = -\alpha_1$, it can be seen that, when $t = \dfrac{2a}{(m-1)\,\alpha_1}$, the expansion is infinite, whereas p cancels out. The velocity of the piston is then equal to $-\dfrac{2a}{m-1}$. When it exceeds this limit, the velocity of the end stratum then ceases to increase and a vacuum is formed in front of the piston.

The formulae giving the velocity and the expansion in terms of x and t can be easily deduced from the integrals for §§ 112 and 113, bearing in mind that they must be reduced to the values of v and z below when we make $x = 0$. This gives us

$$v = \alpha\left[t - \frac{x}{a\left(1 + \dfrac{m-1}{2a}v\right)^{\frac{m+1}{m-1}}}\right],$$

$$1 + z = \left\{1 + \frac{m-1}{2a}\,\alpha\left[t - \frac{x}{a}(1+z)^{\frac{m+1}{2}}\right]\right\}^{-\frac{2}{m-1}}.$$

If we replace $1 + z$ in this equation by $\left(\dfrac{p}{p_0}\right)^{-\frac{1}{m}}$ we could easily deduce from this a relationship between p, x and t.

It is evident, according to the first of these formulae, that any velocity v developed in the end stratum propagates uniformly and has moved a distance x after a period of time equal to

$$\frac{x}{a}\left(1 + \frac{m-1}{2a}v\right)^{-\frac{m+1}{m-1}}.$$

Similarly, any expansion z developed in the end stratum propagates unchanged, at a velocity of

$$a\,(1+z)^{-\frac{m+1}{2}},$$

So that we need not consider infinite velocities, we can assume that the velocity of the piston stops increasing, in terms of its absolute value, from time t_1 onwards, thereafter remaining constant and equal to $-\alpha t_1 = v_1$. The above formulae apply until time t_1. Thereafter, the column of gas can be considered, at each time t, as being divided into three parts:

1) One part in which the motion has still to propagate and which, consequently, is in a state of rest;

2) Another part whose motion is represented by the two formulae above; this part is limited to the stratum where x has the value αt_1;

3) A third part in which the velocity is constant, as are the expansion and the pressure. This is limited by the stratum at the following point along the x-axis

$$at_1\left(1+z_1\right)^{-\frac{m+1}{2}},$$

where z_1 is the constant expansion corresponding to the velocity v_1. Furthermore

$$\left(1+z_1\right)^{-\frac{m+1}{2}} = \left(1+\frac{m-1}{2a}\,v_1\right)^{\frac{m+1}{m-1}} = \left[1+\frac{m-1}{2a}\,\alpha t_1\right]^{\frac{m+1}{m-1}}.$$

The point on the x-axis which limits the part of the column where the velocity and expansion are constant, is thus

$$at_1\left[1+\frac{m-1}{2a}\,\alpha t_1\right]^{\frac{m+1}{m-1}}.$$

In this part, the gas is homogeneous and subjected to a uniform translational motion.

118. Since the motion is represented in geometrical form, either by the velocity surface, or by the expansion surface, we may obtain a very clear picture of it by sectioning one of these surfaces through planes perpendicular to Ot, and by continuously varying the distance between these planes and the origin. The sections obtained are similar in shape for both surfaces, and we shall therefore limit ourselves to a consideration of the expansion surface. Each section produced by the plane $t = const.$ represents all the expansions at all points in the fluid.

If we consider the expansions z and z', which came into existence at the end stratum at moments τ and τ', each of the expansions has propagated at a particular velocity $a(1+z)^{-\frac{m+1}{2}}$ in the case of the first, and $a(1+z')^{-\frac{m+1}{2}}$ in the case of the second.At instant t , the first expansion has been propagated during time $t - \tau$, and the second expansion during time $t - \tau'$. The distance separating the points which these expansions z and z' have reached at time t is therefore, assuming $\tau' > \tau$,

$$a(1+z)^{-\frac{m+1}{2}}(t-\tau) - a(1+z')^{-\frac{m+1}{2}}(t-\tau'),$$

or

$$a\left[\tau'(1+z')^{-\frac{m+1}{2}} - \tau(1+z)^{-\frac{m+1}{2}}\right] + at\left[(1+z)^{-\frac{m+1}{2}} - (1+z')^{-\frac{m+1}{2}}\right].$$

We can see that the distance between the two points where the expansions have the given values z and z' vary in direct proportion to time. This distance increases or decreases depending on the sign in front of the coefficient *at*.

It is easy to find out that the sign in front of coefficient *at* is positive, when $z' - z$ is negative and *vice versa*.

Therefore, if the expansion of the end stratum decreases in algebraic terms, the distance separating the points where the expansions have the given values at a given time will increase with time. The opposite takes place when the expansion of the end stratum increases with time in algebraic terms.

If we join together, in the expansion curve at time t, the two points with the ordinates z and z', we obtain a cord at an angle from the x-axis which increases or decreases with t, depending on whether the difference $z' - z$ is positive or negative. The same is true for the tangent to the curve, since there is nothing to prevent us assuming that the values of z and z' are infinitesimally close.

We can conclude from this that, if the expansion of the end stratum continually increases with time, the tangents to the expansion curve form ever steeper angles with the x-axis as time goes on; in other words, the curve becomes increasingly straighter. The opposite happens when the expansion of the stratum continually decreases.

119. Let us return to the specific case where the motion of the end stratum is uniformly varied, we shall first assume that the velocity is constantly negative; the velocity of the end stratum is then $-\alpha_1 t$, and the expansion, zero when $t = 0$, must then be positive and increasing. It is easy to see how the shape of the expansion curve changes over time.

Indeed, let us suppose that the velocity of the piston ceases to increase from time t_1 onwards, we see that at a time t before t_1, the initial ordinate of the curve is OA

$$z = \left[1 - \frac{m-1}{2a}\alpha_1 t\right]^{-\frac{2}{m-1}} - 1.$$

The ordinate then decreases continually until it cancels out when OB $= at$; from this point on it remains constantly equal to zero.

At a time t' after t, but before t_1, the initial ordinate OA' is greater than the previous one. The end point B of the curve has moved and is now at point B_1. Point A_1 on the new curve corresponds to point A on the initial curve with the same ordinate. The points A and A_1, B and B_1 are points on the same expansion and, according to what was said in the previous number, $AA_1 < BB_1$.

If we consider two points from the same expansion on the two curves, M and M_1 for instance, we have $AA_1 < MM_1 < BB_1$. The distance from two points on the curve located on the same horizontal line is inversely proportional to the ordinate. This curve changes shape, becoming increasingly flatter; the tangent at one point whose ordinate is known forms an increasingly smaller angle with the axis Ox.

The curve continues to change shape until time t_1. If t_1 exceeds the value

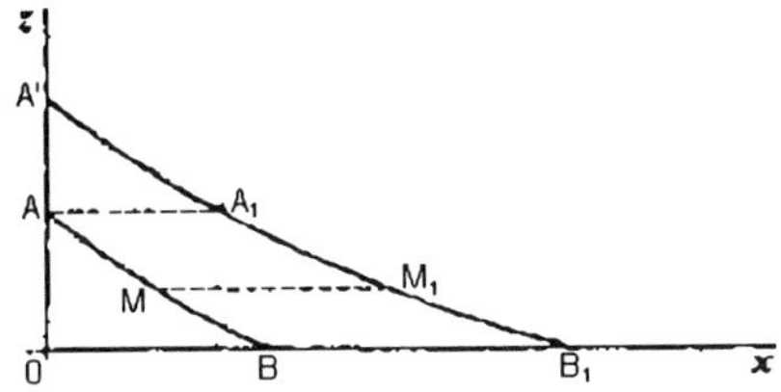

Fig. 1

$\frac{2a}{(m-1)\,\alpha_1}$, the initial ordinate OA continues to increase until time $t'' = \frac{2a}{(m-1)\,\alpha_1}$ At this instant, it becomes infinite; the curve has an infinite branch with the z-axis as its asymptote. We saw that from this moment on, a vacuum formed behind the piston which limited the column of gas; furthermore, the laws of deformation are still the same and for all values above t'' the curve still has the same vertical asymptote.

The curve of velocities follows laws of deformation similar to the expansion laws. However, at time $t'' = \frac{2a}{(m-1)\,\alpha_1}$ the velocity is not infinite; it is equal to $\frac{2a}{m-1}$ and thereafter remains constant. We set out below two states of the expansion curve and the velocity curve which correspond to times after t''.

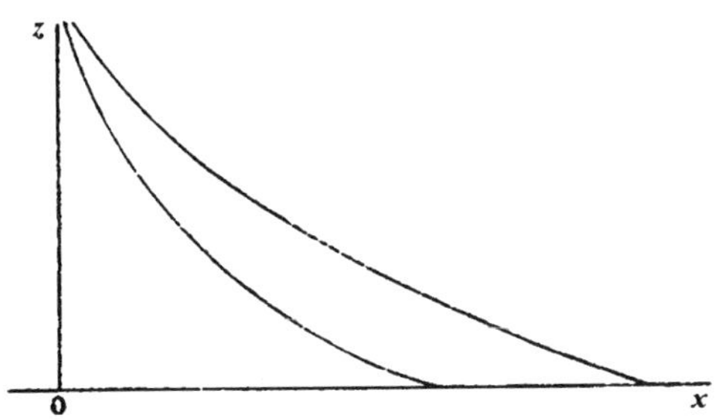

Fig. 2. Expansion curves

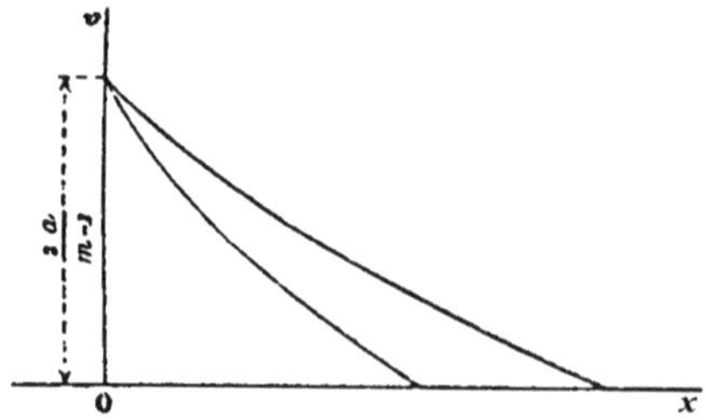

Fig. 3. Velocity curves

For the experiment to be possible, the temperature of the gas would have to drop to absolute zero without altering the properties of the gas, which never happens.

However, this is not what occurs when the velocity of the piston stops increasing after time t_1, before t''. Let z_1 be the expansion corresponding to $v_1 = -\alpha_1 t_1$, AB the expansion curve at this time; at a time t after t_1, the curve will contain three parts, a horizontal line AA_1, whose length is $a\left(t - t_1\right)\left(1 + z_1\right)^{-\frac{m+1}{2}}$, a curvilinear part A_1B_1 and, after point B_1, the x-axis; the distance BB_1 is equal to $a\ (t - t_1)$ and, thereafter, greater than AA_1. At a time t' after t, the curve will have the shape AA_2B_2x, and so on, the horizontal distance separating the end points A_1, B_1 from the curvilinear part increases more dramatically as time goes on.

There is no reason to stop at the velocity curve, which is similar in shape.

If the tube is indefinite on the positive side of the x's, there is nothing to prevent the motion continuing to propagate indefinitely in the same way; the length of the curvilinear part will then increase indefinitely.

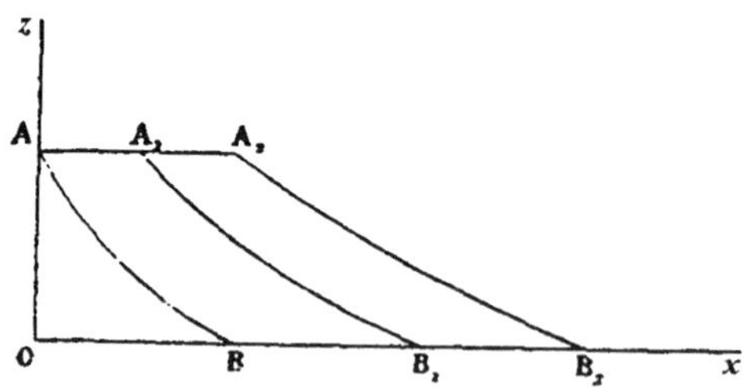

Fig. 4

Conversely, when the tube is closed, there is a moment at which the point B_1 on the expansion curve is located at the end of the tube. At this moment, a phenomenon of reflection occurs, whose nature depends on the condition applied at the end and which will not be studied in this paper. The integral obtained above will then no longer be applicable.

120. The situation is completely different when the velocity of the piston limiting the column is altered to compress the gas. The ordinates of the expansion curve are then negative and always below one. If we consider the curves AB and $A'B_1$, which correspond to the times t and t', and if we join two points on these curves M and M_1 with the same expansion, we obtain

$$BB_1 < MM_1 < AA_1;$$

the distance MM_1 decreases as the ordinate of the point M approaches zero. The tangent to the curve at a particular ordinate point therefore forms an increasingly steeper angle with the x-axis.

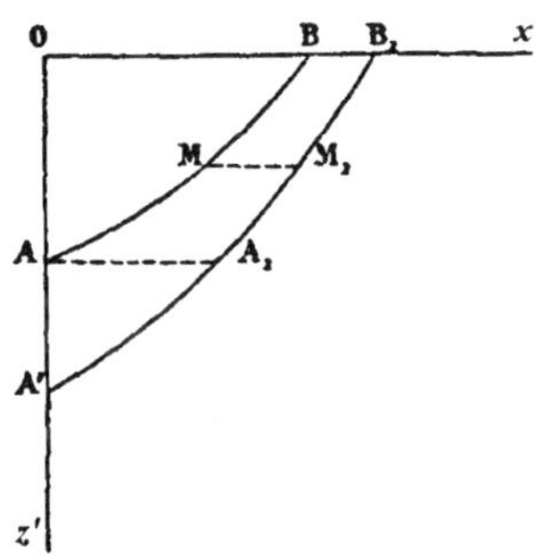

Fig. 5

The ordinate at the origin OA increases with time and would converge towards one if the motion of the piston were still accelerating. However, we can assume that, as from time t_1, the velocity of the piston remains constant; thereafter, for all times after t_1, the curve will contain a horizontal part and a curvilinear part.

Below, we have drawn three positions for the expansion curve, the first AB corresponding to the time t_1, the two others A_1B_1 and A_2B_2 at the times t and t' after t_1.

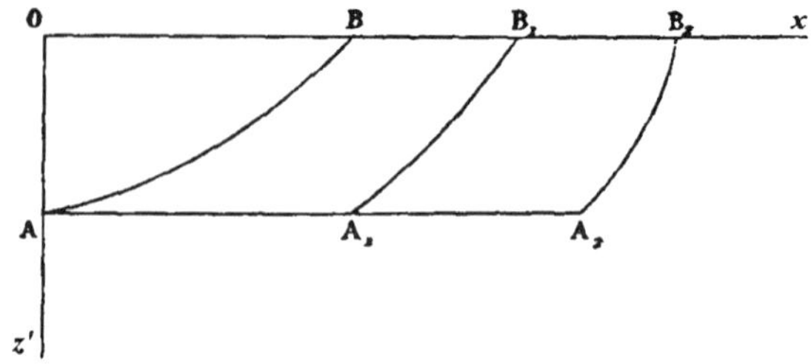

Fig. 6

The difference in the points where the curves cross the x-axis which corresponds to the two ends of the curvilinear part decreases steadily as time goes on; the tangents form angles which approach a right angle with the x-axis.

We saw earlier how the moment when one of the tangents became vertical corresponded to the moment when the motion came into contact with the edge of regression of the developable surface and we know that as of this moment, the integrals generally cease to be applicable.

We must therefore attempt to find the moment at which we come across a vertical tangent in the expansion curve. The moment where this phenomenon occurs is that where the motion comes into contact with the horizontal projection of the edge of regression for the distance surface.

But the co-ordinates X and T of this curve are given by the equations

$$X = \frac{2a^2\left(1 + \frac{m-1}{2a}v\right)^{\frac{2m}{m-1}}}{(m+1)\frac{dv}{dt}},$$

$$T = t + \frac{2a\left(1 + \frac{m-1}{2a}v\right)}{(m+1)\frac{dv}{dt}},$$

where v is the velocity of the end stratum at time t. In the case in question,

$$v = \alpha t, \qquad \frac{dv}{dt} = \alpha,$$

and the value of T becomes

$$T = t + \frac{2a\left(1 + \frac{m - 1}{2a}\alpha t\right)}{(m + 1)\alpha}.$$

The value of T obviously increases with t; T is therefore at its lowest value when $t = 0$.

As a result, the two successive elementary waves which first meet up with one another are those which left at time *0* and time dt. The point on the expansion curve where, following this change, the tangent for the first time becomes perpendicular to Ox is therefore the point whose ordinate z is zero. The moment when this phenomenon occurs is given by the value of T corresponding to $t = 0$, i.e. $\frac{2a}{(m + 1)\alpha}$ The motion has then propagated for a distance equal to $\frac{2a^2}{(m + 1)\alpha}$.

At time $\frac{2a}{(m + 1)\alpha}$ the velocity curve also has a vertical tangent which, as was the case with the expansion curve, corresponds to the point whose ordinate is zero.

121. The formulae given above are applicable until the moment $t = \frac{2a}{(m + 1)\alpha}$ or until the motion has propagated for a distance equal to $\frac{2a^2}{(m + 1)\alpha}$. Under normal conditions, this distance is extremely long. Thus, for instance, in the case of air at normal temperature and pressure, $a = 330^{\text{m}}$ approximately, $m = 1.4$. This then gives us

$$\frac{2a^2}{m + 1} = 90750 \qquad \frac{2a^2}{(m + 1)\alpha} = \frac{90750}{\alpha}.$$

The distance to which the motion has propagated at the time when the formulae cease to be applicable can be obtained by dividing 90 750 by the acceleration of the motion of the piston.

If, for example, this acceleration was equal to 10 m per second, the formulae would not cease to be applicable before the motion had succeeded in propagating a distance of over 9 km.

This distance is moreover inversely proportional to the acceleration of the motion and we could make α large enough so that the distance would be lower than any given length.

This is where the most interesting question is raised: what happens from the moment that the elementary waves, which left the end stratum at different times, begin to meet up? What would happen to the integrals when we go beyond the edge of regression for the distance surface?

This is a new and extremely complex problem, whose general solution is no concern of this paper.

However, very simple logic shows that in general the formulae which represent the motion until this moment must cease to be applicable after the moment when this phenomenon occurs.

Since the motion of the piston which limits the stratum is accelerated in a uniform manner, there comes a moment when its velocity is greater than the velocity a at which the motion propagates into the fluid, as long as there have been no discontinuities. Let us suppose that, since the piston has reached a velocity a_1 at time t_1, its velocity thereafter remains constant; at this time t_1, let l be the length of the column of fluid affected by the motion.

If there is no discontinuity in the velocities, at the point at which the part of the fluid at rest is separated from the fluid in motion, the length of the part in motion would extend in front of this point at a velocity a and would decrease in length at its rear at a constant velocity v_1. At a time t after t_1, the length of this part in motion would have become

$$l-(t-t_1)(v_1-a)\,,$$

such that this length would cancel out when

$$t = t_1 + \frac{1}{v_1 - a}\,.$$

For this, all of the strata affected thereafter by the motion would have to undergo a shortening by as much as their initial length, which is impossible. This absurdity would become even more apparent if t were given values greater than $t_1 + \dfrac{l}{v_1 - a}$, because the length of the part affected by the motion would then be negative, which is completely meaningless.

To avoid physical impossibilities, it is absolutely essential that a discontinuity arise in the velocities and the expansions at the point at which the two portions of fluid separate, because it is only in these conditions that the velocity of propagation, in a fluid at rest, can be modified. However, the considerations used to draw up the partial differential equation governing the motion of the gases cease to be accurate in the presence of discontinuities. The problem is not therefore to find new integrals for the initial equation, but above all to find the new equation governing the motion.

We shall return to this topic in Chapter V.

122. The results we have just obtained remain unchanged when the gases obey Mariotte's law and when the coefficient m becomes equal to one. However, when $m = 1$, as we have already shown, there is no limit velocity of flow, such that, if the piston limiting the column of gas is moved at a velocity sufficient to expand the gas, this velocity could increase indefinitely unless a vacuum forms at the rear of the piston.

Irrespective of the law governing the motion given by the piston to the end stratum, we will always obtain similar results to these if the velocity increases continually in terms of its absolute value. Only when the velocity is altered to compress the gas, does it become possible that in certain cases the first tangent of the expansion curve which becomes vertical is not the one which corresponds to the point whose ordinate is zero. It will be necessary in each particular case to discuss this topic in order to ascertain just how long the integrals obtained provide the answer to the problem.

We can, however, make the following remark, which applies not only to the motion of perfect gases, but also to the motion of all fluids which were considered in this Chapter.

Let us consider the expansion curve and let A and A' be two forms of this curve which correspond to the times t and t'. Let M and M' be two points with the same ordinate which correspond to each other.

Let t'' be another moment in the motion; if we assume that

$$\frac{t'' - t}{t' - t''} = K,$$

the expansion curve which corresponds to the time t'' is the locus of the points M'' which divide the horizontal line MM' into two segments such that

$$\frac{MM''}{t' - t''} = K .$$

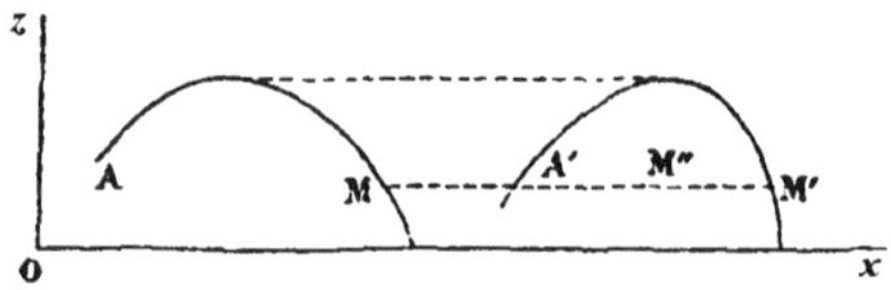

Fig. 7

We need only ascertain two of the expansion curves to obtain all the others using a very simple geometric formula. The same considerations are applicable to the velocity curve.

VI. - Propagation of a periodic vibrational motion

123. In the normal theory of the propagation of acoustic vibrations in a cylindrical tube, as it shows in the Physics courses, we assume that the tube is closed at its end by a vibrating shutter, whose velocity is represented by an expression of the form

$$v = \alpha \cos \beta t ,$$

where α and β are constants, and whose motion is transferred to the end stratum to propagate into the gas. The partial differential equation governing the motion is simplified and reduced to the form

$$\frac{\partial^2 u}{\partial t^2} = a^2 \frac{\partial^2 u}{\partial x^2} ,$$

it can be seen that, at a given time t, the expansion and velocity curves are in fact the sine waves which, as the time t varies, move in a uniform manner at a velocity a.

It is interesting to try to find how the above results are modified in reality.

Owing to the relationship which exists between the velocity and the expansion at a particular point in the body, the expansion of the end stratum is

$$z = \left[1 + \frac{m-1}{2a} \alpha \sin \beta t\right]^{-\frac{2}{m-1}} - 1 .$$

Each individual wave leaves the end stratum at time t then propagates at a constant velocity, represented by

$$a\,(1+z)^{-\frac{m+1}{2}}$$

in other words by

$$a\left[1+\frac{m-1}{2a}\,\alpha\,\sin\,\beta t\right]^{\frac{m+1}{m-1}}.$$

It is thus easy to formulate the velocity or expansion curves for any time t.

124. We shall focus our attention more particularly on the velocity curve and will first formulate that curve which corresponds to the end of the first period of motion for the end stratum. The length of this period is $2\pi/\beta$.

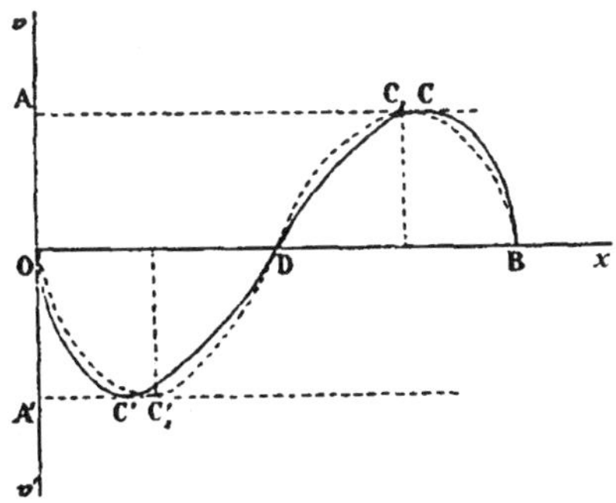

Fig. 8

The individual wave which began at the origin of the motion corresponds to an infinitesimally small velocity and has propagated a distance of $\frac{2\pi a}{\beta}$, which we shall represent by OB.

The individual wave which left at time $\frac{\pi}{2\beta}$ corresponds to a velocity α; its velocity of propagation is

$$a\left(1+\frac{m-1}{2a}\,\alpha\right)^{\frac{m+1}{m-1}};$$

and, since α is assumed to be positive, it is greater than a; the distance to which it has propagated is

$$a\left(1+\frac{m-1}{2a}\alpha\right)^{\frac{m+1}{m-1}}\frac{3\pi}{2\beta} > \frac{3}{4}\,\text{OB}.$$

If we therefore consider a length OA = α along O*v* and also, taken from A, a length parallel to O*x*

$$\text{AC} = a\left(1+\frac{m-1}{2a}\alpha\right)^{\frac{m+1}{m-1}}\frac{3\pi}{2\beta},$$

we shall obtain the maximum of the velocity curve.

The individual wave which left at time $\frac{\pi}{\beta}$ corresponds to a zero velocity; its velocity of propagation is a; therefore, at time $\frac{2\pi}{\beta}$ is has propagated a distance $\text{OD} = \frac{a\pi}{\beta} = \frac{1}{2}\,\text{OB}$.

Finally, the individual wave which left at time $\frac{3\pi}{2\beta}$ corresponds to a velocity of $-\alpha$; it therefore propagated at a velocity of

$$a\left(1-\frac{m-1}{2a}\alpha\right)^{\frac{m+1}{m-1}},$$

which is consequently less than a, and the distance it has travelled at time $\frac{2\pi}{\beta}$ is

$$a\left(1-\frac{m-1}{2a}\alpha\right)^{\frac{m+1}{m-1}}\frac{\pi}{\beta} < \frac{\text{OB}}{4}.$$

If we therefore consider a length OA' = α along O*v*' and, from A', a length parallel to O*x*

$$\text{A}'\text{C}' = a\left(1-\frac{m-1}{2a}\alpha\right)^{\frac{m+1}{m-1}}\frac{\pi}{\beta},$$

we shall obtain the point C' on the curve which corresponds to the y-axis minimum.

Similar formulations could be carried out for all points on the curve; however, the four points obtained C', D, C and B, together with the point O, are enough to provide an idea of its shape to compare it with the sine wave OC'_1DC_1B which would be obtained by normal theory and which is shown by a dotted line on the figure.

The real curve shares three points with the sine wave, which are located on the x-axis; furthermore, the two curves intersect at two other points. The value of x for C, the highest point on the real curve, is greater than that for C_1, the highest point on the sine wave; the opposite applies at the lowest point.

125. In order to formulate the velocity curve corresponding to the end of the second period of the end stratum, we shall consider that this curve initially consists of a part identical to the previous one, from the elementary waves which left the origin between times $\frac{2\pi}{\beta}$ and $\frac{4\pi}{\beta}$; then another part which can be easily deduced from the first by ascertaining the velocities of propagation of the different elementary waves.

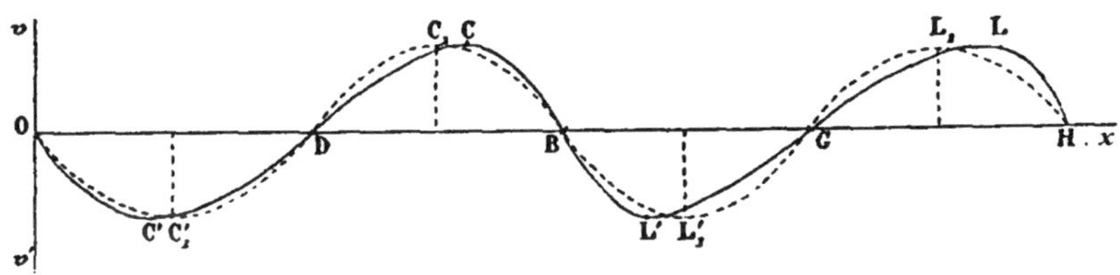

Fig. 9

If we transpose the first sine wave parallel to Ox, by a length equal to OB, we shall obtain a curve BL'_1GL_1H, which would represent the velocity in the portion BH of the gas, if the individual waves all propagated at the same velocity. To obtain the maximum from this, we shall consider a length L_1L along the horizontal L_1, twice the length of C_1C, along the positive x-axis; we shall obtain the minimum in a similar manner by considering a horizontal line from the point L'_1 in the direction of the negative x-axis, of length L'_1L' equal to twice $C_1C'_1$.

We can continue in the same way and plot successive curves which correspond to the times $\frac{6\pi}{\beta}$, $\frac{8\pi}{\beta}$ etc.

It is obvious that, as we travel away from the origin, the maxima move further away from the corresponding points on the sine wave in the direction of the positive x-axis; the distances increase as a series of natural numbers. Similarly, the minima move further away from the corresponding points on the sine wave, but in the opposite direction; the distances again increase like a series of natural numbers.

The points located on the x-axis are common to both curves.

126. Therefore, although according to normal theory the velocity curve should advance in a uniform manner without changing shape, it can be seen that in reality this curve does move in a uniform manner but changes shape continuously. The gradient of the curve decreases continually in the part ranging between one of the points such as L' and the following point L; the gradient increases, however, between a point such as C and the next point L'.

The propagation continues until the edge of regression of the developable motion for the distance is reached by the motion, in other words until the moment when one of the tangents to the velocity curve becomes perpendicular to the axis Ox.

Let us consider the curve and a projection along the horizontal plane xOt of the edge of regression, the co-ordinates X and T of one point are

$$X = \frac{2a^2\left(1 + \dfrac{m-1}{2a}v\right)^{\frac{2m}{m-1}}}{(m+1)\dfrac{dv}{dt}},$$

$$T = t + \frac{2a\left(1 + \dfrac{m-1}{2a}v\right)}{(m+1)\dfrac{dv}{dt}}.$$

In this case,

$$v = \alpha \sin \beta t, \qquad \frac{dv}{dt} = \alpha\beta \cos \beta t,$$

and the value of T becomes

$$T = t + \frac{2a\left(1 + \dfrac{m-1}{2a}\,\alpha \sin \beta t\right)}{(m+1)\,\alpha\beta \cos \beta t}.$$

We must find the value of t corresponding to the minimum value of T. The second term is made up of two parts, one of which t is increasing, the other periodic. We must therefore find the value of t, ranging between zero and $\frac{2\pi}{\beta}$, for which the value of T is a minimum.

We must first disregard the case where we were able to cancel out the numerator of the periodic part; for this, the velocity $\alpha \sin \beta t$ had to become negative and equal to $\frac{2a}{m-1}$ in terms of its absolute value, in other words the limit value for the flow rate velocity.

Having postulated this, we shall show that if α, in other words the maximum value of the velocity of the end stratum, is much smaller than a, as is always the case with acoustic vibrations, the smallest value of T is obtained when $t = 0$ and, consequently, is equal to

$$\frac{2a}{(m+1)\,\alpha\beta}.$$

This is the time in which the motion propagates according to the rules set out above. At that time, it has travelled a distance of

$$X = \frac{2\,a^2}{(m+1)\,\alpha\beta}.$$

After this time, a new phenomenon comes about, which usually introduces discontinuities in the velocities.

127. We can substitute in real values to gain some idea of this distance X.

Let N equal the number of full vibrations produced in a second by the cnd stratum $x = 0$, and 2λ the total amplitude of the vibration; it is easy to see that

$$\beta = 2N\pi, \qquad \alpha = 2\pi N\lambda,$$

such that the value of X becomes

$$X = \frac{a^2}{2\,(m+1)\,N^2\,\pi^2\,\lambda}\,.$$

Let us suppose, to organise our thoughts, that we are dealing with air at room temperature and pressure, $a = 330^{m}$, $m = 1.4$ approximately. If we put $N = 500$, $\lambda = 0^{m},001$, we find that X has a value of less than 10^{m}.

We can see from this example that the normal theories governing the propagation of acoustic vibrations in the gas are soon found lacking, even in the case of small amplitude motions.

More particularly, in the case of periodic motion, the undulating curve which represents the velocity or expansion of a stratum at a given moment, changes shape as it propagates. However, as long as the motion does not reach the edge of regression of the developable surface, the velocity of propagation remains unchanged; and from this point of view, the normal theory provides exact results.

But this edge of regression is reached by the motion after travelling a distance which, in normal circumstances, is not that great; then new phenomena come about.

The distance X, after which the formulae obtained are no longer applicable, can be arranged in a different form. As usual, we call the product $\frac{a}{N}$ of the duration of a full vibration and the speed of sound, the *wavelength.* If we take this wavelength to be L, we obtain

$$X = \frac{L^2}{2\,(m+1)\,\pi^2\,\lambda}\,.$$

Thus the distance X is proportional to the square of the wavelength and inversely proportional to the amplitude of the vibrational motion given to the end stratum.

A study of the expansion curve would have provided us with results which were quite similar to those obtained by considering the velocity curve. In addition, the results would in no way have changed if, with the gas obeying Mariotte's law, the coefficient m had become equal to one.

VII - A study of a specific case of the propagation of motion

128. We showed in § 107 that, for any fluid in a state of rest, we can determine the condition applied to the end stratum for all elementary waves leaving the origin at different moments during the motion, and which propagate at different velocities and meet up at the same time at the same distance from the origin.

Let X be this distance. Since the velocity of propagation corresponding to the expansion z is represented by $\phi'(z)$, the expansion of the end stratum must satisfy the condition

$$\phi'(z) = \frac{X\,\phi'(0)}{X - t\,\phi'(0)}.$$

In the case of perfect gases

$$\psi'(0) = a, \qquad \psi'(z) = a(1+z)^{-\frac{m+1}{2}},$$

such that the previous formula becomes

$$(1+z)^{-\frac{m+1}{2}} = \frac{X}{X - at};$$

and hence

$$1 + z = \left(1 - \frac{at}{X}\right)^{\frac{2}{m+1}}.$$

Since the expansion z of the end stratum is given by this formula, we can easily obtain the velocity which must be applied to this stratum using the relationship which exists between the velocity and the expansion. We therefore find that

$$\text{(1)} \qquad v = \frac{2a}{m-1}\left(1 - \frac{at}{X}\right)^{-\frac{m-1}{m+1}} - \frac{2a}{m-1},$$

a formula which could have been obtained directly by integrating the equation

$$X = \frac{2a^2 \left(1 + \frac{m-1}{2a} v\right)^{\frac{2m}{m-1}}}{(m+1)\frac{dv}{dt}}$$

in which X should be considered as a constant.

When the gas obeys Mariotte's law, the expressions for the expansion z of the end stratum and its velocity v become

$$1 + z = 1 - \frac{at}{X},$$

(2) $$v = -a \log\left(1 - \frac{at}{X}\right).$$

129. To satisfy these conditions, we need just delimit the column of gas by a piston and to give a motion to this piston which is defined by equation (1) or equation (2).

The velocity of the piston, initially zero, continuously increases; it reaches infinity at time $\frac{X}{a}$, it is thus easy to imagine it *a priori*, since the velocity of propagation of the wave which left the end of the column at this last moment, must equal infinity.

The acceleration of the motion of the piston is

$$\frac{dv}{dt} = \frac{2a^2}{X(m+1)}\left(1 - \frac{at}{X}\right)^{-\frac{2m}{m+1}}$$

and we can see that it too continues to increase and reaches infinity when $t = \frac{x}{a}$.

It would be impossible in practice to make the velocity of the piston increase indefinitely; however, there is nothing to prevent us supposing that, after altering the velocity in accordance with the formula in (1), or the formula in (2) if the gas obeys Mariotte's law, for a certain time t_1, we can then maintain this velocity constant.

Let v_1 be the velocity attained at time t_1, z_1 the corresponding expansion; for all values of t greater than t_1, the gas will be divided into three parts, each subject to a different motion.

After the point at on the x-axis, the gas will be at rest; before the point $a\left(1 + z_1\right)^{-\frac{m+1}{2}}\left(t - t_1\right)$ on the x-axis, the expansion will be equal to z_1 and the velocity equal to v_1 at all points; between the two points we have just mentioned, the velocity and expansion will vary from one point to the next.

All individual waves which leave the origin between t and t_1 will meet at distance X and time $\mathrm{T} = \dfrac{\mathrm{X}}{a}$; as a result, the middle section disappears at this point, such that the gas is divided into two parts: one is in a state of rest with zero expansion; the other is travelling at a velocity of v_1 and has an expansion z_1.

130. The propagation of motion is more clearly shown by studying the velocity and expansion curves.

Let us, for instance, consider the velocity curve; let AB be its form at time t_1, OD the distance X.

If we extend the ordinate DC and a series of horizontal lines such as MN, limited by the curve AB and this ordinate, we know that the locus of the points which divide the straight lines such as MN, in any given ratio, is another expansion curve corresponding to one of the moments ranging between t_1 and T.

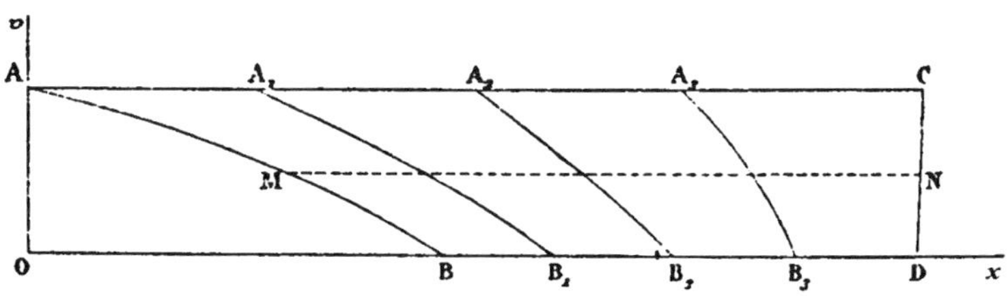

Fig. 10

Let us, for instance, divide all these horizontal lines into four equal parts; this gives us three intermediate curves A_1B_1, A_2B_2 and A_3B_3, which represent the distribution of velocities in the strata at times

$$t_1 + \frac{T - t_1}{4}, \qquad t_1 + \frac{T - t_1}{2} \qquad t_1 + \frac{3(T - t_1)}{4}.$$

We observe that the curvilinear part of the curve becomes increasingly straighter as the motion propagates, and suddenly transforms into a vertical straight line at time T.

If we give a continuous motion to the end stratum, we manage to create a discontinuity in the column which, as was shown in Chapter V, can not thereafter disappear. When we make the transition from the stratum X to the next infinitesimally close stratum, the velocity passes from v_1 to zero and the expansion from z_1 to zero.

The values of X and of v_1 can be taken at random; as for the expansion z_1, it is related to v_1 by the equation

$$v_I = \frac{2a}{m-1}\left(1 + z_1\right)^{-\frac{m-1}{2}} - \frac{2a}{m-1}.$$

The time t_1 needed to communicate the velocity to the end stratum v_1 can be deduced from equation (1) which gives

$$t_I = \frac{X}{a}\left[1 - \left(1 + \frac{m-1}{2a}v_1\right)^{-\frac{m+1}{m-1}}\right].$$

This time is proportional to the distance X.

The maximum for the acceleration of the motion of the piston takes place when $t = t_1$; it therefore has the value

$$\frac{2a^2}{(m+1)X}\left[1 + \frac{m-1}{2a}v_1\right]^{\frac{2m}{m-1}},$$

which, when the velocity v_1 is much smaller than a, is significantly reduced to

$$\frac{2a^2}{(m+1)X};$$

if we took $X = 100^m$, we would find, assuming that the gas in question is air at room temperature and pressure, that this acceleration is a little greater than 900 m per second.

VIII. - Motion of a mobile object subjected to the pressure of a column of gas

131. A gas is enclosed in a cylinder. One of the bases is immovable, the other is closed by a piston of mass M. The external pressure acting on the piston is disregarded. With the system in a state of rest and abandoned to its own devices, we must determine how the piston and the gas will move.

This problem has already been dealt with in papers by several geometrists, among whom we must in particular mention Lagrange, who dealt with this problem in a special paper, initially unpublished, and which Poisson later had published in the *Journal de l'Ecole Polytechnique.*

It is in attempting to use analytical means to determine the motion of projectiles in the barrels of guns that Lagrange was led to consider this problem. Let us indeed suppose that the powder has all burnt by the time the projectile has begun to move, the combustion products form a fluid which we can consider as being at rest at the outset, the pressure remaining constant and the velocity nil at all points.

Since the mass of the barrel is far larger than that of the shot, the gun can be considered, without making too serious an error, immovable. The shot is then set into motion by the pressure of the gaseous fluid alone.

Lagrange realised that the pressure of the fluid was, during the process of expansion, inversely proportional to the power m of the motion; under these conditions, the partial differential equation which governs it is in fact the equation corresponding to perfect gases. Lagrange attempted to determine the integral which matched the particularities of the problem. He seems, moreover, to have been fairly unsatisfied with the results obtained, because he decided that it was unwise to publish the results of his calculations.

More recently, Piobert turned his attention to the same problem in his papers. Like Lagrange, he accepted that the pressure of the powder's combustion products is inversely proportional to the power m of the volume they occupy. This hypothesis also seemed perfectly justified by Rumford's experiments. The partial differential equation governing the motion of the fluid was, again for Piobert, the equation which corresponded to the perfect gases.

Piobert complicated his problem somewhat by attempting to account for the motion of the barrel. Using very long and cumbersome calculations, he managed to arrive at formulae which are incapable of representing the real phenomenon, since they are not in agreement with the partial differential equation.

The failure of Lagrange's and Piobert's attempts seems to have discouraged the geometricians. The problem also lost a great deal of the interest it held for the gunners. The hypothesis of instantaneous combustion of the powder was far from the truth even at the time when Lagrange and Piobert were carrying out their research. However, at that time only old fashioned gunpowder, with small-sized particles, was used. Today, the gunners use large-particle powder which burns slowly and has not finished burning when the projectile leaves the barrel. The hypothesis of instantaneous combustion could therefore only lead to serious errors.

Furthermore, the relationship which exists during the expansion between the pressure and the volume of the powder's combustion products is far from being as simple as Lagrange and Piobert supposed. The partial differential equation governing the motion would therefore seem to differ from that for perfect gases, even if we ignored the time needed to ensure the complete combustion of the powder charge.

The question is still very interesting from the speculative and historical viewpoints. In the following, we shall provide part of the solution; however, it is worthwhile showing the reasons for Lagrange's and Piobert's failures before we go any further.

132. Let us assume that the end stratum of the gas in contact with the piston or projectile is located at point $x = 0$ on the x-axis; let us take λ to be the point on the x-axis representing the other end, and assume that it is in a permanent state of rest.

The partial differential equation governing the motion is

$$\frac{\partial^2 u}{\partial t^2} = a^2 \left(1 + \frac{\partial u}{\partial x}\right)^{-m-1} \frac{\partial^2 u}{\partial x^2}$$

where $a^2 = \dfrac{mp_0}{\rho_0}$, p_0 being the initial pressure and ρ_0 the initial density, which are the same for all points in the fluid.

The initial conditions are $u = 0$, $\dfrac{\partial u}{\partial t} = 0$, since the gas is assumed to be in a state of rest.

If M designates the mass of the piston in contact with the end $x = 0$, it can only be being moved by the pressure of the end stratum; the conditions applied to the ends are

$$M\frac{\partial^2 u}{\partial t^2} = -\omega p \text{ when } x = 0 \qquad \text{and} \qquad u = 0 \text{ when } x = \lambda,$$

irrespective of the value of t, with ω designating the cross-section of the cylinder.

Lagrange and Piobert set out to determine a single integral for the partial differential equation which would satisfy the two initial conditions and the end conditions. It was shown, in Chapter III, that the problem obtained in this manner usually cannot be solved; it is impossible to represent the entire phenomenon using a single analytical expression. We can only obtain an answer to the question by calculating an infinite number of integrals, each representing the motion at a certain time.

At the very beginning, an integral u_1 comes into existence at the end $x = 0$, compatible with the state of rest and satisfying the condition imposed at this end. This integral propagates at a velocity a and arrives at the end λ after a period of time $t_1 = \frac{\lambda}{a}$ At this moment, another integral u_2 comes into existence at the end λ; it must satisfy the condition imposed on this end and also be compatible with the integral u_1, which is sufficient to determine it.

The integral u_2 propagates into u_1 at a particular velocity which changes continuously and reaches the end $x = 0$ at a particular moment t_2, when a new integral u_3 comes into existence in the stratum $x = 0$ which is compatible with u_2 and satisfies the condition imposed on the end $x = 0$.

Successive integrals come into existence alternately at the two ends. If the tube which contains the gas is indefinite, in the direction of the negative x's, the piston will always be subjected to the pressure of the end stratum, irrespective of the distance it has travelled. Its motion will be described, between times zero and t_2 by the integral u_1; between times t_2 and t_4 by the integral u_3, and so on, but never by the integrals u_2, u_4 etc., which come into existence at the end $x = \lambda$. To determine the motion in full, we must know the entire series of integrals.

However, in the ballistics problem that Lagrange and Piobert intended to solve, the tube was of finite length, in fact that of the barrel of a gun. Let this length be L. When the projectile has travelled the distance $L - \lambda$, it is outside the tube; the gas enters the atmosphere and the pressure exerted at the rear of the mobile part becomes rapidly insignificant, such that when gravity and the resistance of air are disregarded, its motion would be to all extents and purposes uniform.

Let u_{2n+1} be the integral which represents the motion of the projectile at the moment when it has travelled a distance $L - \lambda$; it is totally pointless determining subsequent integrals, since the problem can be solved by determining a finite number of integrals.

When it is desired to account for the displacement of the gun, as Piobert attempted to do, the problem becomes even more complex. Let M_1 be the mass of the gun, the friction of the gas against its envelope is considered to be insignificant; the gun is moved by the pressure of the extreme layer which is located at point λ on the x-axis. The condition $u = 0$, applied at this end, must therefore be replaced by the following

$$M_1 \frac{\partial^2 u}{\partial t^2} = \omega p,$$

where $x = \lambda$, irrespective of the value of t.

At the beginning, an integral comes into existence at each end. Let these two integrals be u_1 and u'_1, both compatible with the state of rest, one of them satisfying the condition applied at the end $x = 0$, and the other satisfying the condition applied at the end λ.

The two integrals u_1 and u'_1 propagate into the column in opposite directions, at the same velocity a, and meet at the middle of the column at time $t_1 = \frac{\lambda}{2a}$. A new integral u_2 develops at this point, compatible with the previous integrals; the new integral develops at the expense of the old ones, such that it arrives at the end $x = 0$ after a period of time t_2 and at end λ at time t'_2.

At time t_2 a new integral comes into existence at the end $x = 0$; similarly, at time t'_2 an integral u'_2 comes into existence at the end λ. They propagate into u_2 and end up meeting at a point in the column where an integral u_4 comes into existence and so on.

The motion of the projectile M is successively represented by the integrals u_1, u_3 etc., the motion of the gun by the integrals u'_1, u'_3 etc. As in the previous case, we need just determine these integrals up to the moment when the projectile leaves the gun.

133. It is not our concern here to solve the problem in full; we shall only determine the motion of the mobile part M during the time it is represented by the first integral. We shall assume, to begin with, that the end λ is immovable.

Since the condition applied at the end $x = 0$ is

$$M \frac{\partial^2 u}{\partial t^2} = - \omega p,$$

bearing in mind that

$$\frac{\partial^2 u}{\partial t^2} = \frac{dv}{dt}, \qquad p = p_0 (1 + z)^{-m},$$

can be written as

$$M \frac{dv}{dt} = - \omega p_0 (1 + z)^{-m},$$

where v is the velocity of the mobile part and z is the expansion of the end stratum.

Since the integral which comes into existence on contact with the piston is compatible with the state of rest, it is represented in geometrical terms by a developable surface, and the following relationship exists between the velocity v and the expansion z of a stratum

$$1 + z = \left(1 + \frac{m - 1}{2a} v\right)^{-\frac{2}{m - 1}},$$

hence the previous equation becomes

$$M \frac{dv}{dt} = - \omega p_0 \left(1 + \frac{m - 1}{2a} v\right)^{-\frac{2m}{m - 1}}.$$

This is the differential equation for the motion of the piston. The force which drives it is thus a function of its velocity.

If we integrate this and take into account the fact that the velocity v is zero when $t = 0$, we obtain the following

$$\left(1 + \frac{m-1}{2a} v\right)^{-\frac{m+1}{m-1}} = \frac{m+1}{2a\,\mathrm{M}} \omega p_0 t + 1$$

or

$$v = -\frac{2a}{m-1}\left[1 - \left(1 + \frac{m+1}{2a\,\mathrm{M}} \omega p_0 t\right)^{-\frac{m-1}{m+1}}\right].$$

This formula determines the velocity of the piston as represented by the first integral. But the first integral represents the motion of the piston until the moment when the integral u_2, which comes into existence at the end λ through reflection of the first one, has reached the origin.

The first integral reaches the end λ after a period of time $\frac{\lambda}{a}$ since it propagates at a velocity a. The integral u_2, which comes into existence at the end λ, propagates in the opposite direction at a changing velocity $a\,(1 + z)^{-\frac{m+1}{2}}$: this velocity is less than a, because z is positive, as can easily be confirmed, for all points in the gas where the first integral applies.

Thus the previous formula represents the motion of the piston for a period of time greater than $\frac{2\lambda}{a}$ which we could calculate exactly simply by considering the first integral, since it would be sufficient to determine the horizontal projection of the characteristic of the second system which travels through the point with the co-ordinates $t = \frac{\lambda}{a}$, $x = \lambda$.

134. We could easily obtain the analytical expression for the integral, but it would be of no use in determining the velocities and expansions for the different strata of the gas at each moment. It is easier to construct the curves representing them point by point, relying on the fact that the velocities or expansions developed in the end stratum $x = 0$ propagate unchanged at a constant velocity which, for an expansion z, is $a\,(1 + z)^{-\frac{m+1}{2}}$.

For each moment t, ranging between zero and $\frac{\lambda}{a}$, the gas is divided in two by the stratum which lies at point $\xi = at$ on the x-axis; one of the parts is still at rest, while the motion of the other part is represented by the developable integral.

In the last part, the velocities are always negative while the expansions are positive. The ordinate for the velocity curve or the ordinate for the expansion curve decreases regularly, in terms of its absolute value, as values increase along the x-axis. Each of the curves changes shape with time such that, for a given point of the ordinate, the angular coefficient of the tangent decreases constantly in terms of its absolute value. As a result, even if the length λ were infinite, the edge of regression of the developable surface could not be affected by the motion. All of its points correspond to negative values of t.

For each moment between $\frac{\lambda}{a}$ and $\frac{2\lambda}{a}$, the column of gas is again divided into two parts by a point which moves continuously further from λ. In the part closest to the origin, the motion is represented by the developable integral; in the other part, it is given by the integral u_2 which, at time $\frac{\lambda}{a}$, has come into existence at the end λ.

135. Let us now return to the motion of the mobile part, we can see that its velocity, zero in the beginning, increases constantly in terms of its absolute value, as time goes on. If the length λ of the column is infinite and if the tube is indefinite on the side of the negative x's, such that the gas can never spread outside, then no matter what distance the piston has travelled, the velocity will always be represented by the formula in § 133, irrespective of the length of time t.

When this variable increases indefinitely, the velocity of the mobile part converges to the limit

$$-\frac{2a}{m-1};$$

which has already been encountered several times. This limit is, as we can see, independent of the weight of the mobile body. However, the smaller the mass M, the sooner the velocity reaches this limit.

Let M and M' be the masses of two different pistons. The formula shows that the velocities applied to these two mobile parts by the same gas would be the same at times t and t', such that

$$\frac{t}{t'}=\frac{M}{M'} \quad .$$

When the velocities are equal, the corresponding times are therefore proportional to the masses of the pistons.

We can easily obtain the distance travelled by the piston at a particular moment. Let u be this distance, then

$$u = \frac{2\,a^2\,\mathrm{M}}{(m-1)\,\omega p_0}\left[\left(1 + \frac{m+1}{2a\,\mathrm{M}}\,\omega p_0\,t\right)^{\frac{2}{m-1}} - 1\right] - \frac{2\,at}{m-1};$$

It is evident that the velocity applied to the piston by a column of gas with a length λ, whose pressure is p_0, and whose density is ρ_0, cannot decrease when λ is increasing. However, the limit to which this velocity converges when λ is equal to infinity is $\frac{2a}{m-1}$ Thereafter, this value represents the greatest possible velocity that the gas can apply to a mobile body.

The above results can be generalised and extended to cover any fluid with negligible conductivity; however, an examination of this more general case would take far too long. We shall therefore limit our discussion to the following theorem, which can be easily proven:

When a non-conducting fluid enclosed in a cylinder is limited by an immovable base, and the other end of the column is formed by a piston which can move without friction in the tube, which is assumed to be extended to infinity, the fluid applies a gradually increasing velocity to the piston, a velocity which may not exceed a certain limit, depending on the nature of the fluid and its initial state, but independent of the mass of the piston.

When, for instance, we use a certain charge of gunpowder to set a projectile into motion, it is evident that the velocity applied to the projectile at the moment it leaves the muzzle could only be increased if the initial volume occupied by the charge were reduced and if this charge were completely burnt before the shot began to move. It would increase further if the friction could be eliminated and if the barrel of the gun were extended indefinitely. Nonetheless, there is a top velocity which the projectile can never exceed.

If the gunpowder in the gun were replaced by another type of explosive, the limit would be different, but there would still be one.

136. Matters are not quite the same when the different strata of the fluid in motion can receive heat from an external source. If the fluid is a perfect gas and if the envelope acts as a source of heat, keeping the gas constantly at the same temperature, then the expansion would obey Mariotte's law, such that, in the case of the developable integral, the relationship between the expansion z and the velocity v for a stratum is

$$1 + z = e^{-\frac{v}{a}},$$

Since the equation for the motion of the piston is

$$M\frac{dv}{dt} = -\omega p,$$

we have

$$p = p_0 \ (1 + z)^{-1} = p_0 \, e^{-\frac{v}{a}},$$

such that this equation becomes

$$M \ \frac{dv}{dt} = -\ \omega p_0 \, e^{-\frac{v}{a}}.$$

If we integrate the above and take into account that the velocity cancels out when $t = 0$, we have

$$v = -\ a \ \log \left(1 + \frac{\omega p_0}{Ma} \ t\right).$$

If the initial length of the column, λ, is infinite, in which case the integral is applicable for all values of the time, we can see that the velocity v, zero when $t = 0$, is always negative and increases indefinitely with t. Thus, in this case, the velocity increases beyond any limit.

137. If, in addition to the pressure of the gas, the mobile part was subjected to the action of a known force F, parallel to the generatrices of the tube, the differential equation for its motion would be

$$M \ \frac{dv}{dt} = -\ \omega p_0 \left(1 + \frac{m-1}{2a} \ v\right)^{\frac{2m}{m-1}} + F;$$

and we would have to integrate this to obtain the law governing the motion.

If, for instance, the force F is constant, which is the case when friction is taken into consideration (it is assumed to be independent of the velocity), or with the cylinder inclined, there is reason for us to consider the effect of gravity; this gives us

$$t = \int_0^v \frac{Mdv}{F - \omega p_0\left(1 + \frac{m-1}{2a} v\right)^{\frac{2m}{m-1}}} .$$

138. We have assumed, until now, that an absolute vacuum exists at the rear of the piston. This condition is not possible in practice. In the case of a gun, for instance, the projectile comes up against the pressure of the air in the tube in front of it, and we can attempt to take this circumstance into account.

To do this, we must modify the conditions of the problem and assume that the piston is located between two columns of gas, one on the side of the positive x's, and the other on the opposite side.

Under these conditions, the equation for the motion will be

$$\frac{M}{\omega} \frac{dv}{dt} = p' - p,$$

where p and p' are the pressures corresponding to the end strata of the two gases.

The integrals which come into existence in the two columns are both developable; in the case of the first gas, as before we have

$$p = p_0 \left(1 + \frac{m - 1}{2a} v\right)^{\frac{2\,m}{m - 1}}.$$

In the case of the second gas, we could also find fairly easily that

$$p' = p'_0 \left(1 + \frac{m' - 1}{2a'} v\right)^{\frac{2\,m'}{m' - 1}}.$$

The equation for the motion of the piston is therefore

$$\frac{M}{\omega} \frac{dv}{dt} = p'_0 \left(1 - \frac{m' - 1}{2\,a'} v\right)^{\frac{2\,m'}{m' - 1}} - p_0 \left(1 + \frac{m - 1}{2a} v\right)^{\frac{2\,m}{m - 1}}.$$

The integration operation involves a simple quadrature.

Finally, we can also point out that if the end stratum λ of the cylinder is not immovable, but limited, like the other, by a piston with mass M_1, we may obtain the motion of this second piston, and also that for the piston with mass M, which is located at the end $x = 0$. Indeed, if v is taken to designate the velocity of the piston, and z the expansion of stratum λ,

$$M_l \frac{dv}{dt} = \omega p, \qquad p = p_0 (1 + z)^{-m}.$$

Furthermore, the integral which comes into existence at the end λ is developable, and we can see quite clearly that

$$l + z = \left(1 - \frac{m-1}{2a} v\right)^{-\frac{2}{m-1}}.$$

The equation for the motion therefore becomes

$$M_l \frac{dv}{dt} = \omega p_0 \left(1 - \frac{m-1}{2a} v\right)^{\frac{2m}{m-1}},$$

and we would integrate as before. We would then have the motions for the two pistons M and M_1, but the formulae would only be applicable here for a period a little greater than $\frac{\lambda}{a}$.

The above conditions are those adopted by Piobert. We can see that his problem is easy to solve with some thoroughness as regards the first moments of the motion. The following integrals are much more complex; the scope of this paper does not allow me to indicate how they can be determined.

Chapter V: Concerning the Discontinuities Apparent in the Propagation of Motion

I. - General considerations

139. The propagation of motion in a body gives rise to discontinuities when, in an infinitesimally small period of time, the velocity or the expansion of a stratum undergoes a finite variation or when, at a given moment, the velocity or the expansion of two infinitesimally close strata differ by a finite amount.

The discontinuities can be introduced in several different ways:

1) They may already exist in the initial state. This is a case which we shall never consider in what follows.

2) They can be introduced under the conditions imposed at the ends. If, for instance, an elastic prism-shaped solid is knocked at its end by a body

travelling at a finite velocity V, the end stratum, initially at rest, suddenly acquires this velocity V, whereas the next infinitesimally close stratum is still in a state of rest.

3) In the two cases we have just indicated, it is not surprising that we find a discontinuity in the motion, given that we introduced it in the limit conditions. However, we were able to see, in the previous section that, for certain bodies, discontinuities can come into existence in the motion, even when the initial conditions and those applied to the ends are perfectly continuous. These discontinuities, which are by far the most curious, result from the specific form of the partial differential equation which governs the motion.

We may ask ourselves whether the discontinuities we encounter in the theory of the propagation of motion are pure analytical fiction, or whether they do actually correspond to a physical reality. This is a question which it is difficult to answer given the current state of Science.

If we were only dealing with the discontinuities introduced by the conditions imposed at the ends, the answer would not be suspect. The discontinuous analytical form given to these conditions does not correspond to the actual situation. Thus, in the example quoted of one body colliding with the end of an elastic rod, the action of the body on this end begins even before the two come into contact, owing to the action exerted at a distance by the last particles of the material. This action, initially negligible, increases very rapidly when the distance between the body and the rod becomes very slight and the end of this rod only reaches the velocity V after first passing through all the intermediate velocities. For greater simplicity, this very rapid variation in velocity-the law of which is unknown-is replaced by the hypothesis of abrupt variation.

It is much more difficult to make a judgement when, with continuous conditions being imposed at the ends, and also at the initial state, we can see a discontinuity come suddenly into existence in the propagation of the motion; it even seems, at first sight, that this discontinuity must actually occur in reality. However, we must point out that the partial differential equations used to study the motion are only based on very approximate hypotheses, which would moreover lead to quite different equations. This is why, to establish the equation for the motion of the gases, it was assumed that the specific conductivity of the fluid was strictly nil. However, it is well known that this conductivity, while being very small, does all the same have a perfectly perceptible value; the partial differential equation used is therefore only an approximation.

Irrespective of the cause, discontinuities are bound to enter into the problems of pure mathematics. Whether we consider them as having a real existence, or if we just admit them as a simple approximation, we must examine the effect they have on the phenomena of propagation of motion.

140. To be more general, we shall consider two bodies A and A', which are being moved in strata perpendicular to a straight line Ox. The two bodies remain in contact along a plane B perpendicular to Ox. Each of the bodies is subject to a motion represented by an integral with its own second order partial differential equation. At the moment in question, the velocity and the pressure are not the same in the adjacent strata. The task facing us is to determine the phenomenon which will arise in the neighbourhood of the common section B.

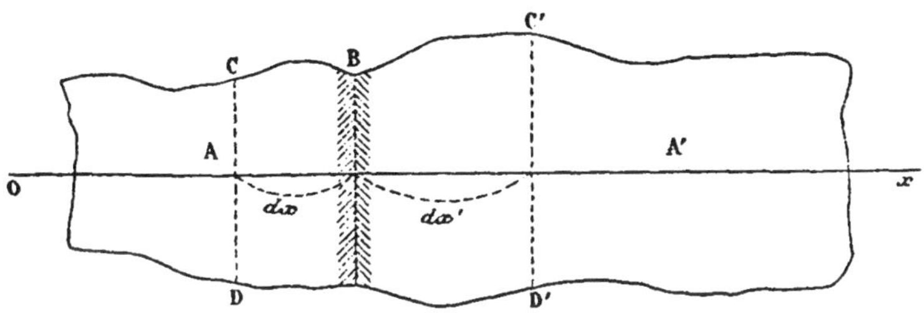

Fig. 11

Let v be the velocity, z the expansion and p the pressure of the stratum in contact with A' in the case of the body A, and v', z' and p' the corresponding values for the stratum from body A' which is in contact with A.

Owing to the effect that they exert on each other, the states of the touching strata will be modified, such that after a time dt, a certain portion of the body A, with length dx, will have reached a velocity of v_1, will have an expansion z_1 and a pressure p_1.

Similarly, at this moment dt, a certain portion of the body A' with a length of dx' will have attained a velocity v'_1, will have an expansion z'_1 and a pressure p'_1.

Our task is to determine the six values v_1, v'_1, z_1, z'_1, p_1, p'_1 and also the lengths dx and dx', or rather the ratios $\frac{dx}{dt}$, $\frac{dx'}{dt}$. This gives us eight unknowns and the problem facing us is to find the eight equations we need to determine the values of these variables.

141. Since the two bodies are forced to remain in contact, the velocities of the two touching strata must always be the same, such that if, at a given moment, there is a finite difference between them, this difference must disappear instantaneously. This provides us with a first equation

(1) $$v_1 = v'_1 .$$

However, for these two values to remain equal, the pressures of the touching strata must also be constantly equal, and hence

(2) $$p_1 = p'_1 .$$

Let us now apply the theorem for balance of momentum in stratum dx of body A; we can see that the increase in its momentum, during time dt, is $\rho_0 \omega \, dx$ $(v_1 - v)$, where ρ_0 is the initial density of the stratum, and ω is its section. The elementary impulse of the external forces, and the force corresponding to the friction and the viscosity, is infinitesimally small compared with the previous one: therefore

(3) $$\rho_0 (v_1 - v) \frac{dx}{dt} = p - p_1 .$$

Similarly, we find when we consider the stratum dx' of the body A' that

(4) $$\rho'_0 (v'_1 - v') \frac{dx'}{dt} = p'_1 - p' .$$

One of the bases of the stratum dx is travelling at a velocity of v for a period of time dt, the other by a velocity of v_1; it thus, for this period of time, undergoes an increase in length equal to $(v_1 - v)\, dt$. The initial length was $dx\,(1 + z)$; at the end of the period dt, it is represented by $dx\,(1 + z_1)$; this gives us

(5) $$(z_1 - z) \frac{dx}{dt} = v_1 - v .$$

Similarly, we would find, when we considered the stratum dx', that

(6) $$(z'_1 - z') \frac{dx'}{dt} = v' - v'_1 .$$

II - Case where the relationship between the pressure and the density is independent of the transformations undergone by the body.

141.[1] The six equations we have just obtained are totally independent of the nature of the bodies; however, the same does not apply for the last two, which

---

[1] Eds. Note. The original paper has two sections numbered 141.

must complete the determination of unknown values. We must make a distinction between two separate cases.

It may be possible that, in the case of the two bodies, the pressure is related to the expansion by means of an invariable equation, independent of the transformation undergone by the body and which is unaltered even when the velocities are subjected to sudden alterations at certain moments. The last two equations can then be obtained immediately. Let $\varphi(z)$ and $\phi(z)$ be the functions for the expansion which represent the pressure in each of the bodies: we then have

$$p_1 = \varphi(z_1), \tag{7}$$

$$p'_1 = \phi(z'_1), \tag{8}$$

and the problem has been solved in full.

Among the bodies whose motion was studied in Chapter III, there are a certain number of them to which the above relationships apply. Firstly the elastic solids, for which the pressure can always be obtained by multiplying the expansion z, taken with the opposite sign, by Young's modulus of elasticity E, such that

$$p = -\mathrm{E}z,$$

a relation which is independent of the transformations undergone by the body earlier.

The same applies for a gas which obeys Mariotte's law, whose temperature therefore remains constant throughout the duration of the motion. In this case,

$$p = p_0 (1 + z)^{-1}.$$

The envelope of the gas is then assumed to act as a perfect source of heat, maintaining the temperature of each stratum unchanged.

However, as we shall see soon, this is not what happens with a gas whose strata cannot receive heat, either from the envelope or from the neighbouring strata.

142. Still keeping to the case where the relationship between the pressure and the expansion is independent of earlier transformations, we shall assume that the two juxtaposed bodies are identical, which does not prevent the above theory from being applicable. The two functions φ and ϕ are then identical; and, since $p_1 = p'_1$, equations (7) and (8) give

$$z'_1 = z_1 .$$

Equations (3) and (4) then become, bearing in mind that $\rho_0 = \rho'_0$,

$$\rho_0 (v_1 - v) \frac{dx}{dt} = \varphi (z) - \varphi (z_1) ,$$

$$\rho_0 (v_1 - v') \frac{dx'}{dt} = \varphi (z_1) - \varphi (z') .$$

Equations (5) and (6) in turn give

$$(z_1 - z) \frac{dx}{dt} = v_1 - v ,$$

$$(z_1 - z') \frac{dx'}{dt} = v' - v_1 .$$

These equations contain only four unknown variables, v_1, z_1, $\frac{dx}{dt}$ and $\frac{dx'}{dt}$. We can attempt to find the condition for which one of the velocities of propagation, $\frac{dx'}{dt}$ for instance, cancels out. We must first have

(9) $$v_1 = v',$$

(10) $$\varphi (z_1) = \varphi (z') ;$$

and hence

$$z_1 = z'.$$

Let us substitute these values of v_1 and z_1 into the two other equations; dividing term by term will provide us with the equation we require

(11) $$\rho_0 (v' - v)^2 = [\varphi (z) - \varphi (z')] (z' - z) .$$

When this equation is satisfied, the phenomenon is much simpler than in the normal case.

143. Let us come back to general case: the two strata of lengths dx and dx' which have attained equal velocities, and thereafter equal pressures, are subjected to the same motion. The body is therefore, at time dt, divided into three parts: one, to the left of CD, is subjected to a motion represented by a particular integral of the partial differential equation which, furthermore, is already known; similarly, the part to the right of C'D' is represented by another integral which too is known. Between these two parts there is another part which has an infinitesimally small length $dx + dx'$, in which the velocity is v_1 and the expansion z_1. The velocity and the expansion therefore contain sudden discontinuities in the sections CD and C'D'; the initial discontinuity, which came into existence in section B, has given rise to two further discontinuities.

However, when one of these values $\frac{dx}{dt}$ or $\frac{dx'}{dt}$ is nil, the middle part CDC'D' disappears. The initial discontinuity has only changed position and the entire motion of the body is again represented by two unique integrals, as at time zero, and these two integrals support the continued existence at their meeting point of a discontinuity in the velocities and the expansions.

Thus, when condition (11) is satisfied, the two integrals which alone represented the motion of the body at the initial moment still represent it at time dt. Only the point common to the two integrals has moved a distance of dx, in other words, one of the integrals has propagated into the other at a velocity $\frac{dx}{dt}$, without giving rise to any new phenomenon. When this is the case, we shall take the two integrals to be mutually *compatible*.

The notion of compatibility is therefore extended to the case where, at the point common to the two integrals, the velocity and the expansion exhibit a discontinuity.

For the two integrals u and u' in the partial differential equation governing the motion of the body to be mutually compatible, the condition in (11) must always be satisfied for the values v, v', z and z' which correspond to the two adjacent strata. The area of contact is difficult to determine since the distance travelled has to be the same for both strata, giving us the following for the adjacent strata

$$u = u',$$

an equation which, for all values of t, determines the points along the x-axis marking the section common to the two integrals.

144. The compatibility condition can be arranged into yet another form. If we multiply the equations containing $\frac{dx}{dt}$ term by term, after first replacing z_1 with z' and v_1 with v', we obtain

$$\rho_0 \left(\frac{dx}{dt}\right)^2 = \frac{\varphi(z) - \varphi(z')}{z' - z}. \tag{12}$$

For the integrals to be compatible, the value of $\frac{dx}{dt}$ obtained by differentiating the equation $u - u' = 0$, where x is considered to be a function of t, must satisfy equation (12) for all values of t.

This equation (12) can also be written as follows

$$\rho_0 \left(\frac{dx}{dt}\right)^2 = \frac{p - p'}{z' - z};$$

in these two forms, it provides the value of the velocity at which one integral propagates into another.

145. When the compatible integrals are such that, at the common point, the velocity and expansion are constantly equal, in other words when the two corresponding integral surfaces connect along the line of intersection, this line forms a characteristic common to the two surfaces, as shown in Chapter III, such that its horizontal projection is known in advance if we know only one of the two surfaces. In other words, we need only know one of the integrals to ascertain the velocity of propagation.

This does not, however, apply when a discontinuity exists in the velocities and expansions for the point common to the two integrals. Equation (12) shows that the velocity of propagation $\frac{dx}{dt}$ then depends on both integrals at once, since the corresponding values of the expansions z and z' are also involved. The representative surfaces then intersect along a curve which depends on both surfaces at once and which cannot be ascertained when only one of these surfaces is given. These curves, which are beyond the scope of this study, must be considered as the generalisation of the characteristics.

We can easily verify that, if the discontinuity present in the expansions for the point common to the two integrals converges towards zero, the curve of intersection moves ever closer to the characteristic. If we replace z' by $z + \delta z$ in equation (12), it becomes

$$\rho_0 \left(\frac{dx}{dt}\right)^2 = \frac{\varphi(z) - \varphi(z + \delta z)}{\delta z} = - \varphi'(z).$$

Since the pressure is represented by $\varphi(z)$, the terms in the partial differential equation which are given to the second order derivatives are

$$\rho_0 \frac{\partial^2 u}{\partial t^2} + \varphi'(z) \frac{\partial^2 u}{\partial x^2}.$$

The value of $\frac{dx}{dt}$ is therefore actually the angular coefficient of the tangent to the horizontal projection of the characteristic.

146. There is, however, a case where the velocity of propagation is not altered by the presence of discontinuities: this is the case when the coefficients of the second order derivatives in the partial differential equation governing the motion are independent of the first derivatives $\frac{\partial u}{\partial x} = z$ and $\frac{\partial u}{\partial t} = v$.

In this case, the terms assigned to the second order derivatives in the partial differential equation are

$$\rho_0 + A\frac{\partial^2 u}{\partial x^2},$$

and, in the case where there is no discontinuity, the velocity of propagation is given by

$$\rho_0 \left(\frac{dx}{dt}\right)^2 = - A.$$

In the opposite case, we must use equation (12); but that then becomes

$$\rho_0 \left(\frac{dx}{dt}\right)^2 = \frac{Az - Az'}{z' - z} = - A$$

since the values of A are the same for both integrals.

The expression for the velocity of propagation is therefore the same, whether there are discontinuities or not, for the point common to the two integrals.

We can therefore ascertain the velocity of propagation in a given integral for all possible cases of compatibility, without having to find the integral which is being propagated.

This is what happens in particular when the second order derivatives in the partial differential equation have coefficients which only depend on x; the velocity of propagation no longer even depends on the integral being considered; it is only a function of the value of x along the x-axis for the stratum.

Let us assume that the partial differential equation is perfectly linear, like the equation governing the motion of an elastic solid by strata; from the above we can see that two integral surfaces which intersect along a characteristic represent two compatible motions. This result agrees with those obtained in Chapter I, where it was shown that two integral surfaces for a linear equation which intersect along a characteristic can, from certain viewpoints, be considered as forming a single integral surface.

147. We still need to examine what happens when the two integrals, which represent the motion of the body at a given moment and which exhibit a discontinuity at the common point, are mutually incompatible. We saw that at time dt, a stratum expanding on both sides of the common section, to distances of dx and dx', was travelling at a velocity v_1 and with an expansion z_1 such that, instead of a single section of discontinuity, there were now two, i.e. those limiting the stratum with length $dx + dx'$.

The motion of this stratum with length $dx + dx'$ is obviously represented by a new integral of the equation governing the motion since there are no discontinuities in the velocities and expansions for the stratum; however, this integral is only extending at the moment in question over an infinitesimally small distance.

It is easy to see that this new integral is compatible with the two initial integrals, because the values of $\frac{dx}{dt}$ and $\frac{dx'}{dt}$ satisfy the following equations, which can be deduced quite easily from the equations in § 142:

$$\rho_0 \left(\frac{dx}{dt}\right)^2 = \frac{\varphi(z) - \varphi(z_1)}{z_1 - z},$$

$$\rho_0 \left(\frac{dx'}{dt}\right)^2 = \frac{\varphi(z_1) - \varphi(z')}{z' - z_1};$$

these are exactly the same as the compatibility conditions.

The new integral will therefore propagate purely and simply into the first two without giving rise to any new phenomena.

148. The above analysis shows that if, in a body, the relationship between the pressure and the expansion of a stratum is independent of the transformations undergone by this stratum, the mechanism of the propagation of the motion is the same, whether there are discontinuities or not.

When two integrals have a common point, they may be compatible; one of them propagates and expands at the expense of the other without giving rise to any new phenomena.

When these two integrals are incompatible, a new integral comes into existence at the common point; it is defined by the condition of being compatible with the first two, into which it propagates, expanding at their expense.

Since the body is subjected to a motion, represented by a given integral, if we apply a certain condition to its end, a new integral comes into existence at this end which is compatible with the first and which satisfies the condition, whether it introduces discontinuities or not.

However, when there are discontinuities at the point common to the two compatible integrals, the velocity of propagation is usually changed; furthermore, there is an exception to the rule when the coefficients of the second order derivatives are independent of u and its first order derivatives; the definition of compatible integrals, given for the case where there are no discontinuities, is still applicable even when these discontinuities arise.

This is the case with the equation

$$\frac{\partial^2 u}{\partial t^2} = a^2 \frac{\partial^2 u}{\partial x^2},$$

which governs the motion of a solid elastic cylinder, the necessary and sufficient condition for the two integrals to be compatible is that they have one of the arbitrary functions in common, whether there are discontinuities or not.

III - Case where the relationship between the pressure and the density depends on the transformations undergone by the body

149. In the majority of bodies considered in the earlier chapters, the relationship between the pressure and the expansion of a stratum cannot be considered to be independent of the transformations it has undergone.

In order to simplify our discussion, in what follows we shall consider only perfect gases which are assumed to be enclosed in a cylindrical tube; however, it is easy to apply the theory to the general case and to all non-conductive fluids, provided the envelope is considered to be impermeable to heat.

We know that in any transformation undergone by a mass particle of a gas not involving the gain or loss of heat, the work of external forces is equal to the increase in total energy, in other words the sum of the internal energy and the active force.

Let us consider two columns of gas A and A', juxtaposed at time t in a cylindrical tube so that they come into contact at section BB', where the velocities v and v', the expansions z and z' and the pressures p and p' are different.

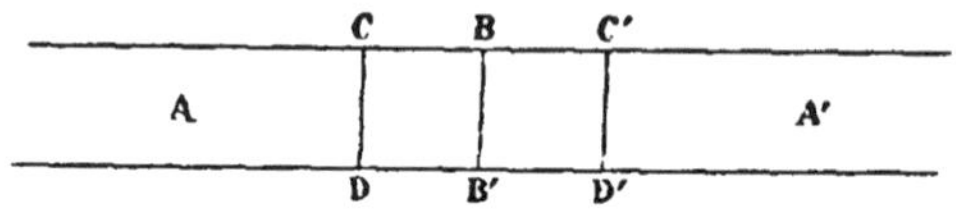

Fig. 12

At time $t + dt$, the two touching strata, whose lengths are dx and dx', will have acquired velocities of v_1 and v'_1, expansions of z_1 and z'_1, and pressures of p_1 and p'_1. We are faced with the task of determining the eight unknowns v_1, v'_1, z_1, $z'_1, p_1, p'_1, \dfrac{dx}{dt}, \dfrac{dx'}{dt}$.

The six equations from §140 are applicable, but we still need to find the last two.

To this end, we shall assume that the variation in total energy for each of the strata is equal to the work of the external forces.

Let us in particular consider the stratum CDB'B which has a length of dx and a mass of $\rho_0\omega\, dx$, and with a total energy at time t of

$$\frac{\rho_0\omega dx}{2}v^2 + \frac{p\omega dx(1+z)}{m-1} \quad ;$$

at time $t + dt$ this has become

$$\frac{\rho_0\omega dx}{2}v_1^2 + \frac{p_1\omega dx(1+z)}{m-1} \quad .$$

The work of the forces is obviously

$$\omega\, pv\, dt - \omega\, p_1 v_1\, dt,$$

because the stratum is subjected at its bases to pressures of p and p_1. This gives us the following equation

$$(13)\qquad \left[\frac{\rho_0}{2}\left(v_1^2 - v^2\right) + \frac{p_1\left(1 + z_1\right) - p\left(1 + z\right)}{m - 1}\right]\frac{dx}{dt} = pv - p_1 v_1 .$$

Similarly, we would find, when we considered the stratum BB'C'D' with a length of dx', that

$$(14)\qquad \left[\frac{\rho'_0}{2}\left(v_1'^2 - v'^2\right) + \frac{p'_1\left(1 + z'_1\right) - p'\left(1 + z'\right)}{m' - 1}\right]\frac{dx'}{dt} = p'_1 v'_1 - p' v' .$$

These two equations, together with the six already obtained, will allow us to ascertain all of the unknowns.

150. For each of the strata dx and dx', we can use these equations to deduce the relationship between the pressure, the initial expansion of the stratum and its final expansion.

Let us, for instance, consider the stratum dx; from equations (3), (5) and (13) we know that

$$pv - p_1 v_1 = \frac{\rho_0}{2}\left(v_1 - v\right)\left(v_1 + v\right)\frac{dx}{dt} + \frac{p_1 - p}{m - 1}\frac{dx}{dt} + \frac{p_1 z_1 - pz}{m - 1}\frac{dx}{dt}$$

$$pv - p_1 v_1 = \frac{v_1 + v}{2}\left(p - p_1\right) + \frac{p_1 - p}{m - 1}\frac{v_1 - v}{z_1 - z} + \frac{p_1 z_1 - pz}{m - 1}\frac{v_1 - v}{z_1 - z};$$

hence, if we bear in mind that

$$pv - p_1 v_1 - \frac{p - p_1}{2}\left(v_1 + v\right) = \frac{p + p_1}{2}\left(v - v_1\right)$$

and eliminate the common factor $v - v_1$,

$$\frac{p + p_1}{2} + \frac{p_1 - p}{m - 1}\frac{1}{z_1 - z} + \frac{p_1 z_1 - pz}{m - 1}\frac{1}{z_1 - z} .$$

If we solve this with respect to p_1, we have

$$(15)\qquad p_1 = p\,\frac{2\,(1 + z) - (m - 1)\,(z_1 - z)}{2\,(1 + z) + (m + 1)\,(z_1 - z)}$$

Similarly, we would find that

$$(16) \qquad p'_1 = p' \frac{2(1+z') - (m'-1)(z'_1 - z')}{2(1+z') + (m'+1)(z'_1 - z')}$$

Since the pressures p_1 and p'_1 must be equal, we can also deduce from this that

$$(17) \qquad p \frac{2(1+z) - (m-1)(z_1 - z)}{2(1+z) + (m+1)(z_1 - z)} = p' \frac{2(1+z') - (m'-1)(z'_1 - z')}{2(1+z') + (m'+1)(z'_1 - z')}$$

151. The above results are still the same when the two gases juxtaposed along the stratum BB' are identical. In this case $m' = m$, and we can also assume that $\rho'_0 = \rho_0$, having first given z' a suitable value.

The gas which, at time t, was divided into two by the stratum BB', is, at time $t + dt$, divided in four by the strata CD, BB' and C'D'. The two middle sections have infinitesimally small lengths; there are discontinuities in CD and C'D' as regards the velocities, expansions and pressures; in BB', the pressures and velocities are equal, but the same does not apply for the expansions.

We may attempt to find the condition for which one of the strata CDBB' or BB'C'D' disappears, for instance the second. For this, $\frac{dx'}{dt}$ has to cancel out. Equation (4) then becomes

$$v'_1 = v'$$

and hence, by virtue of equation (1),

$$v_1 = v'.$$

Equation (6) then gives

$$\rho'_1 = \rho'$$

and hence, by virtue of equation (2),

$$\rho_1 = \rho',$$

leaving us with the equations

$$\rho_0 (v' - v) \frac{dx}{dt} = p - p',$$

$$(z_1 - z)\frac{dx}{dt} = v' - v,$$

$$\left[\frac{\rho_0}{2}\left(v'^2 - v^2\right) + \frac{p'\left(1 + z_1\right) - p\left(1 + z\right)}{m - 1}\right]\frac{dx}{dt} = pv - p'v',$$

from which we need only eliminate $\frac{dx}{dt}$ and z_1 to obtain the required condition.

When this condition is satisfied, the gas is divided at time $t + dt$ into three by the strata CD and BB'. In CD there is a discontinuity for the velocities, pressures and expansions; in BB', however, the discontinuity affects only the expansions.

In that part of the gas which is to the left of CD, there are no discontinuities. Let us assume that the friction and the viscosity are negligible, the corresponding motion is therefore represented by an integral of the equation

$$\rho_0 \frac{\partial^2 u}{\partial t^2} = m p_0 \left(1 + \frac{\partial u}{\partial x}\right)^{-m-1} \frac{\partial^2 u}{\partial x^2},$$

or, more generally, by an integral from the equation

$$\rho_0 \frac{\partial^2 u}{\partial t^2} = m f(x) \left(1 + \frac{\partial u}{\partial x}\right)^{-m-1} \frac{\partial^2 u}{\partial x^2} + f'(x)\left(1 + \frac{\partial u}{\partial x}\right)^{-m},$$

obtained in Chapter II and which we must return to, if, at the time taken as the initial moment, the pressure is not uniform in the column.

Similarly, the motion in the column of gas to the right of BB' is represented by another integral from the same equation.

As for the middle part, its motion is represented by neither of the integrals. However, if the value of z_1 was equal to that for z', the motion of the middle part would be represented by the same integral as the part to the right of BB', provided there are no further discontinuities after BB'.

When this is the case, at time $t + dt$, the gas is divided, as in the moment before t, in two only, and each part is subjected to a motion represented by an integral of the above equation; only the section of separation where discontinuities arise has moved an infinitesimally small distance dx during time dt.

We now have three equations

$$\rho_0\,(v' - v)\,\frac{dx}{dt} = p - p',$$

$$(z' - z)\,\frac{dx}{dt} = v' - v,$$

$$\left[\frac{\rho_0}{2}\left(v'^2 - v^2\right) + \frac{p'\,(1 + z') - p\,(1 + z)}{m - 1}\right]\frac{dx}{dt} = pv - p'v',$$

which only contain the unknown $\frac{dx}{dt}$, apart from the known data. Eliminating this would provide two conditional equations, i.e.

$$\rho_0\,(v' - v)^2 = (p - p')\,(z' - z),$$

$$p' = p\,\frac{2\,(1 + z) - (m - 1)\,(z' - z')}{2\,(1 + z) + (m + 1)\,(z' - z')}\,.$$

Assuming these conditions are satisfied, we now have the following equation to solve $\frac{dx}{dt}$,

$$\rho_0\left(\frac{dx}{dt}\right)^2 = \frac{p - p'}{z' - z},$$

which is analogous to the equation arrived at above (§ 144).

152. In the particular case we have just examined, since the body was divided in two at time t, each subject to its own motion, the same applies at time $t + dt$; one of the motions propagates at the expense of the other, without giving rise to any new phenomenon. We shall also say that the two motions are mutually *compatible*.

Let us now return to the general case where the above conditions are not satisfied, we saw that the two strata dx and dx' in contact with BB' for a period of time dt, were subjected to specific motions for which the velocities, pressures and expansions were calculated. It is now easy to see that the motion of the stratum CDBB' is compatible with the motion which exists in part A of the body, because the conditions are obviously satisfied. Similarly, the motion which came into existence in stratum BB'C'D' is compatible with the motion already existing in part A'. This means that at the following moment the state of the gas will still be exactly alike; only the planes CD and C'D' will have moved, one to the left, the other to the right.

The motions which came into existence in the two strata touching the initial section of discontinuity thus propagate in opposite directions at the expense of the initial motions.

Moreover, no further new motion comes into existence in the strata neighbouring BB', because the values of the velocities and pressures are the same in this section for the two new motions. It is evident that, in spite of the inequality of the expansions, the eight equations arrived at above are then satisfied when $\frac{dx}{dt} = 0$, and when $\frac{dx'}{dt} = 0$.

We can therefore sum up the phenomenon arising when two columns of gas are in contact at a section where the velocities, expansions and pressures are different in the following manner.

In each of the adjacent strata, a new motion develops which is compatible with the motion in the corresponding column of gas. Each of these motions propagates at the expense of the initial motion, at a certain velocity which varies continuously, and supports discontinuities at the common point in the velocities, expansions and pressures. In the stratum where the two gases come into contact, the velocity and the pressure are always equal, but the expansions are normally different.

The velocity of propagation for a motion compatible with another motion can be written as follows

$$\frac{dx}{dt} = \frac{v' - v}{z' - z} ;$$

it is equal to the quotient of the difference in the velocities at the common point over the difference in the expansions, or rather to this quotient with the sign changed, because the value of dx is negative in the figure as it has been constructed.

153. The phenomenon is again the same when the two adjacent columns are of the same gas. We then see that any discontinuity which forms at a given moment in a gas will usually give rise to three others; two of them propagate either side of the stratum in which the initial discontinuity was located; the other remains in this same stratum, but only affects the expansion.

One of the end discontinuities disappears when a certain relationship exists between the velocities and the pressures of the two touching columns; the end discontinuities would both disappear if the initial discontinuity only affected the expansions.

Finally, when the motions affecting the two columns are compatible, in other words when the two conditions are satisfied, the initial discontinuity moves purely and simply in the body without giving rise to other discontinuities.

But, in no case does the initial discontinuity disappear.

The discontinuities can come into existence in the column, even when the ends are only subjected to perfectly continuous motions: this was proved in the previous chapter; but they can also come into existence at the ends, through the very nature of the conditions applied to them. If, for instance, the column was closed by a piston, and if the piston was driven at a certain finite velocity, this would cause a discontinuity. The same would occur if a sudden change was brought to bear on the pressure applied to this end.

The phenomenon is also the same in the case where the discontinuity comes into existence in the middle of the body; but we need only concern ourselves with one of the motions which comes into existence, i.e. the one which propagates into the initial motion. This motion must satisfy the condition applied at the end; at the same time, it must be compatible with the initial motion. These are the only conditions which may be imposed on it, and they must suffice for determining it, seeing that only one motion is possible.

154. It was shown in Chapter II that, as long as a stratum of the gas did not undergo a sudden variation in velocity or expansion, its pressure was related to its expansion by the adiabatic relationship

$$p\,(1+z)^m = const.$$

This is not, however, the case when it undergoes a sudden variation in its velocity or expansion.

Let us consider a discontinuity which propagates at a velocity of $\dfrac{dx}{dt}$, regarded as either positive or negative, depending on whether it is driven towards the positive or negative x's; if we designate the velocity, expansion and the pressure of the stratum before it was affected by the discontinuity by v, z and p, and term the same variables after the discontinuity has travelled through the stratum v_1, z_1, and p_1, we obtain the three relationships

$$\rho_0\,(v_1 - v)\frac{dx}{dt} = p_1 - p,$$

$$(z_1 - z)\frac{dx}{dt} = v - v_1,$$

$$\left[\frac{\rho_0}{2}\left(v_1^2 - v^2\right) + \frac{p_1\,(1+z_1) - p\,(1+z)}{m-1}\right]\frac{dx}{dt} = p_1 v_1 - p v.$$

which are in fact the relationships from the preceding paragraphs, with the sign in front of $\frac{dx}{dt}$ changed.

The second of these relationships shows that $v - v_1$ and $z_1 - z$ are finite at the same time. Therefore, any sudden variation in the velocity would bring about a sudden change in expansion for the stratum.

From these relationships we may also deduce that

$$p_1 = p\,\frac{2\,(1 + z) - (m - 1)\,(z_1 - z)}{2\,(1 + z) + (m - 1)\,(z_1 - z)}\,,$$

a formula which is of prime importance since it can be used to calculate the final pressure using the initial data and the variation which occurred in the expansion.

The formula can be written in yet another form. Let ρ and ρ_1 be the densities corresponding to the expansions z and z_1; it is evident that

$$p_1 = p\,\frac{(m + 1)\,\rho_1 - (m - 1)\,\rho}{(m + 1)\,\rho - (m - 1)\,\rho_1}$$

or

$$p_1 = p\,\frac{(m + 1)\,\dfrac{\rho_1}{\rho} - (m - 1)}{(m + 1) - (m - 1)\,\dfrac{\rho_1}{\rho}}\,.$$

The ratio of the pressures is thus a rational function of the ratio of the densities.

The pressure p_1 is mostly positive, which sets the limits for the values which the ratio $\frac{\rho_1}{\rho}$ can adopt.

Let us first assume that this ratio is greater than one, which happens when a sudden variation compresses the stratum; the numerator is always positive; the same must apply for the denominator.

Therefore

$$\frac{\rho_1}{\rho} < \frac{m + 1}{m - 1}\,.$$

If, conversely, the sudden variation expands the stratum, $\frac{\rho_1}{\rho}$ is less than one; the denominator is therefore positive; the numerator must also be positive, which means that the following must be true

$$\frac{\rho_1}{\rho} > \frac{m-1}{m+1} .$$

Thus, when a stratum of gas undergoes a sudden variation in its velocity, expansion and pressure, the ratio of the final density to the initial density must range between $\frac{m-1}{m+1}$ and $\frac{m+1}{m-1}$.

The above limits would then be zero and infinity, if the coefficient m was equal to one, in other words if the gas obeyed Mariotte's law.

155. The above result appears paradoxical at first sight. Yet it is a direct consequence of the principle of the equivalence of heat and work.

When a stratum of gas undergoes a transformation in which its velocity varies continuously, the variation in its active force, as shown in Chapter II, is the same as if the stratum was like an invariable solid. But, if the velocity undergoes a sudden variation, the variation in the active force would bring about a corresponding variation in the internal energy. In other words, if there is an increase in the active force, for instance, some of this increase will be at the expense of the heat contained in the stratum; when, conversely, there is a reduction in the active force, some of the active force lost is used to increase the internal energy or the temperature of the stratum.

For the rest, we may easily calculate the temperature at which the stratum undergoes a sudden variation. Let T_1 be this temperature, and T the initial temperature: the characteristic equation for the gas can then be written as follows

$$\frac{p}{\rho} = RT,$$

this gives us

$$T_l = \frac{1}{R}\frac{p_1}{\rho_l} = \frac{p}{R}\,\frac{(m+1)\frac{\rho_1}{\rho} - (m-1)}{\rho_l\left[(m+1) - (m-1)\frac{\rho_1}{\rho}\right]},$$

or, if we bear in mind that $\frac{p}{R} = \rho T$,

$$T_1 = T\,\frac{m+1-(m-1)\dfrac{\rho}{\rho_1}}{m+1-(m-1)\dfrac{\rho_1}{\rho}}$$

The ratio of absolute, initial and final temperatures for the stratum can be written in terms of a rational function of the ratio of the densities.

The temperature T cancels out when $\dfrac{\rho_1}{\rho} = \dfrac{m-1}{m+1}$ and becomes infinite when $\dfrac{\rho_1}{\rho} = \dfrac{m+1}{m-1}$. This also explains why the pressure p_1 becomes nil or infinity depending on whether or not the ratio $\dfrac{\rho_1}{\rho}$ reaches one of these two limits.

156. If the transformation which changes the density of the stratum from ρ to ρ_1 had been continuous, the value of $\dfrac{p}{\rho^m}$ would have proven constant, such that, if the new pressure corresponding to the density ρ_1 was written p'_1, we would have

$$\frac{p'_1}{\rho_1^m} = \frac{p}{\rho^m}, \qquad p'_1 = p\left(\frac{\rho_1}{\rho}\right)^m.$$

It is natural to compare the pressures p_1 and p'_1. For this reason, we must determine the difference

$$p_1 - p'_1 = p\left[\frac{(m+1)\dfrac{\rho_1}{\rho} - (m-1)}{m+1-(m-1)\dfrac{\rho_1}{\rho}} - \left(\frac{\rho_1}{\rho}\right)^m\right],$$

for the values of $\dfrac{\rho_1}{\rho}$ ranging between $\dfrac{m-1}{m+1}$ and $\dfrac{m+1}{m-1}$.

It is easy to prove that this difference is positive or negative according to whether or not $\dfrac{\rho_1}{\rho}$ is less than or greater than one. So, in the case of a given variation in the density, the variation in pressure will be larger when it takes place suddenly than when it takes place very gradually.

157. The results obtained in the previous numbers lead us to draw some important conclusions.

We showed in Chapter II how, if the external forces, friction and viscosity were disregarded, the partial differential equation governing the motion of a perfect gas would be

$$\rho_0 \frac{\partial_2 u}{\partial t^2} = m f(x)\left(1 + \frac{\partial u}{\partial x}\right)^{-m-1} \frac{\partial^2 u}{\partial x^2} + f'(x)\left(1 + \frac{\partial u}{\partial x}\right)^{-m},$$

or, if a suitable initial state were chosen,

$$\rho_0 \frac{\partial^2 u}{\partial t^2} = m p_0 \left(1 + \frac{\partial u}{\partial x}\right)^{-m-1} \frac{\partial^2 u}{\partial x^2}.$$

The pressure p_0 is constant, but the density ρ_0 is variable in the gas, such that the temperature at the initial moment varies from one stratum to the next; the gas can still be in equilibrium in this state, since the conductivity is regarded as being strictly nil.

This equation is suitable as long as there are no discontinuities in the motion; the function of x designated by ρ_0 remains the same. We shall see that this is no longer the case when discontinuities are involved.

Indeed, the above equation supposes that the adiabatic relationship between the density ρ of a stratum and its pressure p is always

$$\frac{p}{\rho^m} = \text{const.} = \frac{p_0}{\rho_0^m}.$$

This is what actually happens when this stratum is transformed continuously.

However, when the density ρ undergoes a sudden variation, such that it changes from ρ to ρ_1, the equation $\frac{p}{\rho^m} = \frac{p_1}{\rho_1^m}$ is no longer satisfactory; it is replaced by the following

$$p_1 = p\,\frac{(m+1)\dfrac{\rho_1}{\rho} - (m-1)}{m+1-(m-1)\dfrac{\rho_1}{\rho}},$$

which cannot be exactly the same as the previous one, unless $m = 1$.

When m is not equal to one, the value of $\frac{p_1}{\rho_1^m}$ is different from $\frac{p}{\rho^m}$ and thus also from $\frac{p_0}{\rho_0^m}$. As long as the stratum does not undergo any sudden variations, the value of $\frac{p}{\rho^m}$ will again become a constant, but the value of this constant will have changed and become different from the first.

We can determine a value ρ'_0 such that

$$\frac{p_0}{\left(\rho'_0\right)^m} = \frac{p_1}{\rho_1^m}.$$

In all the states that the stratum adopts after the sudden variation, the relationship between its pressure and density will obviously be the same as if, at the initial moment, the pressure were equal to p_0 as before, and if its density were ρ'_0.

The discontinuity whose effect on a particular stratum has just been determined, will propagate, as shown, into the column of gas, such that it takes place gradually in an infinite number of strata. Let us assume that it came into existence at a point x_1 and that at a certain moment it had reached another point x_2, all the strata ranging between x_1 and x_2 will have undergone similar variations.

In their later transformations, supposed to be continuous, the relationship between the pressure and the density will be the same as if the density of each unit, with a pressure of p_0 in the initial state, was no longer represented by ρ_0, but by another function of x, ρ'_0.

Thus, the part of the column traversed by a discontinuity no longer satisfies the initial second order equation, but does satisfy another equation

$$\rho'_0 \frac{\partial^2 u}{\partial t^2} = m\, p_0 \left(1 + \frac{\partial u}{\partial x}\right)^{-m-1} \frac{\partial^2 u}{\partial x^2}.$$

The transition from one discontinuity into a column of gas therefore modifies the second order partial differential equation governing its motion.

158. If the discontinuities were produced in the gas only by the conditions applied at the end, it would still be possible, just by applying continuous motions to these ends, to always arrive at the same partial differential equation, but this is not the case. We actually know that they must generally come into existence within the column, even when the conditions applied at the ends are perfectly continuous at the time when the motion comes into contact with the edges of regression of the integral surfaces. It is therefore impossible, in a general study of the motion of a gas with a known initial state, to always retain the same second order partial differential equation.

To obtain an equation which will always be satisfied, we must eliminate the function of x, ρ_0, from this second order equation. This can be achieved by writing the equation in the following manner

$$\frac{\dfrac{\partial^2 u}{\partial t^2}}{\left(1 + \dfrac{\partial u}{\partial x}\right)^{-m-1} \dfrac{\partial^2 u}{\partial x^2}} = \frac{m\, p_0}{\rho_0}$$

and by differentiating with respect to t, which gives

$$\frac{\partial^2 u}{\partial x^2}\frac{\partial^3 u}{\partial x^3} - \frac{\partial^2 u}{\partial t^2}\frac{\partial^3 u}{\partial x^2\,\partial t} + (m+1)\,\frac{\dfrac{\partial^2 u}{\partial t^2}\dfrac{\partial^2 u}{\partial x^2}\dfrac{\partial^2 u}{\partial x\,\partial t}}{1 + \dfrac{\partial u}{\partial x}} = 0,$$

a third order differential equation, which is the only one to remain constant throughout the motion; it is, as we can see, independent of the initial state and remains the same for all gases which have the same specific heat ratio.

The partial differential equation which governs the motion of a perfect gas enclosed in a cylindrical tube is therefore a third order equation and not a second order equation, as had always been postulated until now.

This is the equation we shall have to use to analyse some of the phenomena of the motion, especially the way in which the discontinuities develop when the motion comes up against the edges of regression of the integrals of the various second order equations, considered until now, and which are the specific integrals of the third order equation.

This delicate study will be dealt with in another paper. We shall limit ourselves here to considering a few specific cases of discontinuities which can be analysed in full without having to use the third order equation.

Nonetheless, we are now in a position to make a comment concerning the characteristics of this equation. Their horizontal projections are given by the differential equation

$$\frac{\partial^2 u}{\partial x^2}\left(\frac{dx}{dt}\right)^3 - \frac{\partial^2 u}{\partial t^2}\frac{dx}{dt} = 0;$$

from which we obtain three values for $\frac{dx}{dt}$

$$\frac{dx}{dt} = 0, \qquad \frac{dx}{dt} = \pm\sqrt{\frac{\frac{\partial^2 u}{\partial t^2}}{\frac{\partial^2 u}{\partial x^2}}},$$

which are easy to calculate when the specific integral which represents the motion is known. Let ρ_0 and p_0 be the values corresponding to the second order equation satisfied by the particular integral in question; this gives us

$$\frac{\frac{\partial^2 u}{\partial t^2}}{\frac{\partial^2 u}{\partial x^2}} = \frac{m\, p_0}{\rho_0}\left(1 + \frac{\partial u}{\partial x}\right)^{-m-1}.$$

The last two values of $\frac{dx}{dt}$ are therefore the ones which correspond to the velocities of propagation of the motion in this particular integral. The first value of $\frac{dx}{dt}$ seems to correspond to a nil velocity of propagation, which finds its own application. It has indeed been shown that if a discontinuity were to come into existence within the gas, it would give rise to three other discontinuities, i.e. two which propagate on opposite sides of the initial discontinuity, and a third which only affects the expansion and always remains in the same point. The velocity of propagation for this last discontinuity is therefore nil and corresponds to the third characteristic.

The role of these remarkable curves is therefore just as important for the third order equation as for the second order equation.

IV. -Propagation of a uniform shock wave in a gas at rest

159. A column of gas is in a state of rest. One of its ends, corresponding to the value of $x = 0$ on the x-axis is closed by a piston, to which a velocity V is suddenly given, and which is thereafter maintained at this velocity. It is required to determine the motion of the gas.

Colonel Sébert and I have already dealt with this problem in a memorandum inserted in the *Comptes rendus de l'Académie des Sciences* (session of 25 February, 1884). To solve it, we shall assume, as we always have done until now, that for each stratum of the gas, the ratio $\frac{p}{\rho^m}$ remains constant throughout the motion, even when the density undergoes sudden variations. The solution we gave is not strictly exact and we need to modify it.

At the very beginning of the motion, a discontinuity comes into existence in the stratum which is in contact with the piston; this discontinuity propagates through the column, dividing it in two at time t; one part is still in a state of rest, while the other is subjected to a certain motion which we must determine.

But the entire part of the column which is in motion has undergone the effect of the discontinuity; consequently, this motion is not represented by an integral of the initial partial differential equation

$$\rho_0 \frac{\partial^2 u}{\partial t^2} = m\, p_0 \left(1 + \frac{\partial u}{\partial x}\right)^{-m-1} \frac{\partial^2 u}{\partial x^2},$$

where ρ_0 is the density and p_0 is the pressure in the gas at rest.

The motion we are looking for is an integral of another equation of the same form, in which the function ρ_0 is different. We shall show that this integral is linear.

To do so, let us make

$$u = \mathrm{A}x + \mathrm{B}t + \mathrm{C},$$

we need just show that the three constants A, B and C can be determined in such a way that the function u will satisfy the condition applied at the end $x = 0$.

Firstly, when $x = 0$, the value of $\frac{\partial u}{\partial t}$ must always equal V, which implies that B = V, and in addition that u is nil when $x = 0$, $t = 0$, which then implies that C = 0. The condition applied at the end therefore leads us to postulate

$$u = \mathrm{A}x + \mathrm{V}t,$$

where the coefficient A is the only arbitrary variable to be found.

160. We now must try to find whether we can obtain a value for the coefficient A such that the previous motion will be compatible with the state of rest defined by $u = 0$.

The compatibility conditions were given in § 151. In this case, if we take p_1 to be the pressure developed in the gas in motion, they become

$$\rho_0 \mathrm{V} \frac{dx}{dt} = p_1 - p_0, \tag{1}$$

$$\mathrm{A} \frac{dx}{dt} = - \mathrm{V}, \tag{2}$$

$$\left[\frac{\rho_0}{2} \mathrm{V}^2 + \frac{p_1 (1 + \mathrm{A}) - p_0}{m - 1} \right] \frac{dx}{dt} = p_1 \mathrm{V}.$$

As we know, the last equation can be replaced by the following one

$$p_1 = p_0 \frac{2 - (m - 1) \mathrm{A}}{2 + (m + 1) \mathrm{A}} \tag{3}$$

Equations (1), (2) and (3) contain three arbitrary variables, i.e. A, p_1 and $\frac{dx}{dt}$. We need just determine their values using these three equations for the compatibility conditions to be satisfied.

The motion we are looking for is indeed one which is represented by a linear integral; at all points, the velocity is the same and equal to that of the piston V, the expansion is constant and equal to A, the pressure and density are also constant.

161. Eliminating $\frac{dx}{dt}$ from the first two equations gives

$$p_1 = p_0 - \frac{\rho_0 \mathrm{V}^2}{\mathrm{A}}$$

and, if we substitute this into the third, we obtain

$$\mathrm{A}^2 - \frac{m + 1}{2m} \frac{\rho_0}{p_0} \mathrm{V}^2 \mathrm{A} - \frac{1}{m} \frac{\rho_0}{p_0} \mathrm{V}^2 = 0, \tag{4}$$

the binomial equation which describes A.

If we replace A with $-\frac{V}{\frac{dx}{dt}}$, we find

$$(5) \qquad \left(\frac{dx}{dt}\right)^2 - \frac{m+1}{2} V \frac{dx}{dt} - \frac{m p_0}{\rho_0} = 0.$$

This equation is also binomial and gives the value of the velocity of propagation for the motion.

The roots of equation (5) are always real: one is positive and the other negative; let us assume, to settle our ideas, that V is positive, and the positive root is greater than the negative root. One of the roots corresponds to the case where the column of gas is alongside the positive x's, the other to the case where this column is alongside the negative x's.

We can also interpret the two roots in another way, assuming that the gas expands in both directions at once and the piston is infinitesimally thin. If we suddenly apply a velocity V to the piston, the motion propagates in both directions at the velocities represented by the two roots of equation (5) respectively. The product of these two velocities is equal to $\frac{m p_0}{\rho_0}$, in other words to the square of the normal velocity of sound, which is thus the proportional average of the values of $\frac{dx}{dt}$.

The positive root of equation (5) is

$$\frac{dx}{dt} = \frac{m+1}{4} V + \sqrt{\left(\frac{m+1}{4}\right)^2 V^2 + \frac{m p_0}{\rho_0}}$$

or, if we use a to represent the normal velocity of sound $\sqrt{\frac{m p_0}{\rho_0}}$

$$\frac{dx}{dt} = \frac{m+1}{4} V + \sqrt{\left(\frac{m+1}{4}\right)^2 V^2 + a^2}.$$

The velocity of propagation increases with V. When V is cancelled out, this velocity equals a, a result which is in agreement with the findings of the previous sections. As V increases even more, and the ratio $\frac{a}{V}$ becomes insignificant, the velocity of propagation can be represented approximately by $\frac{m+1}{2}$ V and thus becomes proportional to the velocity of the piston.

We must still consider the negative root which corresponds to the case where the column of gas has negative values of x; however, we arrive at the same result if we assume that the column is arranged as above and if we also assume that V is negative. We then need only replace V in the above formula by $-V_1$; the equation becomes

$$\frac{dx}{dt} = \sqrt{\left(\frac{m+1}{4}\right)^2 V_1^2 + a^2} - \frac{m+1}{4} V_1.$$

The motion of the piston is then driven to expand the gas. The velocity of propagation, equal to a when $V_1 = 0$, decreases as V_1 increases and would converge towards zero if V_1 could increase indefinitely; we shall see later that V_1 cannot exceed a finite limit beyond which the gas would no longer follow the piston, such that a vacuum would form behind it.

162. We know that

$$\frac{dx}{dt} = \frac{p_1 - p_0}{\rho_0 V}.$$

If we replace $\frac{dx}{dt}$ in equation (5) with this value, we have the following binomial equation for determining the pressure p_1 in the gas in motion

$$(6) \qquad (p_1 - p_0)^2 - \frac{m+1}{2} \rho_0 V^2 (p_1 - p_0) - m \rho_0 p_0 V^2 = 0.$$

The roots of this equation are always real; one is positive and gives $p_1 > p_0$; it corresponds to the case where the motion of the piston is driven to compress the gas; the other is negative and corresponds to the case where the motion is driven in the opposite direction.

Solving this equation gives the following value for the positive root,

$$p_1 = p_0 + \frac{m-1}{4} \rho_0 V^2 - \rho_0 V \sqrt{\left(\frac{m+1}{4}\right)^2 V^2 + a^2}.$$

The pressure p_1, equal to p_0; when $V = 0$, increases with V and increases indefinitely at the same time as the velocity. When the velocity is large enough for the ratio $\frac{a}{V}$ to be negligible in front of $\frac{m+1}{4}$, the difference $p_1 - p_0$ is very nearly proportional to the square of the velocity V.

The second root can be obtained by replacing V with $-V_1$, this gives

$$p_1 = p_0 + \frac{m-1}{4} \rho_0 V_1^2 + \rho_0 V_1 \sqrt{\left(\frac{m+1}{4}\right)^2 V_1^2 + a^2}\,.$$

It is equal to p_0 when V_1 is nil; however, it decreases as V_1 increases. However, we must bear in mind that the pressure p_1 cannot become negative and this sets a limit on V_1 that the velocity cannot exceed. This limit can be obtained by putting the second term of the above equation equal to zero; it is then easy to obtain the following

$$V_l = \sqrt{\frac{2\ p_0}{(m-1)\,\rho_0}}$$

or

$$V_l = a\sqrt{\frac{2}{m\,(m-1)}}\,.$$

This is the greatest velocity that a gas at rest can suddenly assume. This limit would be infinite if m was equal to one.

It is true that there is nothing to prevent a velocity greater than V_1 being suddenly applied to the piston, driving it in a manner to expand the gas. But this would only take a velocity equal to $a\sqrt{\frac{2}{m\,(m-1)}}$, and a vacuum would be created behind the piston.

When the value of V_1 equals this limit, the velocity of propagation for the motion becomes

$$\frac{dx}{dt} = a\left[\sqrt{1 + \frac{2}{m\,(m-1)}\left(\frac{m+1}{4}\right)^2} - \sqrt{\frac{2}{m\,(m-1)}\left(\frac{m+1}{4}\right)^2}\,\right]$$
$$= a\sqrt{\frac{m-1}{2\ m}}\ .$$

This is the smallest possible value of $\frac{dx}{dt}$ in the case of a sudden discontinuity.

When m is equal to 1.4, as is the case in air, this maximum velocity of propagation is very approximately equal to 0.378 a.

With the pressure p_1 nil, the expansion A is, by virtue of equation (3), equal to $\frac{2}{m-1}$. As a result, the density has become $\frac{\rho_0 (m-1)}{2}$. Finally, it is evident that the temperature of the gas has dropped to absolute zero.

Therefore, if a column of gas is limited by a piston, and this piston is suddenly driven at a velocity of greater than or equal to $a \sqrt{\frac{2}{m(m-1)}}$, in such a way as to release the pressure on the fluid, the temperature of the stratum closest to the piston would immediately drop to absolute zero; its density would not become nil, but would drop in the ratio of 1 to $\frac{m-1}{2}$; this stratum would assume a velocity equal to $a \sqrt{\frac{2}{m(m-1)}}$ and the motion would propagate into the column at a velocity of $a \sqrt{\frac{m-1}{2}}$, all the strata affected by the motion being instantaneously changed to the same state as the first.

The experiment obviously cannot be performed, not only because of the impossibility of suddenly applying a finite velocity to the piston, but also because the properties of the gases change completely when the temperature drops below a certain level. However, the above theory can be used to explain, in a better way than has until now been possible, the large drop in temperature which occurs when a gas is suddenly expanded and which can cause the gas to solidify.

This drop in temperature is in no way due, as is normally claimed, to the strata which are being expanded borrowing the heat needed from the neighbouring strata to produce their active force, but results from the fact that the internal energy of each stratum which is being expanded is used to generate its own active force. We are therefore able to obtain the value for the highest possible flow velocity caused by a sudden expansion.

The internal energy of a stratum with length dx is, indeed, equal to $\frac{p_0 \omega \, dx}{m-1}$ and its mass is $\rho_0 \omega \, dx$. If, when the stratum assumes a velocity V, its active force is acquired at the expense of its own internal energy, the limit for this velocity is given by the equation

$$\frac{p_0\, \omega\, dx}{m-1} = \frac{\rho_0}{2}\, \omega\, dx\, V^2, \quad \text{hence} \quad V^2 = \frac{2}{m-1}\, \frac{p_0}{\rho_0},$$

which is exactly the value we obtained earlier.

163. As we have already observed, it is impossible to suddenly communicate a finite velocity V to the piston, but we can communicate such a velocity to it in a very short period of time, and the above results closely approximate the more complex phenomenon which actually occurs.

However, there does seem to be a contradiction regarding the velocity of propagation of the motion which, while constantly equal to a in the absence of discontinuities, assumes very different values when the velocity is applied suddenly to the piston.

It is worthwhile giving a few explanations concerning this matter; however, we must examine separatly the case where the velocity of the piston is driven to compress the gas in a separate section, and the case where it increases its volume.

In the first case, if the velocity of the piston increases continuously until it attains the value V, the motion is first represented by a developable integral which propagates through the part at rest at a velocity of a. But the phenomenon is modified as soon as the motion comes into contact with the edge of regression of the integral surface and we saw, in the previous chapter, that the sooner this edge of regression is encountered, the quicker the rate at which the velocity varies.

From this moment on, the individual waves which left the origin in succession meet, thereby creating a discontinuity in the last stratum in motion. Through this discontinuity, the velocity of propagation is modified and undoubtedly converges towards the value which would be obtained by assuming that the velocity V is applied suddenly to the piston.

The hypothesis of the velocity being suddenly applied to the piston therefore only eliminates the first stage of the phenomenon, which now lasts only a very short time, as is the case in real time, to communicate the velocity V to the piston.

We may therefore assume, when the velocity is communicated very rapidly to the piston, that this velocity was attained instantaneously, without inducing any serious error.

Let us now consider the case where the velocity of the piston is controlled to increase the volume of the fluid, if we assume that it is acquired by a continuous variation, the motion is at all times represented by the developable integral, since the edge of regression does not come into play in these conditions, as shown earlier. The velocity of propagation of the motion is therefore always represented by a, whereas it would be less than a if the piston attained the velocity instantaneously. The phenomena are therefore markedly different, depending on whether the piston attains its velocity suddenly or over a finite period of time, however small it may be. This difference is highlighted in the diagrams below which compare the velocity curves at different instants for both hypotheses.

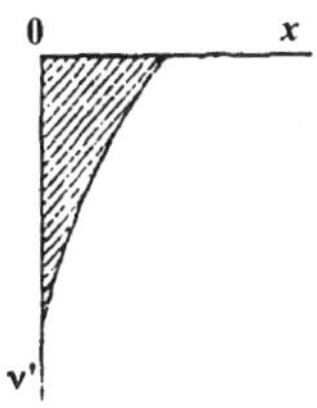

Fig. 13

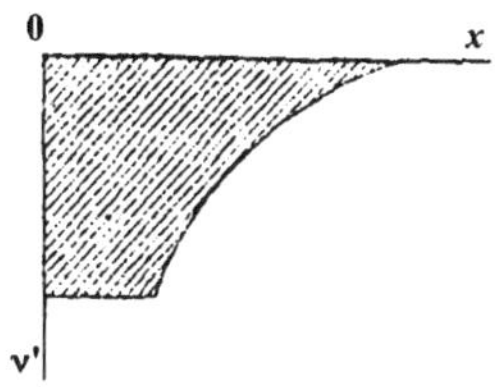

Fig. 14

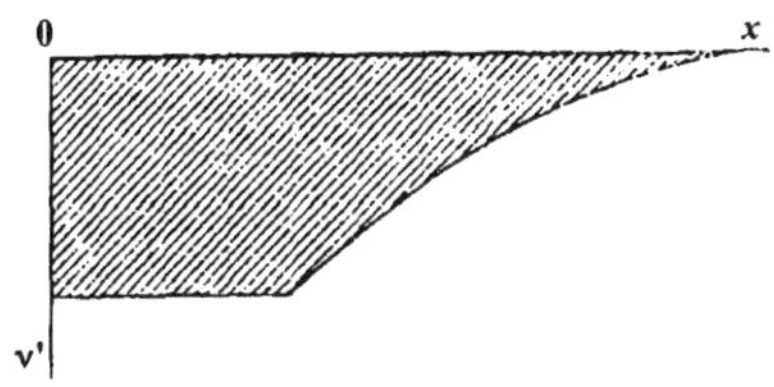

Fig. 15

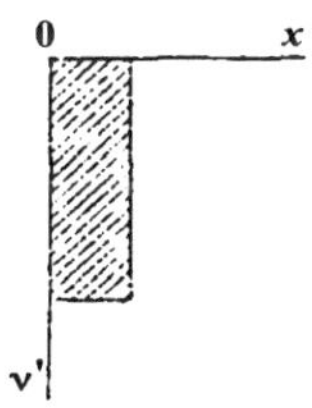

Fig. 16

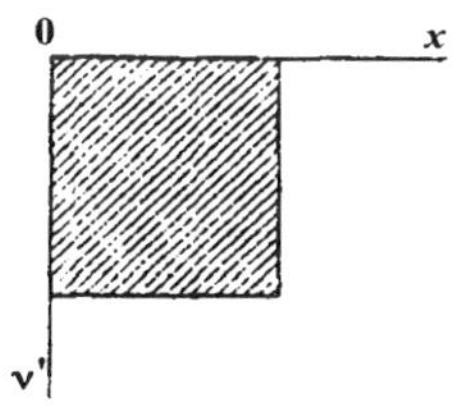

Fig. 17

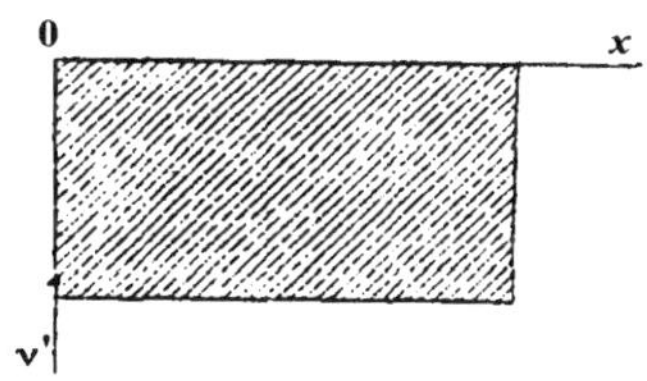

Fig. 18

One of these curves is always a horizontal straight line, while the other contains a horizontal part and a curvilinear part. This last part moves further and further away, such that, if the difference between the two diagrams can almost be ignored in the first moments, this is no longer the case after a certain amount of time has elapsed.

The state of the gas expanded by a piston to which a finite velocity has suddenly been applied can, in certain circumstances, be considered unstable, because if the time taken by the piston to attain this velocity is not exactly nil, the state of the gas is increasingly different from the state which corresponds to a sudden variation.

Our findings would have been similar if we had considered the expansion curves.

The above can be used to account for another apparent contradiction related to the maximum flow velocities. It was shown in Chapter IV that if, with a gas limited by a piston, a motion was applied to this piston with constant acceleration and if this motion was controlled to increase the initial volume of the gas, it would no longer follow the piston once its velocity had exceeded the limit $\frac{2a}{m-1}$, which therefore is the maximum flow rate velocity of the gas. The end stratum of the gas then has an infinite expansion and its temperature has dropped to absolute zero.

However, when the velocity is communicated to the piston in a sudden manner, a vacuum produces at its rear as soon as it goes beyond $a\sqrt{\frac{2}{m(m-1)}}$, which is substantially lower than the previous value. The maximum flow rate velocity in this case is therefore much lower than in the first case. The ratio of the first velocity to the second is equal to $\sqrt{\frac{2m}{(m-1)}}$, in other words $\sqrt{7}$ when we make $m = 1.4$.

The phenomenon is considerably different in the two cases; indeed, in the first, only the end stratum has the velocity $\frac{2a}{m-1}$; the velocity of the following strata becomes increasingly lower until the end point reached by the motion, where this velocity cancels out; similarly, the expansion varies between infinity and zero. Conversely, in the second case, all the strata affected by the motion have the same velocity; the expansion is therefore constant and equal to $\frac{2}{m-1}$. Eventually, the temperature at all points has dropped to absolute zero, although the expansion is not infinite.

V. - Application to air resistance

164. Assuming that a mobile piston separates two columns of the same gas which, at the initial moment have the same temperature and the same density, let us determine the resistance that the gas will apply to the motion of the piston.

This determination is of some interest because of the indications it can provide on the way in which the resistance that the air applies to the motion of the projectiles varies with the velocity. However, the conditions are far from being the same because, since the projectile moves in an unlimited volume of air, the fluid can flow in all directions.

If the velocity of the piston, initially in a state of rest, was to vary in tiny fractions of degrees, the following relationship would always exist between the pressure p in front of the piston and the velocity V

$$p = p_0 \left(1 + \frac{m-1}{2a} \mathrm{V}\right)^{\frac{2m}{m-1}}$$

(Chapter IV, § 116); similarly, the pressure p' at the rear would be

$$p' = p_0 \left(1 - \frac{m-1}{2a} \mathrm{V}\right)^{\frac{2m}{m-1}}.$$

The resistance R applied to the piston by the fluid, reported to the surface unit, is $p - p'$; this therefore gives us

$$\mathrm{R} = p_0 \left[\left(1 + \frac{m-1}{2a} \mathrm{V}\right)^{\frac{2m}{m-1}} - \left(1 - \frac{m-1}{2a} \mathrm{V}\right)^{\frac{2m}{m-1}}\right],$$

an expression which applies as long as there are no discontinuities in the motion.When the velocity V is very low, the above expression is considerably reduced to

$$\frac{2mp_0 a}{a} V = \frac{2mp_0 a}{a^2} = 2a\rho_0 V .$$

The resistance is thus proportional to the velocity and can be obtained by multiplying the velocity by the mass of a column of fluid of length equal to the velocity of sound. This is the known value, given by elementary theory. We can see here why it is no longer suitable when the velocity V is not very low.

The value of R increases with V; however, it must be pointed out that, when V becomes greater than $\frac{2a}{m-1}$, the pressure p' to the rear of the piston cancels out, such that the resistance can be represented just using the formula

$$\mathrm{R} = p_0 \left(1 + \frac{m-1}{2a} \mathrm{V}\right)^{\frac{2m}{m-1}}.$$

We must therefore use two formulae: one for when V ranges between zero and $\frac{2a}{m-1}$, and the other for when V ranges between $\frac{2a}{m-1}$ and infinity.

165. The formulae are quite different when we assume that the velocity V is communicated to the piston in a sudden manner. The pressures p_1 and p'_1 at the front and rear of the piston are then roots of equation (6) in § 162, such that

$$p_1 = p_0 + \frac{m+1}{4}\rho_0 V^2 + \rho_0 V\sqrt{\left(\frac{m+1}{4}\right)^2 V^2 + a^2},$$

$$p'_1 = p_0 + \frac{m+1}{4}\rho_0 V^2 - \rho_0 V\sqrt{\left(\frac{m+1}{4}\right)^2 V^2 + a^2}.$$

The resistance, reported to the surface unit, is $p_1 - p'_1$; thus

$$R = 2\rho_0 V\sqrt{\left(\frac{m+1}{4}\right)^2 V^2 + a^2}.$$

When the velocity V is very low, we again find the simplified expression from the previous section that the resistance is proportional to the velocity. When V increases, so does R. There is, however, a moment when the formula ceases to apply. We have shown that the pressure p'_1 cancels out when V exceeds the limit $a\sqrt{\frac{2}{m(m-1)}}$. The formula therefore only applies when V ranges between zero and $a\sqrt{\frac{2}{m(m-1)}}$. When the velocity V increases beyond $a\sqrt{\frac{2}{m(m-1)}}$, we must make $R = p_1$, in other words

$$R = p_0 + \frac{m+1}{4}\rho_0 V^2 + \rho_0 V\sqrt{\left(\frac{m+1}{4}\right)^2 V^2 + a^2}.$$

When the velocity V is very large, such that the ratio $\frac{a}{V}$ is negligible compared with the unit, we in fact have

$$R = \frac{m+1}{2}\rho_0 V^2.$$

The resistance would therefore become proportional to the square of the velocity and to the density; in addition, it would be independent of the temperature, which is not the case with the full formula, since the pressure p and the value of a^2 are proportional to the absolute temperature for a given density ρ_0.

166. Let us now return to the case where the velocity of the piston, at first equal to zero, varies rapidly and without discontinuity, until it reaches a value V and thereafter remains constant. It is obvious that during the first few moments, the resistance felt by the mobile piston is represented by the formulae in § 164, in which the velocity V is given the same value as for the piston at each moment.

However, since the length of the column of gas is assumed to be indefinite in both directions, it is not long before a discontinuity arises at the front of the piston and the state of the gas rapidly approaches the condition it would be in if the velocity V had been applied to the piston in a sudden manner. Conversely, at the rear of the piston, the state of the fluid is at all times represented by the developable integral.

Let us therefore ignore the first few moments; the pressure at the front of the piston must be considered as equal to p_1 and the pressure at the rear as equal to p' or zero, depending on whether the velocity V is above or below $\dfrac{2a}{m-1}$.

The resistance R can therefore be obtained from one of the following two formulae

$$\mathrm{R} = p_0 + \frac{m+1}{4}\rho_0 \mathrm{V}^2 + \rho_0 \mathrm{V}\sqrt{\left(\frac{m+1}{4}\right)^2 \mathrm{V}^2 + a^2} - p_0\left(1 - \frac{m-1}{2a}\mathrm{V}\right)^{\frac{2m}{m-1}},$$

$$\mathrm{R} = p_0 + \frac{m+1}{4}\rho_0 \mathrm{V}^2 + \rho_0\sqrt{\left(\frac{m+1}{4}\right)^2 \mathrm{V}^2 + a^2},$$

the first equation applying when V is below $\dfrac{2a}{m-1}$, and the second when V is greater than $\dfrac{2a}{m-1}$.

When the velocity V is very low, the first expression is reduced substantially to $2a\,\rho_0\mathrm{V}$. When the velocity V is high enough for the ratio $\dfrac{a}{\mathrm{V}}$ to be negligible when compared with one, the second expression gives the following approximate value again

$$R = \frac{m+1}{2} \rho_0 V^2.$$

167. In the case of ballistic projectiles, the velocity V is communicated to the body over an extremely short period of time, barely 1 or 2 hundredths of a second. The resistance that the air opposes to their motion would therefore be represented by the formulae in the previous number, if we could liken this phenomenon to what happens in an indefinite cylindrical tube. It is very clear that such a comparison is in no way acceptable because, in reality, the gas displaced by the projectile can flow sideways. The way in which the gas flows also depends on the shape of the front of the bodies and the same applies to air resistance.

However, one of the results provided by the above theory has been borne out by experiments carried out on projectiles. These experiments proved conclusively that the resistance opposed by air to the motion of the projectiles was very nearly proportional to the square of the velocity, when the velocity was in excess of 400 m per second or 500 m per second (for more information on this, consult the *Traité de Balistique expérimentale*, vol. II, p. 177 and 238[1]).

This paper considers the acceleration r due to air resistance which, if ϖ represents the mass of the shot and d its diameter, is related to the resistance R, reported to the surface unit, by the formula

$$r = \frac{\pi d^2}{4} R \frac{g}{\varpi}.$$

If we replace R with the value provided in the previous theory, when, at very large values of V, the resistance is very nearly proportional to the square of the velocity, we find that

$$r = \frac{\pi}{4} \frac{m+1}{2} \frac{\rho_0 g d^2}{\varpi} V^2.$$

or, if we take Δ to be the mass of one cubic meter of air,

$$r = \frac{\pi}{4} \frac{m+1}{2} \frac{\Delta d^2}{\varpi} V^2.$$

which, when we assume that $m = 1.4$, becomes

[1] *Traité de Balistique expérimentale*, M. Hélie, 2e édition; Paris, Gauthier-Villars, 1884.

$$r = 0.943 \frac{\Delta d^2}{\varpi} V^2 .$$

Experiments carried out with cylindrical shot have shown that at very high velocities, we actually have

$$r = 0.580 \frac{\Delta d^2}{\varpi} V^2 ,$$

a value which is far lower than the previous one, as we could expect since the projectile is moving in an indefinite environment and not in a tube which prevents the sideways flow of the gas.

168. Instead of assuming that the column of gas is limited by a piston to which we give a certain velocity, it is easy to imagine that a finite difference in pressure is formed instantaneously at the end of the column, or rather that the pressure of this end stratum is constantly maintained at a value p_1, which is different from the initial pressure p_0.

In this case, the motion of the gas is again represented by a linear expression

$$u = Ax + Vt,$$

because this motion will at all times be compatible with the state of rest if the following three equations are satisfied

$$\rho_0 V \frac{dx}{dt} = p_1 - p_0 ,$$

$$A \frac{dx}{dt} = -V,$$

$$p_1 = p_0 \frac{2 - (m-1) A}{2 + (m+1)A}$$

which contain three arbitrary variables, V, A and $\frac{dx}{dt}$. The last one gives

$$A = \frac{2 (p_0 - p_1)}{(m+1) p_1 + (m-1) p_0} .$$

It is evident that the value of A always ranges between $-\frac{2}{m+1}$ and $\frac{2}{m-1}$, this last value corresponding to the case when $p_1 = 0$.

The first two equations then give

$$V^2 = \frac{2}{\rho_0} \frac{(p_0 - p_1)^2}{(m+1)p_1 + (m-1)p_0},$$

$$\left(\frac{dx}{dt}\right)^2 = \frac{(m+1)\,p_1 + (m-1)\,p_0}{2\,\rho_0}.$$

We shall not devote any of our time to discussing these formulae; we need just point out that if p_1 cancels out, we have

$$V = \sqrt{\frac{2\,p_0}{(m-1)\,\rho_0}} = -a\sqrt{\frac{2}{m\,(m-1)}},$$

$$\frac{dx}{dt} = a\sqrt{\frac{m-1}{2m}},$$

which is consistent with the results obtained earlier.

VI. - Reflection of a uniform shock wave

169. In the preceding analysis, the column of gas was assumed to be indefinite, such that we did not need to consider the reflection phenomena which occur at the other end of the tube. But, when the motion which comes into existence at the end $x = 0$ is represented by a linear expression of the form

(1) $$u = Ax + Bt,$$

we can easily study this reflection phenomenon, provided that all the conditions applied at the other end are one of the following: 1) this end is fixed; 2) the end stratum is obliged to retain a constant pressure. These two specific cases correspond to the ones which are studied in the Laws of Physics under the name of *reflection with change of sign* and *reflection without change in sign*.

The reflection occurs at the moment when the motion represented by formula (1) has reached the end $x = \lambda$. The entire column is then at a velocity B; applying a velocity equal and opposite to B would in no way alter the nature of the phenomenon. The column is brought back to a state of rest; the expansion has a constant value A and the pressure also has a constant value p_1. If the end d is kept fixed, the general motion applied to the system communicates a velocity of – B to this end, and the problem is the same as that dealt with above. If, however, the pressure of the end stratum is kept constant and equal to p_0, the problem is the same as that in § 168.

The new motion which comes into existence at the end λ is therefore still represented by a linear formula whose coefficients can be determined using the compatibility conditions.

We may even go one stage further and determine the successive reflections of the motion at the two ends of the column; we will still obtain linear integrals whose coefficients can be obtained using the compatibility conditions. The calculations were carried out without difficulty.

170. As a possible application, we shall give the solution of the following problem

A gas is at rest in a cylindrical tube limited on one side by an immovable base; the tube is closed at the other end by a piston which is suddenly accelerated to a velocity V, controlled in such a way as to compress the gas. The state of the fluid at each moment must be ascertained.

Let

p_0 be the pressure;
ρ_0 be the density of the gas;
λ be the initial length of the column.

The motion which comes into existence at the origin at the very beginning of the phenomenon is represented, as we know, by a linear expression of the form

$$u_1 = \mathrm{A}_1 x + \mathrm{B}_1 t + \mathrm{C}_1.$$

When this motion, which propagates at a velocity $\left(\frac{dx}{dt}\right)_1$, has reached the end λ, a reflection occurs, in other words a new motion u_2 comes into existence at this end; it is also represented by a linear formula where

$$u_2 = \mathrm{A}_2 x + \mathrm{B}_2 t + \mathrm{C}_2.$$

The motion u_2 propagates into u_1 at a velocity $\left(\frac{dx}{dt}\right)_2$, and ends by returning to the origin, where a new motion comes into existence similar to the previous one and so on.

We shall represent one of these motions in general by

$$u_n = \mathrm{A}_n x + \mathrm{B}_n t + \mathrm{C}_n.$$

And we shall use p_n and ρ_n to represent the corresponding pressure and density, and lastly, we shall take $\left(\frac{dx}{dt}\right)_n$ to be its velocity of propagation. The problem will obviously be solved if we find formulae which will enable us to calculate the constants A_n, B_n, C_n, ρ_n, p_n and $\left(\frac{dx}{dt}\right)_n$ for all values of n.

When n is odd, the motion in question comes into existence on contact with the piston, and the formula must represent the motion of this mobile body which has a velocity which always equals V; thus $B_n = V$. When n is even, however, the formula must represent the motion of the stratum λ, which is always immobile; therefore $B_n = 0$. The coefficient B_n is therefore alternately equal to V and zero.

Let t_{n-1} be the moment when the motion with the subscript n comes into existence, and t_n the moment when it reflects; it is obvious that

$$t_n = t_{n-1} + \frac{\lambda}{\left(\frac{dx}{dt}\right)_n}.$$

When n is odd, the distance travelled by the piston at the time when the motion with subscript n comes into existence is represented by the formula with subscript $n-1$ as well as by the formula with subscript n. So $u_n = u_{n-1}$ when $x = 0$ and $t = t_{n-1}$, which gives

$$B_n t_{n-1} + C_n = B_{n-1} t_{n-1} + C_{n-1}$$

or, since n is odd, $B_{n-1} = 0$, $B_n = V$,

$$C_n = C_{n-1} - Vt_{n-1}.$$

When n is even, the distance travelled from the end λ must always be nil, and since $B_n = 0$,

$$C_n = -\lambda A_n.$$

Lastly, irrespective of the value of n,

$$\rho_n = \frac{\rho_0}{1 + A_n}.$$

We now only need to obtain the formulae for calculating p_n, $\left(\frac{dx}{dt}\right)_n$ and A_n.

171. To this end, we must use formulae which express the fact that the motion with subscript n is compatible with the motion with subscript $n-1$.

When n is odd, these formulae are

(2) $$\rho_0 V \left(\frac{dx}{dt}\right)_n = p_n - p_{n-1} ,$$

(3) $$(A_n - A_{n-1}) \left(\frac{dx}{dt}\right)_n = -V ,$$

(4) $$p_n = p_{n-1} \frac{2\left(1 + A_{n-1}\right) - (m-1)\left(A_n - A_{n-1}\right)}{2\left(1 + A_{n-1}\right) + (m+1)\left(A_n - A_{n-1}\right)} .$$

When n is odd, equations (2) and (3) are replaced by the following

$$-\rho_0 V \left(\frac{dx}{dt}\right)_n = p_n - p_{n-1} , \qquad (A_n - A_{n-1}) \left(\frac{dx}{dt}\right)_n = -V .$$

But, in the first case, $\left(\frac{dx}{dt}\right)_n$ is positive, while it is negative in the second. Therefore, if we limit our study to a consideration of the absolute values of $\left(\frac{dx}{dt}\right)_n$, we can always use formulae (2), (3) and (4).

Eliminating A_n and $\left(\frac{dx}{dt}\right)_n$ from these last relationships will provide us with a binomial equation in p_n; we must take the root which has a plus sign in front of its radical, because, according to the direction of the velocity, p_n is obviously greater than p_{n-1}. This therefore gives us

$$p_n = p_{n-1} + \frac{m+1}{4} \frac{\rho_0}{1 + A_{n-1}} V^2 +$$
$$\frac{\rho_0}{1 + A_{n-1}} V \sqrt{\left(\frac{m+1}{4}\right)^2 V^2 + \frac{m\, p_{n-1}\left(1 + A_{n-1}\right)}{\rho_0}}$$

or, if we take into account the fact that $\rho_{n-1} = \dfrac{\rho_0}{1 + A_{n-1}}$,

$$p_n = p_{n-1} + \frac{m+1}{4} \rho_{n-1} V^2 + \rho_{n-1} V \sqrt{\left(\frac{m+1}{4}\right)^2 V^2 + \frac{m p_{n-1}}{\rho_{n-1}}}.$$

The value of p_n is therefore expressed using the values with the subscript $n-1$.

We can draw the following from equations (2) and (3)

$$A_n = A_{n-1} - \frac{\rho_0 V^2}{p_n p_{n1}}.$$

The value of $\left(\frac{dx}{dt}\right)_n$ can in turn be obtained from either equation (2) or (3).

The solution to the problem is now complete since all of the values with subscript n can be unambiguously deduced from the values with subscript $n-1$, and the zero order values are known.

The motion of the gas is quite particular. The fluid is always divided in two: one part, that closest to the piston, is subjected to a motion of uniform velocity V, the other is in a state of rest. The point of separation moves alternately in both directions by a uniform motion of increasing velocity, as can easily be seen at each change of direction. At certain times, the gas is travelling at a velocity V; at other times, all of the gas is in a state of rest. The relationship between the pressure and the density is not the same for the two parts of the gas; it undergoes a sudden variation in each stratum, each time the particular stratum contains a discontinuity.

VII. - Concerning the motions which come into existence at the meeting point between two homogeneous gases

172. We shall now consider two columns of gas which touch at a stratum; these columns are homogeneous; the pressure and the velocity are the same at each point in these columns; however, there is a discontinuity at the transition from one column into the other, such that they affect each other. It was shown earlier, that a new motion comes into existence in each of the two strata which then propagates into the corresponding column. It is evident that the two new motions are represented by linear expressions whose coefficients have to be found.

The two columns present are homogeneous; this means that we can read off the values on the x-axis for the strata from their initial position, such that the expansion must be considered nil. In addition, if we apply a particular velocity to the entire system which is equal and opposite to that of one of the gases, the gas will come to a state of rest without in any way modifying the phenomenon.

Once we have determined the motions which come into existence at the meeting point, we need only increase the velocities by a constant to obtain the real phenomenon.

Let us assume, for instance, that the stratum of contact is located at the origin of the co-ordinates and that the column which is on the side of the positive values of x has a nil velocity; we shall let p be its pressure, ρ its density and m the value of the ratio of specific heat. Let p', ρ' and m' be the corresponding values for the second gas, and let V be its velocity.

The two motions which come into existence at the stratum of contact can be represented using the formulae

$$u_1 = \mathrm{A}x + \mathrm{B}t, \qquad u'_1 = \mathrm{A}'x + \mathrm{B}'t$$

and we have to determine the constants A, B, A' and B' which represent the expansions and the velocities of the new motions, together with their pressures p_1 and p'_1. Lastly, we still have to find the velocities of propagation $\frac{dx}{dt}$ and $\frac{dx'}{dt}$, but we shall limit our consideration to their absolute values.

As was proven in §§ 149 and 150, the following relationships exist between these eight variables.

$$\mathrm{B} = \mathrm{B}', \quad p_1 = p'_1,$$

$$\rho\mathrm{B}\frac{dx}{dt} = p_1 - p, \quad \rho'(\mathrm{B}' - \mathrm{V})\frac{dx'}{dt} = p' - p'_1,$$

$$\mathrm{A}\frac{dx}{dt} = -\mathrm{B}, \qquad \mathrm{A}'\frac{dx'}{dt} = \mathrm{B}' - \mathrm{V},$$

$$p_1 = p\,\frac{2-(m-1)A}{2+(m+1)A'}, \qquad p'_1 = p'\,\frac{2-(m'-1)A'}{2+(m'+1)A'}$$

which, by an obvious series of eliminations, can be reduced to the following four equations

(1) $$\rho\mathrm{B}^2 = \mathrm{A}\,(p - p_1),$$

(2) $$\rho'\,(\mathrm{B} - \mathrm{V})^2 = \mathrm{A}'\,(p' - p_1),$$

(3) $$2\,(p_1 - p) + \mathrm{A}\,[(m+1)\,p_1 + (m-1)\,p] = 0,$$

(4) $$2\,(p_1 - p') + A'\,[(m'+1)\,p_1 + (m'-1)\,p'] = 0,$$

which contain only four unknowns A, A', B and p_1.

173. It is easy to see that if p_1 were known, it would be possible to obtain the other unknowns unambiguously. A and A' would then be given by linear equations. The same applies for B^2. Furthermore, since the values of $\frac{dx}{dt}$ and $\frac{dx'}{dt}$ must be positive, the sign of front of B is that of $p_1 - p$.

We must, however, calculate p_1. If we eliminate A and A', we have

$$B^2 = \frac{2\,(p - p_1)^2}{\rho\left[(m+1)\,p_1 + (m-1)\,p\right]},$$

$$(B - V)^2 = \frac{2\,(p' - p_1)^2}{\rho'\left[(m'+1)\,p_1 + (m'-1)\,p'\right]}.$$

If we represent the second terms from the previous equations by H and H' and eliminate B, we have

(5) $$V^4 - 2V^2\,(H - H') + (H - H')^2 = 0$$

which is an equation containing p_1 to the sixth power.

We shall not limit ourselves to a general discussion of this equation, which would be too lengthy. There are usually two roots for a single value of V^2, one corresponds to the case where V is positive, and the other to the case where V is negative. We shall determine the equation for the roots which answers the question posed, bearing in mind that the equations

$$\rho B\frac{dx}{dt} = p_1 - p, \quad \rho'\,(B - V)\frac{dx'}{dt} = p' - p_1,$$

must give positive values for $\frac{dx}{dt}$ and $\frac{dx'}{dt}$.

174. It was shown in Chapter IV, that if, when a gas in a state of rest is limited by a piston and an increasing velocity is applied to this piston, given by the formula

$$v = \frac{2a}{m-1}\left(1 - \frac{at}{X}\right)^{-\frac{m-1}{m+1}} - \frac{2a}{m-1}$$

and that from a certain moment onwards, the velocity of the piston is kept constant, all the individual waves join up at the same time at a distance X at time $\frac{X}{a}$.

The column of gas is then split in two parts: one is still in a state of rest, the other is travelling at a uniform velocity V and has the same pressure and expansion at all its points.

It is now possible to determine the subsequent motion of the gas; we need just apply the method used in the previous sections, attributing suitable values to the variables p, p', ρ, ρ' and by making $m' = m$. We can see that two motions come into existence at point X, one of which continues to propagate into the gas at rest, while the other doubles back on itself. The original shock wave is therefore reflected before it reaches the end of the tube.

The new motions are represented by linear expressions; let us in particular consider the motion which doubles back on itself to meet the piston; at the moment of the meeting, it is reflected to give birth to a new motion, again represented by a linear integral. This will propagate unchanged through the gas in the direction of the positive x's until it has travelled a distance of X, where a phenomena similar to the first will occur.

All the circumstances of this unusual phenomenon can be determined with no other difficulty than that of performing a few lengthy numerical calculations.

175. The calculations arrived at using the general method can be simplified a little, when the touching columns of gas are both in a state of rest, but at different pressures.

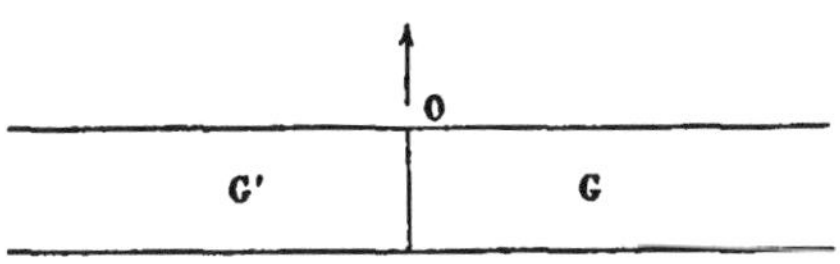

Fig. 19

This state could be produced by introducing two gases G and G', at different pressures, into the same tube, separated into two parts by a shutter O, which is then suddenly removed. The motion which then occurs is interesting to consider from the point of view of the flow of the gases before the permanent state is reached.

We must make V = 0 in the formulae in §§ 172 and 173. Equation (5) is reduced to

$$\mathrm{H} - \mathrm{H}' = 0$$

and we can easily see that it is a quadratic equation. The equation has one root in the range between p and p', because the values of H and H' are always positive; but H cancels out when $p_1 = p$. For the rest, if we assume that $p > p'$, H is a decreasing function and H' is an increasing function, when p_1 lies in the range p' to p. The root in question is therefore unique.

It is also the only root which can be used in the question, because in the two equations

$$\rho \mathrm{B} \frac{dx}{dt} = p_1 - p, \quad \rho' \mathrm{B} \frac{dx'}{dt} = p' - p_1,$$

$\frac{\mathrm{dx}}{\mathrm{dt}}$ and $\frac{\mathrm{dx'}}{\mathrm{dt}}$ must be positive, therefore $\mathrm{p}_1 - \mathrm{p}$ and $\mathrm{p}' - \mathrm{p}_1$ must have the same sign.

The pressure p_1 can therefore be determined without any ambiguity. We can then use it to deduce the values of A, A' and B^2, using equations (1), (3) and (4). It is obvious that A is positive and A' is negative; furthermore, the value of B is negative.

If the pressure p' decreases indefinitely, the same applies for ρ'; since the value of H is finite, the equation H – H' implies that p' must converge towards zero. This then gives us $\mathrm{A} = \frac{2}{m-1}$, and hence

$$\mathrm{B}^2 = \frac{2p}{\rho\,(m-1)}, \qquad \mathrm{B} = \sqrt{\frac{2p}{\rho\,(m-1)}}.$$

This is the maximum flow rate velocity of a gas when there is a sudden change in the pressure.

176. It was assumed in the above calculations that the values of m and m' were not equal to one. The conclusions do not change when $m = m' = 1$, in other words when the two gases present obey Mariotte's law. However, when these two columns consist of the same gas at the same temperature, the two motions which come into existence at the common point are identical and can hence be represented using the same linear equation, as has already been stated.

We already know that in these two motions, the pressure and the velocity are the same; for the motions to be identical, the densities ρ_1 and ρ'_1 must also be equal.

This then gives us

$$A = \frac{\rho}{\rho_1} - 1 , \qquad A' = \frac{\rho'}{\rho'_1} - 1 ,$$

and equations (3) and (4) from § 172 become

$$\frac{\rho}{\rho_1} = \frac{p}{p_1} , \qquad \frac{\rho'}{\rho'_1} = \frac{p'}{p_1} ,$$

hence

$$\frac{\rho_1}{\rho'_1} = \frac{p'}{\rho'} \frac{\rho}{p} .$$

But the two adjacent columns belong to a single gas and assuming that the temperatures are the same, the pressures are proportional to the densities; the second term of the equation is therefore equal to one; therefore $\rho_1 = \rho'_1$.

VIII - Concerning certain discontinuous motions represented by developable integrals

177. Until now, the determination of the motion of a gas from the moment that discontinuities arise has only been carried out for very specific cases, i.e. those where the new motions are represented by linear expressions of the form

$$u = Ax + Bt + C,$$

which, even though the relationship between the pressure and the density has been changed, still satisfy the second order equation usually considered, i.e.

$$(1) \qquad \rho_0 \frac{\partial^2 u}{\partial t^2} = m\, p_0 \left(1 + \frac{\partial u}{\partial x}\right)^{-m-1} \frac{\partial^2 u}{\partial x^2} .$$

This last equation is satisfied for any linear function in x and t.

We shall now study a very general case of motion, for which the integrals can still be obtained but no longer satisfy the above equation.

For this, we must reiterate a comment made in § 96.

When a homogeneous fluid is subjected to a motion represented by a linear integral

$$u = Ax + Bt + C,$$

the integrals compatible with the above along the positive angular coefficient characteristics are developable surfaces which satisfy the first order equation

$$\frac{\partial u}{\partial t} = B + \psi(A) - \psi\left(\frac{\partial u}{\partial x}\right).$$

In the case of perfect gases,

$$\psi\left(\frac{\partial u}{\partial x}\right) = -\frac{2a}{m-1}\left(1+\frac{\partial u}{\partial x}\right)^{-\frac{m-1}{2}};$$

therefore the developable surfaces in question are defined by the first order equation

$$\frac{\partial u}{\partial x} = \frac{2a}{m-1}\left(1+\frac{\partial u}{\partial x}\right)^{-\frac{m-1}{2}} - \frac{2a}{m-1}(1+A)^{-\frac{m-1}{2}} + B.$$

178. We shall now consider a homogeneous gas at rest at a pressure p_0 and with a density ρ_0. The end $x = 0$ is limited by a piston which is suddenly accelerated to a finite velocity V, the velocity of this piston is then made to vary continuously in accordance with an arbitrary law. It is our task to determine the motion which propagates in the gas.

If there were no discontinuities, we know that the motion would be represented by one of the developable integrals in equation (1). However, since the initial velocity was applied suddenly, a discontinuity develops in the end stratum and then propagates into the column.

We may first assume that the velocity V remains constant for a certain time τ, after which it varies continuously. During this time τ, the motion is the same as that studied in § 159 *et sequens*. It is represented by an expression of the form

$$u = Ax + Vt,$$

in which

$$A = \frac{m+1}{4}\frac{V^2}{a^2} - \frac{V}{a}\sqrt{1 + \left(\frac{m+1}{4}\right)^2 \frac{V^2}{a^2}};$$

moreover, the pressure p_1 is

$$p_1 = p_0 + \frac{m+1}{4}\rho_0 V^2 + \rho_0 V \sqrt{\left(\frac{m+1}{4}\right)^2 V^2 + a^2}.$$

The relationship between the pressure and the expansion is no longer as it is postulated in equation (1), because the value of $p_1\,(1 + A)^{-m}$ is different from p_0.

However, we can state that

$$p'_0 = p_1\,(1 + A)^{-m}\ .$$

The gas in motion is then in the same state as if its initial pressure were p'_0 and its density ρ_0, and each stratum expanded according to the adiabatic law; consequently, as long as there is no new discontinuity, the motion of this column will be governed by the equation

$$(2) \qquad \rho_0 \frac{\partial^2 u}{\partial t^2} = m\,p'_0 \left(1 + \frac{\partial u}{\partial x}\right)^{-m-1} \frac{\partial^2 u}{\partial x^2}.$$

Therefore, when, at time τ, the velocity of the piston starts to vary continuously, the new motion which will come into existence in the already disrupted column will be an integral of equation (2).

179. Thus, according to the comment in § 177, this motion will be given by one of the developable integrals of equation (2) which, if we make $a_1^2 = \dfrac{m\,p'}{\rho_0}$, are defined by the first order equation

$$\frac{\partial u}{\partial t} = \frac{2a_1}{m-1}\left(1 + \frac{\partial u}{\partial x}\right)^{-\frac{m-1}{2}} - \frac{2a_1}{m-1}(1 + A)^{-\frac{m-1}{2}} + V.$$

The arbitrary function introduced by the integration will be determined to satisfy the motion of the piston.

The integrals of the above equation, apart from the linear one, do not in any way satisfy equation (1), because the developable integrals of this equation are defined by the following

$$\frac{\partial u}{\partial t} = \alpha \pm \frac{2a}{m-1}\left(1 + \frac{\partial u}{\partial x}\right)^{-\frac{m-1}{2}},$$

where α is an arbitrary constant.

The above results are obviously independent of the duration τ and remain unchanged when τ is cancelled out.

180. We have just obtained motions which do not in any way satisfy the initial partial differential equations, but they do correspond to the case where a discontinuity is introduced into the condition applied to the end. We shall now show how we can obtain analogous motions by applying only fully known motions to the end of the column.

Let us therefore assume that the gas is at rest and we apply a motion to the piston which is defined by the relationship

$$v = \frac{2a}{m-1}\left(1 - \frac{at}{\mathrm{X}}\right)^{-\frac{m-1}{m+1}} - \frac{2a}{m-1},$$

until a certain time τ, after which the velocity is left at V, the value it has reached; we know that all the individual waves arrive together at a distance X, where a discontinuity comes into existence instantaneously. We also showed in § 174 that two motions come into existence at point X, one of which continues to propagate in the same direction, while the other one returns towards the origin, which it reaches at time τ_1. The part of the column between the origin and X is then subjected to a motion represented by a linear integral which, when it meets the piston at time τ_1, undergoes a reflection which can easily be calculated. After this reflection, the part of the column closest to the origin is subjected to a motion defined by a linear integral and the relationship which had existed until then between the pressure and the density undergoes a change. The continuous motions which can come into existence in this column are therefore represented by the integrals of an equation such as (2).

If, after time $\tau_2 > \tau_1$, the velocity of the piston, until then equal to V, varies in a continuous manner, a motion will come into existence which will propagate into the previous one and will be represented by a developable integral analogous to that in the previous number, which will in no way satisfy the initial equation.

This integral, which could be determined with no difficulty other than performing a few lengthy calculations, will therefore develop in the gas without provoking a discontinuity in the condition imposed at the end, because the piston will not have been subjected to a sudden acceleration.

Rayleigh
courtesy of AIP Emilio Segrè Visual Archives

LORD RAYLEIGH (J. W. STRUTT, THIRD BARON RAYLEIGH)

Rayleigh made important contributions to the theory of sound including the propagation of finite-amplitude waves. Rayleigh was awarded the Nobel Prize in physics in 1904 for the discovery and isolation of argon.

He was born in Langdon Grove, England, on November 12, 1842. He was a peer by inheritance and attended Cambridge University, graduating in 1865. At Cambridge he was a student of the mathematician E. J. Routh and attended lectures by G. G. Stokes. Because of his independent wealth, he was able to establish a personal laboratory at the family home at Terling Place. His famous book on the theory of sound was published in 1877. He became Cavendish Professor of experimental physics at Cambridge in 1879 and remained in that position until 1884. From that time on his headquarters were predominantly at Terling Place. He did accept a professorship at the Royal Institution of Great Britain, and served from 1887 to 1905. This appointment did not require his permanent presence in London, however. From 1908 until his death in 1919 he served as chancellor of Cambridge University.

His contributions to shock compression of condensed matter consist mainly of his recognition that in a steady propagating wave of finite amplitude the pressure is a linear function of the specific volume and that the wave rise time can be calculated from knowledge of the dissipative processes involved. The latter is similar to the contributions made by G. I. Taylor, also in 1910. The professional interaction between Rayleigh and Taylor on this matter is opaque, although it is well known that their names are linked with regard to the unstable growth of perturbations in accelerated fluid layers.

Aerial Plane Waves of Finite Amplitude

Lord Rayleigh, O.M., F.R.S., *Proc. Roy. Soc. Lond.* A84, pp. 247-284 (1910).

Waves of Finite Amplitude Without Dissipation

In the investigations which follow, we are concerned with the motion of an elastic fluid in one dimension, say, parallel to x. It is implied not only that there are no component velocities perpendicular to x, but that the motion is the same in any perpendicular plane, so that it is a function of x and of the time (t) only. If u be the velocity at any point x, p the pressure, ρ the density, X an impressed force, the dynamical equation for an inviscid fluid is

$$\frac{du}{dt}+u\frac{du}{dx}=\mathrm{X}-\frac{1}{\rho}\frac{dp}{dx}. \tag{1}$$

At the same time the "equation of continuity" takes the form

$$\frac{d\rho}{dt}+\frac{d(\rho u)}{dx}=0. \tag{2}$$

The first step, and it was a very important one, in the treatment of waves of finite amplitude is due to Poisson.[1] Under the assumption of Boyle's law, $p=a^2\rho$, he proved that for waves travelling in one direction (positive) the circumstances of the propagation are expressed by

$$u=f\{x-(a+u)t\}, \tag{3}$$

in which f denotes an arbitrary function. When u can be neglected in comparison with a, this reduces to the familiar law of undisturbed propagation applicable to infinitesimal waves.

Poisson does not discuss the significance of (3) further than to show that the boundaries of a continuous wave, limited to a finite range along x, are propagated with the ordinary velocity a, and that accordingly the length of the wave does not alter as it advances. The meaning of (3) is that in general u advances with a velocity equal, not to a, but to $a + u$, and that this might be expected is very

[1] "Mémoire sur la Théorie du Son," 'Journ. de l'École Polytechnique, 1808, vol. 14, p. 319.

easily seen (Earnshaw). From the ordinary theory we know that an infinitely small disturbance is propagated with a certain velocity a, which velocity is relative to the parts of the medium undisturbed by the wave. Let us consider now the case of a wave so long that the variations of velocity and density are insensible for a considerable distance along it, and at a place where the velocity (u) is finite let us imagine a small secondary wave to be superposed. The velocity with which the secondary wave is propagated through the surrounding medium is a, but on account of the local motion of the medium itself the whole velocity of the advance is $a + u$, and depends upon the part of the long wave at which the small wave is placed. What has been said of the secondary wave applies also to the parts of the long wave itself, and thus we see that after a time t the place where a certain velocity u is to be found is in advance of its original position by a distance equal, not to at, but to $(a + u)t$, or, as we may express it, u is propagated with velocity $(a + u)$.

A closer discussion of the solution represented by Poisson's integral was given by Stokes,[2] who pointed out the difficulty which ultimately arises from the motion becoming discontinuous. If we draw a curve to represent the distribution of velocity, taking x for abscissa and u for ordinate, we may find the corresponding curve after the lapse of time t by the following construction:—Through any point on the original curve draw a straight line in the positive direction parallel to x, and of the length equal to $(a + u)t$, or, as we are concerned with the shape of the curve only, equal to ut. The locus of the ends of these lines is the velocity-curve after a time t.

But this law of derivation cannot hold good indefinitely. The crests of the velocity-curve gain continually on the troughs and must at last overtake them. After this the curve would indicate two values for u for one value of x, ceasing to represent anything that could actually take place. In fact we are not at liberty to push the application of the integral beyond the point at which the velocity becomes discontinuous, or the velocity-curve has a vertical tangent. In order to find when this happens, let us take two neighbouring points on any part of the curve which slopes downwards in the positive direction, and inquire after what time this part of the curve becomes vertical. If the difference of abscissæ be dx, the hinder point will overtake the forward point in the time $-dx/du$. Thus the motion, as determined by Poisson's integral, becomes discontinuous after a time equal to the reciprocal, taken positively, of the greatest negative value of du/dx.

For example, let us suppose that

$$u = \mathrm{U} \cos \frac{2\pi}{\lambda}\{x-(a+u)t\}. \tag{4}$$

[2] "On a Difficulty in the Theory of Sound," 'Phil. Mag.,' November, 1848.

where U is the greatest initial velocity. When $t = 0$, the greatest negative value of du/dx is $-2\pi U/\lambda$, so that discontinuity will commence at the time $t = \lambda/2\pi U$.

The only kind of wave travelling in the positive direction which can escape ultimate discontinuity is one which has no forward slope. This is the case of a wave forming the transition between a larger constant value of u when x exceeds a certain value, and a smaller constant value when x falls short of a certain value. As time passes, the slope everywhere becomes easier. We shall see presently that this wave is a wave of rarefaction, in the sense that during its passage the gas passes from a greater to a less density.

It is worthy of remark that, although we may of course conceive a wave of finite disturbance to exist at any moment, there is in general a limit to the duration of its previous independent existence. By drawing lines in the negative instead of in the positive direction we may trace the history of the velocity-curve; and we see that as we push our inquiry further and further into past time the forward slopes become easier and the backward slopes steeper. At a time equal to the greatest positive value of dx/du, antecedent to that at which the curve is first contemplated, the velocity would be discontinuous. The exception is now a wave of condensation, involving a passage always from a less to a greater density.

When discontinuity sets in, a state of things exists to which the usual differential equations are inapplicable; and the subsequent progress of the motion has not been determined. It is probable, as suggested by Stokes, that some sort of reflection would ensue. In regard to this matter we must be careful to keep purely mathematical questions distinct from physical ones. We shall see later how the tendency to discontinuity may be held in check by forces of a dissipative character. But this has nothing directly to do with the mathematical problem of determining what would happen to waves of finite amplitude in a medium, free from viscosity, whose pressure is under all circumstances proportional to the density.[3] To suppose that the problem has no solution would seem to be tantamount to admitting an inherent contradiction in the assumption, usually made in hydrodynamics, of a continuous fluid subject to Boyle's law. It would be strange if the necessity of a molecular constitution for gases could be established by such an argument.

With Poisson's integral (3), showing how the velocity is propagated, there is associated another law connecting the velocity and the density in a positive progressive wave. In the case of a fluid obeying Boyle's law this relation is

$$u - a \log \rho = \text{const.} \tag{5}$$

[3] 'Theory of Sound,' 1878, § 251.

It does not occur explicitly in Poisson's memoir, and Earnshaw considers that Poisson did not discover it. Certainly it is remarkable that he omitted to formulate the law, but at the same time it is difficult to suppose him ignorant of it, seeing that it follows by simple subtraction from two of his equations.[4] A formula equivalent to (5) was given explicitly, so far as I know, for the first time by Airy,[5] who attributes it to De Morgan.

The assumption that in a progressive wave there is a definite relation between u and ρ forms the basis of Earnshaw's investigation.[6] That such a relation is to be expected may be shown by a line of argument analogous to that already employed in connection with Poisson's integral.

Whatever may be the law of pressure as a function of density, the velocity of propagation of small disturbances is according to the usual theory equal to $\sqrt{(dp/d\rho)}$, and in a positive progressive wave the relation between velocity and condensation (s) is

$$u{:}s = \sqrt{(dp/d\rho)}, \tag{6}$$

where $s = \delta\rho/\rho$. If this relation be violated at any point, a wave will emerge, travelling in the negative direction. Let us now picture to ourselves the case of a positive progressive wave in which the changes of velocity and density are very gradual but become important by accumulation, and let us inquire what condition must be satisfied in order to prevent the formation of a negative wave. It is clear that the answer to the question whether or not a negative wave will be generated at any point will depend upon the state of things in the immediate neighbourhood of the point, and not upon the state of things at a distance from it, and will therefore be determined by the criterion applicable to small disturbances. In applying this criterion we are to consider the velocities and condensations, not absolutely, but relatively to those prevailing in the neighbouring parts of the medium, so that the form of (6) proper for the present purpose is

$$du = \sqrt{\left(\frac{dp}{d\rho}\right)} \cdot \frac{d\rho}{\rho}, \tag{7}$$

whence

$$u = \int \sqrt{\left(\frac{dp}{d\rho}\right)} \cdot \frac{d\rho}{\rho} \tag{8}$$

[4] Equations (1), p. 364, and (b), p. 367.

[5] 'Phil. Mag.,' 1849, vol. 34, p. 402. The corresponding formula for long tidal waves of finite amplitude was also given.

[6] 'Roy. Soc. Proc.,' January 6, 1859; 'Phil. Trans.,' 1860, p. 133.

which is the relation between u and ρ generally necessary for a positive progressive wave, as laid down by Earnshaw.[7]

Earnshaw worked with the so-called Lagrangian form of the equations, in which the motions of particular particles are followed, and he obtained complete solutions for a wave progressive in one direction. In the case of Boyle's law the relation between velocity and density is that already given (5), and in the case of the adiabatic law, where

$$\frac{p}{p_0} = \left(\frac{\rho}{\rho_0}\right)^{\gamma}, \tag{9}$$

Earnshaw finds

$$\left(\frac{\rho}{\rho_0}\right)^{\frac{1}{2}(\gamma-1)} = 1 + \frac{(\gamma-1)u}{2a}, \tag{10}$$

where a is the velocity of infinitesimal disturbances under the condition represented by p_0, ρ_0, viz. $a^2 = \gamma p_0/\rho_0$. In (9) γ denotes as usual the ratio of the two specific heats; and in (10), applicable to a *positive* progressive wave, the constant of integration has been so chosen that $u = 0$ corresponds to $\rho = \rho_0$.

The generalised form of Poisson's integral, appropriate when p is *any* given function of ρ, does not appear quite explicitly in Earnshaw's memoir. The line of argument already used shows that it must be

$$u = f[x - \{u + \surd(dp/d\rho)\}t]. \tag{11}$$

In the case of a gas obeying Boyle's law,

$$\surd(dp/d\rho) = \text{const.}, \tag{12}$$

and (11) reduces to Poisson's form.

In the case of the adiabatic law, we have from (9), (10),

$$\surd\left(\frac{dp}{d\rho}\right) = a\left(\frac{\rho}{\rho_0}\right)^{\frac{1}{2}(\gamma-1)} = a + \frac{\gamma-1}{2}u, \tag{13}$$

so that (Earnshaw)

$$u + \surd(dp/d\rho) = a + \tfrac{1}{2}(\gamma+1)u. \tag{14}$$

[7] 'Theory of Sound,' 1878, § 251.

Thus (11) assumes the form

$$u = f[x - \{a + \tfrac{1}{2}(\gamma + 1)u\}t], \tag{15}$$

and this with (10) may be considered to constitute the solution of the problem up to the point where discontinuity sets in. We may fall back upon Boyle's law by putting in (15) $\gamma = 1$.

It appears that whether the relation of pressure to density be isothermal or adiabatic, there is a change of type as the wave advances. There can be no escape from such a change unless $u + \surd(dp/d\rho)$ be constant. Using (8), we may deduce in this case

$$\surd(dp/d\rho) = \mathrm{B}/\rho,$$

B being constant, whence

$$p = \mathrm{A} - \mathrm{B}^2/\rho \tag{16}$$

expresses the only law of pressure under which waves of finite amplitude can be prolonged without undergoing a change of type (Earnshaw). A simpler derivation of (16) will be given presently.

Earnshaw further considers the genesis of disturbance in a gas originally at rest by the motion of a piston (supposed to be contained in a tube), but some of his conclusions appear to need revision. All that is required in these problems is virtually contained in (8), (11). If X denote the position of the piston at time T, the velocity of its motion is $\mathrm{U} = d\mathrm{X}/d\mathrm{T}$, and this velocity is shared by the gas in contact with it. On the positive side the velocity of propagation of U (equal to u) is, by (11),

$$\mathrm{U} + \surd(dp/d\rho);$$

so that if at time t (greater than T), U is to be found at x, we must have

$$x = \mathrm{X} + \{\mathrm{U} + \surd(dp/d\rho)\}(t - \mathrm{T}). \tag{17}$$

Among the problems which naturally suggest themselves would be to determine what happens when the piston originally at rest at $\mathrm{X} = 0$ begins to move at time $\mathrm{T} = 0$ with a constant velocity. But if this velocity be positive, the discontinuity, which immediately ensues, causes the failure of our equations. On the other hand, if the constant velocity be negative, say $-\mathrm{V}$, the initial discontinuity disappears forthwith, and the subsequent motion may be traced.

Take, for example, the case of Boyle's law. We have

$$\mathrm{X} = \mathrm{UT}, \qquad \mathrm{U} + \surd(dp/d\rho) = \mathrm{U} + a;$$

so that

$$x = \mathrm{U}t + a(t - \mathrm{T}) \qquad (t > \mathrm{T}),$$

and we have to consider where the velocities 0 and $-\mathrm{V}$ are to be found at time t. Now $\mathrm{U} = 0$ corresponds to the range of T from $-\infty$ to 0, so that x ranges from at to ∞. Again, $\mathrm{U} = -\mathrm{V}$ corresponds to the range of T for 0 to t, so that x ranges from $(a - \mathrm{V})t$ to $-\mathrm{V}t$. The whole range of x on the positive side of the piston is not accounted for, except the interval for $x = (a - \mathrm{V})t$ to $x = at$. This is occupied by the transition of velocity from 0 to $-\mathrm{V}$, and we infer, from what was said in the discussion of Poisson's integral, that this transition must take place linearly.

Under Boyle's law the relation between velocity and density (5) is such that, however fast the piston may recede (u negative), a complete vacuum can never be formed behind it. It is otherwise under the adiabatic law (10), where $\rho = 0$ corresponds to

$$u = -2a/(\gamma - 1) \quad \text{(Earnshaw)}.$$

It may be of interest to consider further a few examples of (17). Still assuming Boyle's law, let us suppose that the piston is at rest ($\mathrm{X} = 0$) until $\mathrm{T} = 0$, and then moves with uniform acceleration (g), so that (T+)

$$\mathrm{U} = g\mathrm{T}, \qquad \mathrm{X} = \tfrac{1}{2} g\mathrm{T}^2 = \mathrm{U}^2/2g.$$

The use of these in (17) gives

$$x = \frac{a^2}{2g} + (\mathrm{U} + a)t - \frac{(\mathrm{U} + a)^2}{2g}, \tag{18}$$

showing that the relation between x and U is parabolic. In (18) we see that $dx/d\mathrm{U}$ vanishes, when $t = (\mathrm{U} + a)/g$. Thus, if g be positive, *i.e.* if the wave be one of condensation, discontinuity sets in at the front after an interval, reckoned from the beginning of the motion, equal to a/g. But if g be negative, there is no discontinuity, and (18) remains valid for an indefinite time.

In general, as in the last example (g+), the discontinuity sets in locally at one point of the velocity-curve, while other parts are temporarily exempt. It is of interest to inquire under what law the piston must advance so as to generate a linear velocity-curve. For then, under the adiabatic law, which includes Boyle's, a velocity-curve, once linear, remains linear, and if discontinuity enters, it must affect the whole curve simultaneously.

It may be worth while to pause here for a moment to inquire what law of pressure is implied in the permanence of the linear character of the velocity-curve. By (7)

$$\rho\frac{du}{d\rho}=\sqrt{\left(\frac{dp}{d\rho}\right)};$$

so that if $u+\sqrt{(dp/d\rho)}$ is a linear function of u, $du/d\log\rho$ must also be a linear function of u. This requires that

$$\rho du/d\rho = C\rho^n,$$

where C and n are constants, and the most general relation between p and ρ consistent with requirements is

$$p = A + B\rho^\gamma, \tag{19}$$

where A, B, γ, are constants. The relation (19) may be regarded as a kind of generalised adiabatic law; it includes the special law (16) under which a velocity-curve is absolutely permanent in type.

Supposing the motion to commence at T = 0, we have

$$X=\int_0^T U dT = UT - \int_0^U T dU,$$

and hence for (17), under the supposition of Boyle's law,

$$x = -\int T dU - (U+a)t - aT; \tag{20}$$

and the question before us is so to determine T as a function of U that (20) may be linear in U. From (20) when t is constant

$$\frac{dx}{dU} = -T + t - a\frac{dT}{dU} = t_0, \text{ say},$$

where t_0 is a constant, whence

$$T = t - t_0 + He^{-U/a},$$

H being the constant of integration. But, since U - 0 when T = 0, this assumes the form

$$T = T'(1 - e^{-U/a}), \tag{21}$$

T′ being written for $t - t_0$. Or, if we express U in terms of T,

$$U = -a\log\left(1-\frac{T}{T'}\right). \tag{22}$$

In (21), (22), T′ is positive, if U is positive; and U becomes infinite when T = T′. We must therefore regard the law as limited to values of T less than T′.

From (22) we find

$$X = \int_0^T U dT = a\left\{(T' - T)\log\left(1 - \frac{T}{T'}\right) + T\right\}, \tag{23}$$

which completely expresses the motion of the piston. The corresponding velocity-curve at time t may be verified by means of (20), (21). It is expressed by

$$U = \frac{at - x}{T' - t}, \tag{24}$$

exhibiting the linear character of the slope of velocity. Evidently the slope becomes vertical throughout when $t = T'$.

In the above example it is not necessary to suppose the law of motion of the piston continued up to $T = T'$. On the contrary, we may imagine U to increase up to some prescribed finite value and then to remain constant. In this case the slope expressed by (24) forms the transition between $U = 0$ for values of x greater than at and the finite value at which the acceleration of the piston stops.

If the wave be one of rarefaction we must take T' negative, say $-T''$. In this case the analogue of (22) shows that U constantly increases in numerical value but does not become infinite in any finite time. The analogue of (24) is

$$U = \frac{x - at}{t + T''}, \tag{25}$$

representing a slope which ever grows easier as time passes.

The problem also admits of solution when the gas follows the adiabatic law (9). As in (15), (20),

$$x = X + \{a + \tfrac{1}{2}(\gamma + 1)U\}(t - T); \tag{26}$$

and (26) is to satisfy the condition of making dx/dU constant. In this

$$dX/dU = U.dT/dU,$$

so that

$$\frac{dT}{dU}\{(\gamma - 1)U + 2a\} + (\gamma + 1)(T - t) = \text{const.}$$

On integration we obtain, under the condition that U and T vanish together,

$$\left(1 - \frac{T}{T'}\right)\left[1 + \frac{\gamma - 1}{2a}U\right]^{\frac{\gamma+1}{\gamma-1}} = 1, \tag{27}$$

T′ being a constant which, if positive, corresponds to U = ∞, thereby determining U as a function of T.

For the value of $X = \int_0^T U dT$, we have

$$\frac{\gamma-1}{2a}\frac{X}{T'} = -\frac{\gamma+1}{2}\left[\left(1-\frac{T}{T'}\right)^{\frac{2}{\gamma+1}}-1\right]-\frac{T}{T'}$$

$$= \frac{\gamma-1}{2}-\frac{\gamma+1}{2}\left(1+\frac{\lambda-1}{2a}U\right)^{-\frac{2}{\gamma-1}}+\left(1+\frac{\lambda-1}{2a}U\right)^{-\frac{\gamma+1}{\gamma-1}}. \tag{28}$$

It may be observed that, although U is infinite when T = T′, X remains finite. Using this in (26), we get finally, on reduction,

$$x = at - \tfrac{1}{2}(\gamma + 1)U(T' - t), \tag{29}$$

or

$$U = \frac{2}{\gamma+1}\frac{at-x}{T'-t}. \tag{30}$$

If $x < at$, U is a linear function of x, and (T′ being positive) the slope of the velocity curve increases until it becomes vertical when $t = T'$.

If T′ is negative, the wave is one of rarefaction, and (30) applies however great t may be.

By putting $\gamma = 1$ we fall back from (30) to (24), and less simply from (28) to (23).[8]

Riemann's work[9] is of somewhat later date than Earnshaw's, but in one important respect is more general. It may be convenient to recall the principal result.

Taking p a given function of ρ, say $\phi(\rho)$, and putting X = 0, we have from (1) and (2)

$$\frac{du}{dt}+u\frac{du}{dx} = -\phi'(\rho)\frac{d\log\rho}{dx}, \qquad \frac{d\log\rho}{dt}+u\frac{d\log\rho}{dx} = -\frac{du}{dx}.$$

---

[8] I have since found that this problem was successfully treated by Hugoniot.

[9] 'Göttingen Abhandlungen.' 1860, vol. 8.

If the second of these equations be multiplied by $\pm\surd\{\phi'(\rho)\}$ and be added to the first, we find

$$\frac{dr}{dt} = -\{u + \surd\phi'(\rho)\}\frac{dr}{dx}, \qquad \frac{ds}{dt} = -\{u - \surd\phi'(\rho)\}\frac{ds}{dx}, \tag{31}$$

where

$$2r = f(\rho) + u, \qquad 2s = f(\rho) - u \tag{32}$$

and

$$f(\rho) = \int \surd\phi'(\rho).d\log\rho. \tag{33}$$

From these follow

$$dr = \frac{dr}{dx}\left[dx - \{u + \surd\phi'(\rho)\}dt\right]$$
$$ds = \frac{ds}{dx}\left[dx - \{u - \surd\phi'(\rho)\}dt\right]$$

so that r remains constant when x and t change in such a manner that $dx = \{u + \surd\phi'(\rho)\}dt$, and s remains constant when x and t change so that $dx = \{u - \surd\phi'(\rho)\}dt$. In the case of a positive progressive wave $s = 0$, whence $f(\rho) = u$ and also $r = u$. The velocity with which u travels in such a wave is accordingly $u + \surd(dp/d\rho)$, of which fact (11) is merely another form of statement. Riemann's equations are more general than anything previously given, as not limited to a single progressive wave.

Since Riemann's equations do not seem to have been applied in any example of continuous motion, I have thought it worth while to inquire whether they can be satisfied when r and s are both linear functions of x, Boyle's law being assumed, so that in (32), (33)

$$\surd\phi'(\rho) = a, \qquad f(\rho) = a\log\rho.$$

If we suppose

$$r = \mathrm{A}x + \mathrm{B}, \qquad s = \mathrm{C}x + \mathrm{D}, \tag{34}$$

we obtain, on substitution in (31), equations for the determination of A, B, C, D, as functions of the time. In the first instance we find

$$\mathrm{A} - \mathrm{C} = 1/t, \tag{35}$$

in which to t a constant may be added, and further

$$A = \frac{H+1}{2t}, \qquad C = \frac{H-1}{2t}, \tag{36}$$

when H is an arbitrary constant. Also

$$B - D = -aH + L/t, \tag{37}$$

L being another arbitrary constant, and thence

$$B = \tfrac{1}{2}a\left(H^2 - 1\right)\log t + \frac{L(H+1)}{2t} + \tfrac{1}{2}M, \tag{38}$$

$$D = \tfrac{1}{2}a\left(H^2 - 1\right)\log t + \frac{L(H-1)}{2t} + \tfrac{1}{2}N, \tag{39}$$

with

$$M - N = -2aH. \tag{40}$$

If we allow the origin of x to be arbitrary, as well as that of t, we may write

$$2r = (H + 1)x/t + (H^2 - 1)a \log t + M, \tag{41}$$

$$2s = (H - 1)x/t + (H^2 - 1)a \log t + N, \tag{42}$$

$$u = r - s = x/t - aH, \tag{43}$$

$$a \log \rho = r + s = Hx/t + (H^2 - 1)a \log t + \tfrac{1}{2}(M + N). \tag{44}$$

If $H = \pm 1$, the logarithmic term disappears, and either r or s is constant. In these cases we fall back upon single progressive waves.

If $H = 0$, $u = x/t$ in (43), and (44) gives $1/\rho$ proportional to t. The density is thus uniform with respect to x, but the volume of a given mass grows proportionally with t. The uniform expansion occurs in such a manner that the gas remains unmoved at the origin of co-ordinates. Since in this case, $du/dt + u du/dx = 0$, we see that every part of the gas moves with unaccelerated velocity.

Waves of Permanent Regime

When waves are propagated in one dimension without change of type, the circumstances are dynamically the same as in steady motion, as appears at once by impressing on the system a velocity equal and opposite to that of wave-propagation. The problem may conveniently be considered under this form.

From the general equation of continuity (2), by making $d\rho/dt$ equal to zero, or independently, we have

$$\rho u = \rho_0 u_0 = m, \tag{45}$$

where m is a constant which Rankine called the mass-velocity. The dynamical equation (1) reduces to

$$u\frac{du}{dx} + \frac{1}{\rho}\frac{dp}{dx} = \mathrm{X}. \tag{46}$$

If X = 0, (46) may be written

$$\int_{p_0}^{p} \frac{dp}{\rho} = \tfrac{1}{2}u_0^2 - \tfrac{1}{2}u^2, \tag{47}$$

where u_0 is the velocity corresponding to p_0.

Eliminating u, we get

$$\int_{p_0}^{p} \frac{dp}{\rho} = \tfrac{1}{2}u_0^2\left(1 - \frac{\rho_0^2}{\rho^2}\right), \tag{48}$$

determining the law of pressure under which alone it is possible for a stationary wave to maintain itself in fluid moving (outside the wave) with velocity u_0. From (48)

$$\frac{dp}{d\rho} = u_0^2\,\frac{\rho_0^2}{\rho^2}, \tag{49}$$

or

$$p + m^2/\rho = p_0 + m^2/\rho_0 \tag{50}$$

the law found by Earnshaw.

Since, under the adiabatic law, the relation between density and pressure differs from (50), we conclude that a self-maintaining stationary aerial wave is an impossibility, unless it be in virtue of impressed forces, or of viscosity, or other dissipative agencies not now regarded.

When the changes of density concerned are *small*, (50) may be satisfied approximately; and we see from (49) that the velocity of the stream (outside the wave) necessary to keep the wave stationary is given by

$$u_0 = \surd(dp/d\rho),$$

which is the same as the velocity of the wave reckoned relatively to the fluid at a distance.

This way of regarding the subject shows, perhaps more clearly than any other, the nature of the relation between velocity and density. In a stationary wave-form a loss of velocity accompanies an augmented density, according to the principle of energy, and therefore the fluid composing the condensed parts of a wave moves forward more slowly than the disturbed portions. Relatively to the fluid at a distance, the motion of the condensed parts is in the same direction as that in which the waves travel.

By means of (46), we can find what impressed force is required in order to ensure a stationary wave-form when (50) is not satisfied. For example, if $p = a^2\rho$, we find from (45), (46),

$$\mathrm{X} = u\frac{du}{dx} + a^2\frac{d\log\rho}{dx} = \left(u^2 - a^2\right)\frac{d\log u}{dx}, \tag{51}$$

showing that an impressed force is necessary at every place where u is variable and unequal to a. In (51) X is the accelerating force so called. The actual force operative upon the element of mass $\rho\, dx$ is $\mathrm{X}\rho\, dx$. Thus, on integration,

$$\int \mathrm{X}\rho\, dx = m\int\left(1 - \frac{a^2}{u^2}\right)du = m\left(u_2 - u_1\right)\left\{1 - \frac{a^2}{u_1u_2}\right\}, \tag{52}$$

if the range of integration extend from the place where the velocity is u_1 to the place where it becomes equal to u_2. The integral applied force vanishes if the terminal velocities are such that their geometric mean is a. We may apply this to the case of a velocity-curve giving a simple gradual transition from one constant velocity u_1 to another velocity u_2. Under the above condition the integral force vanishes, but finite forces are required at all points of the slope, except the particular point where $u = a$.

It is of some importance to notice that although, under the condition $u_1u_2 = a^2$, the applied forces contribute on the whole no momentum, yet they do contribute energy, positive or negative. To find the work done in unit of time by the forces we have

$$2/m.\int \mathrm{X}\rho\,.\,u\,.\,dx = u_2^2 - u_1^2 - 2u_1u_2\log\left(u_2/u_1\right). \tag{53}$$

The better to interpret this let us suppose that u_1 and u_2 are positive, and in the first instance that $u_2>u_1$, so that the fluid passes from a less to a greater velocity, or by (45) from a greater to a less density. In the case of a wave of rarefaction we have therefore to consider the sign of

$$y^2 - 1 - 2y\log y, \tag{54}$$

when $y>1$. It is not difficult to prove that this sign is always positive. When $y-1$ is small, the approximate value of (54) is $\frac{1}{3}(y-1)^3$, and is therefore positive when $y>1$. Again, if we remove the positive factor y from (54) and then differentiate, we obtain $(1 - 1/y)^2$, which is positive. Hence, when $y>1$, (54) is necessarily positive. The propagation of the wave of rarefaction without change of type requires that the impressed forces, contributing on the whole no momentum, should nevertheless do work upon, *i.e.* communicate energy to, the gas.

In like manner, if the wave be one of condensation, *i.e.*, if the gas passes from a less to a greater density, the operation of the impressed forces is to remove energy from the gas forming the wave. It follows that although dissipative forces, such as those arising from viscosity, may possibly constitute a machinery capable of maintaining the type of a wave of condensation, in no case can they maintain the type of wave of rarefaction.

It is desirable to extend this argument to waves propagated under the adiabatic law. In general, from (45), (46),

$$\mathrm{X}\rho = m\frac{du}{dx} + \frac{dp}{dx};$$

so that $\displaystyle\int \mathrm{X}\rho dx = m(u_2 - u_1) + p_2 - p_1 = m^2/\rho_2 - m^2/\rho_1 + p_2 - p_1.$

As in (50), the condition that on the whole no momentum is communicated is

$$m^2/\rho_2 - m^2/\rho_1 + p_2 - p_1 = 0. \tag{55}$$

Again,

$$m^{-1}.\int \mathrm{X}\rho\,.\,u\,.\,dx = \tfrac{1}{2}\left(u_2^2 - u_1^2\right) + \int_{p_1}^{p_2}\frac{dp}{\rho} = \int_{p_1}^{p_2}\frac{dp}{\rho} - \frac{p_2 - p_1}{2}\left(\frac{1}{\rho_2} + \frac{1}{\rho_1}\right). \tag{56}$$

The question in which we are interested is the sign of (56). If we regard v (the volume of unit mass) viz., $1/\rho$, as the ordinate, and p as the abscissa of a curve, the first term on the right of (56) represents the area of the curve bounded by two ordinates and the axis of p, while the second term is what the area would be if the ordinate retained throughout the mean of the terminal values.

So far the argument is general. If the relation between p and v be adiabatic, and $p_2>p_1$, the expression (56) is negative, since v proportional to $p^{-1/\gamma}$ makes

d^2v/dp^2 positive.[10] The final pressure exceeding the initial pressure denotes a wave of condensation, and we conclude, as before, that maintenance of type in such a wave requires removal of energy from the wave, while in the contrary case of a wave of rarefaction additional energy would need to be supplied.

The problem now under discussion is closely related to one which has given rise to serious difference of opinion. In his paper of 1848 already referred to, Stokes considered the *sudden* transition from one constant velocity to another, and concluded that the necessary conditions for a permanent regime could be satisfied. Results equivalent to his may be deduced from (45) in connection with the condition ($u_1u_2 = a^2$) already found from (52) to express that there is no change of momentum on the whole. Thus,

$$u_1 - a\sqrt{(\rho_2/\rho_1)}, \qquad u_2 - a\sqrt{(\rho_1/\rho_2)}. \tag{57}$$

Similar conclusions were put forward by Riemann in 1860 (*loc. cit.*). Commenting on these results in the "Theory of Sound" (1878), I pointed out that although the conditions of *mass* and *momentum* were satisfied, the condition of *energy* was violated, and that therefore the motion was not possible; and in republishing this paper[11] Stokes admitted the criticism, which had indeed already been made privately by Kelvin. On the other hand, Burton[12] and H. Weber[13] maintain, at least to some extent, the original view.

Inasmuch as they ignored the question of energy, it was natural that Stokes and Riemann made no distinction between the cases where energy is gained or lost. As I understand, Weber abandons Riemann's solution for the discontinuous wave (or bore, as it is sometimes called for brevity) of rarefaction, but still maintains it for the case of the bore of condensation. No doubt there is an important distinction between the two cases; nevertheless, I fail to understand how a loss of energy can be admitted in a motion which is supposed to be subject to the isothermal or adiabatic laws, in which no dissipative action is contemplated. In the present paper the discussion proceeds upon the supposition of a gradual transition between the two velocities or densities. It does not appear how a solution which violates mechanical principles, however rapid the transition, can become valid when the transition is supposed to become absolutely abrupt. All that I am able to admit is that under these circumstances dissipative forces (such as viscosity) that are infinitely small may be competent to produce a finite effect.

---

[10] Compare Lamb's 'Hydrodynamics,' § 280.

[11] 'Collected Works,' vol. 2, p. 55.

[12] 'Phil. Mag.' 1893, vol. 35, p. 316.

[13] 'Die Partiellen Differentialgleichungen der Mathematischen Physik,' Braunschweig, 1901, vol. 2, p. 496.

If we suppose that under the influence of small dissipative forces the bore of Stokes and Riemann can be propagated, at least approximately, we naturally inquire whether it can be regarded as the complete outcome of the simple progressive wave with a straight velocity slope which, as we have found, tends after a definite interval of time to assume the character of a bore. It would seem that the answer must be in the negative. Taking Boyle's law, we recognise from (5) that in the progressive wave, just before the formation of the bore, the relation between the velocities and densities is

$$u_2 - u_1 = a \log \rho_2 - a \log \rho_1,$$

while in (57) the relation is

$$u_2 - u_1 = a\sqrt{(\rho_2/\rho_1)} - a\sqrt{(\rho_1/\rho_2)}.$$

The two functions of ρ on the right, which are independent of any common addition to u_1 and u_2, cannot be identified (unless $\rho_2 = \rho_1$), as we have found already in discussing (54). This incompatibility may be regarded as a confirmation of Stokes' opinion that something of the nature of reflection must ensue.

Permanent Regime Under the Influence of Dissipative Forces

The first investigation to be considered under this head is a very remarkable one by Rankine "On the Thermodynamic Theory of Waves of Finite Longitudinal Disturbance."[14] which (except a limited part expounded by Maxwell in his "Theory of Heat") has been much neglected.[15] Conduction of heat is here for the first time taken into account and although there are one or two serious deficiencies, not to say errors, presently to be noticed, the memoir marks a very definite advance.

The first step is the establishment of an equation equivalent (when the wave is reduced to rest) to (45), and of Earnshaw's relation (50), in which equations we shall usually substitute v, the volume of unit mass, for $1/\rho$. Rankine remarks that "no substance yet known fulfills the condition expressed by (50) between finite limits of disturbance, at a constant temperature, nor in a state of non-conduction of heat (called the *adiabatic* state). In order, then, that permanency of type may be possible in a wave of longitudinal disturbance, there must be both

---

[14] 'Phil. Trans.,' 1870, vol. 160, Part II, p. 277.

[15] I must take my share of the blame. Rankine is referred to by Lamb ('Hydrodynamics,' 1906, p. 466). The body of Rankine's memoir seems to have been composed without acquaintance with the writings of his predecessors; but in a supplement he notices the work of Poisson, Stokes, Airy, and Earnshaw.

change of temperature and conduction of heat during the disturbance." However, we shall see later that even under Boyle's law *viscosity* is competent to endow a wave with permanency.

The question is, how can Earnshaw's law be satisfied? Obviously not (in the absence of viscosity) if the expansions are adiabatic; but if at every stage the right quantity of heat is added or subtracted, the gas may be made to follow any prescribed law. This is the idea underlying Rankine's investigation. For the unit mass of a perfect gas we have, as usual, $pv = \mathrm{R}\theta$, θ denoting absolute temperature. The condition of the gas is defined by any two of the three quantities p, v, θ, and the third may be expressed in terms of them. The relation between simultaneous variations is

$$d\theta/\theta = dp/p + dv/v. \tag{58}$$

In order to effect the changes specified by dp and dv, it is in general necessary to communicate heat to the gas. Calling the necessary quantity of heat $d\mathrm{Q}$, we may write

$$d\mathrm{Q} = \left(\frac{d\mathrm{Q}}{dv}\right)dv + \left(\frac{d\mathrm{Q}}{dp}\right)dp. \tag{59}$$

Suppose now (a) that $dp = 0$. Equations (58), (59), give

$$\frac{d\mathrm{Q}}{d\theta}(p \text{ constant}) = \left(\frac{d\mathrm{Q}}{dv}\right)\frac{v}{\theta},$$

where $d\mathrm{Q}/d\theta$ (p constant) expresses the specific heat of the gas under constant pressure. Denoting this by C, we have

$$\mathrm{C} = \left(\frac{d\mathrm{Q}}{dv}\right)\frac{v}{\theta}.$$

Again, suppose (b) that $dv = 0$. We find in a similar manner that if c denote the specific heat under constant volume,

$$c = \left(\frac{d\mathrm{Q}}{dp}\right)\frac{p}{\theta}.$$

Thus, in general,

$$d\mathrm{Q}/\theta = \mathrm{C}dv/v + c\,dp/p. \tag{60}$$

If between (58) and (60) we eliminate dp, there results

$$d\mathrm{Q} = (\mathrm{C} - c)\frac{pdv}{\mathrm{R}} + c\,d\theta. \tag{61}$$

In (61) $dQ = 0$ corresponds to adiabatic expansion, when according to Mayer's principle the cooling effect $-\, cd\theta$ is equal to the external work done by the gas $p\, dv$. Hence

$$C - c = R, \tag{62}$$

and therefore

$$\gamma = \frac{C}{c} = \frac{C}{C - R}, \tag{63}$$

a relation discovered by Rankine himself in 1850.

Rankine then applies (60) to find what heat must be communicated in order that the gas may follow Earnshaw's law making $dp = -m^2 dv$. With regard to (62), it appears that

$$dQ = \frac{dp}{m^2(\gamma - 1)}\left\{p_0 + m^2 v_0 - (\gamma + 1)p\right\}. \tag{64}$$

It will be understood that under the condition now imposed of Earnshaw's relation, as well as of the ordinary gas law, there remains but one independent variable, and that the state of the gas may be expressed in terms of any *one* of three quantities p, v, θ.

We have next to consider how far the necessary supply of heat defined by (64) can be effected by conduction. If the initial state (distinguished by suffix 1) and final state (with suffix 2) be of uniformity with respect to x, the total quantity of heat received by the gas during its passage must be zero, or $\int dQ = 0$. Hence from (64) Rankine finds

$$p_0 + m^2 v_0 = \tfrac{1}{2}(\gamma + 1)(p_1 + p_2). \tag{65}$$

This is a necessary condition; but of course there is nothing so far to show that it is sufficient.

In (65) p_0, v_0 are any corresponding values of p and v within the wave, and we may identify them with p_1, v_1.[16]

Thus

$$m^2 v_1 = \tfrac{1}{2}(\gamma - 1)p_1 + \tfrac{1}{2}(\gamma + 1)p_2. \tag{66}$$

The velocity u_1, equal to mv_1, is that with which the wave advances relatively to the fluid in state (1). And

---

[16] We may, of course, also identify p_0, v_0 with p_2, v_2.

$$u_1^2 = m^2 v_1^2 = v_1\{\tfrac{1}{2}(\gamma-1)p_1 + \tfrac{1}{2}(\gamma+1)p_2\} \tag{67}$$

gives the square of the velocity of wave-propagation relatively to fluid (1). The velocity of propagation of infinitely small disturbances (p_2 nearly equal to p_1) is given by $u_1{}^2 = \gamma p_1 v_1$; and thus a wave of finite condensation is propagated faster than an infinitesimal wave, and according to (67) a wave of finite rarefaction would be propagated slower than an infinitesimal wave. Moreover, there is no limit to the velocity of a wave of condensation.

Rankine proceeds to express the absolute temperature (θ) at a point where the pressure is p in a wave of permanent type. By Earnshaw's law (50) in combination with (65)

$$\frac{\theta}{\theta_0} = \frac{pv}{p_0 v_0} = \frac{p}{p_0} \cdot \frac{(\gamma+1)(p_1+p_2)-2p}{(\gamma+1)(p_1+p_2)-2p_0}, \tag{68}$$

and for the ratio of terminal temperatures

$$\frac{\theta_2}{\theta_1} = \frac{p_2}{p_1} \cdot \frac{(\gamma+1)p_1+(\gamma-1)p_2}{(\gamma+1)p_2+(\gamma-1)p_1}. \tag{69}$$

The second fraction on the right of (69) obviously represents the ratio of volume v_2/v_1, or of densities ρ_1/ρ_2.

In order to justify (65), it is not necessary that the terminal states be states of absolute uniformity. It will suffice that the temperature be there stationary ($d\theta/dx = 0$), which secures that no conduction of heat takes place there, and a state of stationary temperature usually involves a stationary pressure. To make the most of (65) we must apply it to the smallest ranges, *i.e.* between consecutive places where dp/dx vanishes.

But here a question arises which Rankine does not seem to have considered. In order to secure the necessary transfers of heat by means of conduction it is an indispensable condition that the heat should pass from the hotter to the colder body. If maintenance of type be possible in a particular wave as the result of conduction, a reversal of the motion will give a wave whose type cannot be so maintained. We have seen reason already for the conclusion that a dissipative agency can serve to maintain the type only when the gas passes from a less to a more condensed state. If this be so, the application which Rankine makes to a periodic wave is evidently prohibited.

According to the *second* law of thermodynamics, the criterion whether the transformation is possible as the result of dissipative action is the sign of $\int dQ/\theta$. If this be negative, the transformation is not possible. From (64) with use of (65)

$$dQ = \frac{(\gamma+1)dp}{2m^2(\gamma-1)}(p_1 + p_2 - 2p). \tag{70}$$

In (68) we may give p_0, θ_0 any corresponding values found in the wave. Thus p_0 lies between p_1 and p_2, and (γ>1)

$$(\gamma + 1)(p_1 + p_2) - 2\,p_0 \text{ is positive.}$$

Accordingly $\int dQ/\theta$ takes the same sign as

$$\int_{p_1}^{p_2} \frac{dp(p_1 + p_2 - p)}{p\{(\gamma+1)(p_1 + p_2) - 2p\}}. \tag{71}$$

The integral (71) is evaluated without difficulty. Dropping the factor $1/(\gamma + 1)$, and writing $\varpi = p_2/p_1$, we get

$$\log \varpi + \gamma \log \frac{\gamma + 1 + (\gamma-1)\varpi}{\gamma - 1 + (\gamma+1)\varpi}. \tag{72}$$

It is evident that (72) changes sign when we substitute $1/\varpi$ for ϖ.

If we expand (72) in powers of $(\varpi - 1)$ we find

$$(72) = \frac{(\gamma^2-1)(\varpi-1)^3}{12\gamma^2} + \ldots,$$

the terms in $(\varpi - 1)$, $(\varpi - 1)^2$, disappearing. Thus, when $\varpi - 1$ is positive, (72) begins positive. Differentiating (72) with respect to ϖ, we get

$$\frac{1}{\varpi} + \frac{\gamma(\gamma-1)}{\gamma+1+(\gamma-1)\varpi} - \frac{\gamma(\gamma+1)}{\gamma-1+(\gamma+1)\varpi}. \tag{73}$$

When (73) is reduced to a single fraction, the denominator is positive, and the numerator is

$$(\gamma^2 - 1)(\varpi - 1)^2.$$

We infer that if $\varpi > 1$, (72) is always positive, and that if $\varpi < 1$, (72) is always negative. Hence if $p_2 > p_1$, *i.e.* if the wave be one of condensation, the communications of heat required are such as may arise from conduction; but if the wave be one of rarefaction, its permanency can in no wise be attained as the result of conduction. A wave of condensation here means a wave such that during its progress the gas passes always from a less dense to a more dense state, and the most important case is when the limits are finite, so that the passage constitutes the transition from one uniform density to a greater uniform density.

Rankine proceeds to examine more particularly under what conditions a wave can be permanent. "In order that a particular type of disturbance may be capable of permanence during its propagation, a relation must exist between the temperatures of the particle and their relative positions, such that the conduction of heat between the particles may effect the transfers of heat required by the thermodynamic conditions of permanence of type."

The equation of conduction is readily found. The heat conducted in unit time across a layer of the gas is represented by $k\ d\theta/d$, where k is a coefficient of conductivity which may be a function of the condition of the gas, here dependent on one variable. The equation of conduction is ($u = +$)

$$v\frac{d}{dx}\left(k\frac{d\theta}{dx}\right) = \frac{\mathrm{D}}{\mathrm{D}t}\mathrm{Q} = u\frac{d\mathrm{Q}}{dx} = mv\frac{d\mathrm{Q}}{dx},$$

whence, if we reckon Q from the initial condition of constant pressure p_1,

$$k\frac{d\theta}{dx} = m\mathrm{Q}. \tag{74}$$

And from (70)

$$\mathrm{Q} = \frac{\gamma+1}{2m^2(\gamma-1)}\int_{p_1}^{p}(p_1+p_2-2p)dp = \frac{\gamma+1}{2m^2(\gamma-1)}(p-p_1)(p_2-p). \tag{75}$$

And from (50), (65),

$$p + m^2 v = \tfrac{1}{2}(\gamma+1)(p_1+p_2), \tag{76}$$

whence

$$\theta = \frac{pv}{\mathrm{R}} = \frac{p}{2m^2\mathrm{R}}\{(\gamma+1)(p_1+p_2)-2p\}, \tag{77}$$

and

$$\frac{d\theta}{dp} = \frac{(\gamma+1)(p_1+p_2)-4p}{2m^2\mathrm{R}}. \tag{78}$$

Using these, we find with regard to (62)

$$dx = \frac{k}{mQ}\frac{d\theta}{dp}dp = \frac{k\,dp}{mc(\gamma+1)}\frac{(\gamma+1)(p_1+p_2)-4p}{(p-p_1)(p_2-p)}, \tag{79}$$

by which is determined the distribution of pressure (and thence of density and temperature) along the line of propagation.

On the supposition that k is constant, Rankine integrates (79) in terms of logarithms. Writing

$$p-\tfrac{1}{2}(p_1+p_2)=q, \qquad \tfrac{1}{2}(p_2-p_1)=q_1,$$

he obtains

$$\frac{dx}{dq} = \frac{k}{mc(\gamma+1)}\frac{(\gamma-1)(p_1+p_2)-q}{q_1^2-q^2},$$

and

$$x = \frac{k}{mc(\gamma+1)}\left\{\frac{(\gamma-1)(p_1+p_2)}{2q_1}\log\frac{q_1+q}{q_1-q}+2\log\left(1-\frac{q^2}{q_1^2}\right)\right\}, \tag{80}$$

x being measured from the place where $q = 0$. Mathematically the wave is infinitely long; but practically the transition of pressure is effected in a distance comparable with $k/mc(\gamma + 1)$, which may be small in terms of ordinary standards. It is to be observed that the general character of the result does not depend upon the constancy of k.

Reverting to (79), we see that the denominator on the right is positive, and that the numerator is also positive for that part of the wave where p is nearly equal to $\tfrac{1}{2}(p_1+p_2)$. Thus, for this part of the wave at any rate, p and x increase together; or, since u is positive, the gas passes to a condition of greater density—the wave must be one of condensation. This consideration, as we have seen, Rankine overlooked. And a further limitation presents itself: since there cannot be two pressures in one place, it is evident that dx/dp must not change sign. The numerator in (79) must be positive over its whole range from p_1 to p_2, and this will not be the case if p_2/p_1 exceed $(\gamma + 1)(3 - \gamma)$, equal for common gases to 1·61. The conclusion is that the only kind of wave, involving a transition from one uniform pressure to another, which can be maintained with the aid of conduction is a wave of condensation, and then only when the ratio of pressures does not exceed a moderate value.

The next contribution to the subject upon which I have to comment is contained in a long and ably written memoir by Hugoniot.[17] This author, though he covers to a great extent the same ground, makes no reference to Stokes, Earnshaw, Riemann, or Rankine, and but a very slight one to Poisson—a circumstance which increases the difficulty of comparison. Since Hugoniot uses the Lagrangian form of equation, his investigation runs naturally on the same lines as Earnshaw's, whose general solution for a single progressive wave is reproduced. I have already alluded to the solution of special problems relating to the propagation of a wave of variable type.

The most original part of Hugoniot's work has been supposed to be his treatment of discontinuous waves involving a sudden change of pressure, with respect to which he formulated a law often called after his name by French writers. But a little examination reveals that this law is *precisely the same* as that given 15 years earlier by Rankine, a fact which is the more surprising inasmuch as the two authors start from quite different points of view. Rankine's investigation, as we have seen, is expressly based upon conduction of heat in the gas, but Hugoniot supposes his gas to be non-conducting. A question of some delicacy is here involved, which will repay careful examination. It will be convenient to give a paraphrase of Hugoniot's argument.[18]

The argument depends upon an application of the principle of energy to a region bounded by two fixed planes, including the place of discontinuity. The work done by the fluid as it emerges with volume v_2 against the pressure p_2 is p_2v_2. On the whole, therefore, the external work done by the passage of the unit of mass is $p_2v_2 - p_1v_1$. The increase of kinetic energy of the fluid is

$$\tfrac{1}{2}\left(u_2^2 - u_1^2\right) = \tfrac{1}{2}m^2\left(v_2^2 - v_1^2\right) = \tfrac{1}{2}\left(v_2 + v_1\right)\left(p_1 - p_2\right),$$

in virtue of (50), which requires that

$$p_1 - p_2 + m^2\left(v_1 - v_2\right) = 0. \tag{81}$$

The sum of these is

$$p_2v_2 - p_1v_1 + \tfrac{1}{2}\left(p_1 - p_2\right)\left(v_2 + v_1\right) = \tfrac{1}{2}\left(v_2 - v_1\right)\left(p_2 + p_1\right). \tag{82}$$

We have next to consider the internal energy of unit of mass in the initial and final states. For this purpose we suppose the gas to expand adiabatically from its actual volume v to an infinite volume. In this expansion the work done by the gas is

---

17 'Journal de l'École Polytechnique,' 1887, 1889.

18 Compare Lamb's 'Hydrodynamics,' 1906, § 280.

$$\int_v^\infty p dv = \frac{pv}{\gamma - 1}, \tag{83}$$

so that the difference of internal energy in the two states is

$$\frac{p_2 v_2}{\gamma - 1} - \frac{p_1 v_1}{\gamma - 1}. \tag{84}$$

The principle of energy requires that the sum of this and (82) be zero, whence

$$\gamma = \frac{(p_2 - p_1)(v_1 + v_2)}{(p_1 + p_2)(v_1 - v_2)} \tag{85}$$

is the relation between the pressures and volumes in the two states. The result thus found by Hugoniot is the same as Rankine's. From Rankine's equation (65)

$$p_1 + m^2 v_1 = p_2 + m^2 v_2 = \tfrac{1}{2}(\gamma + 1)(p_1 + p_2) = \tfrac{1}{2}(p_1 + p_2) + \tfrac{1}{2} m^2 (v_1 + v_2),$$

it follows that

$$m^2 = \gamma \frac{p_1 + p_2}{v_1 + v_2} = \frac{p_1 - p_2}{v_1 - v_2}, \tag{86}$$

which is identical with (85).

The first remark that I will make is that, although Hugoniot assumes that the transition between the two states is sudden, there is nothing in his argument which requires this, all that is really necessary being that the *régime* is permanent. The next remark is that, however valid (85) may be, its fulfillment does not secure that the wave so defined is possible. As a matter of fact, a whole class of such waves is certainly impossible, and I would maintain, further, that a wave of the kind is never possible under the conditions, laid down by Hugoniot, of no viscosity or heat-conduction.

A closer examination of the process by which (85) was obtained will show that while the first law of thermodynamics has been observed, the second law has been disregarded. The crux of the matter lies in the comparison of the internal energies of the incoming and outgoing gas expressed in (84). If (p_2, v_2) and (p_1, v_1) lie upon the same adiabatic, the work corresponding to the passage from one state to the other is given without ambiguity by (84). But in the present case the two states do not lie upon the same adiabatic, and the work required is deduced upon the assumption that nothing is involved in the passage at $v = \infty$ from one adiabatic to the other. What is actually there required is the communication (positive or negative) of an infinitesimal quantity of heat. From the point of view of the first law the infinitesimal quantity of heat may be neglected, but not so from the point of view of the second law, since the transfer is supposed to take place at the zero of temperature. When heat and work are distinguished, infinitesimal heat at zero may have a finite value. The imaginary passage to infinity has the advantage of leading rapidly to the required conclusion, but it rather tends to obscure the real nature of the process. While all the other items of the account are mechanical work, the passage from one adiabatic to the other (which may take place at constant finite volume) is a question of *heat* as distinguished from work. If during a complete cycle work would be lost and corresponding heat gained, the operation is dissipative and there need be no contradiction if viscosity or heat-conduction enter, but the opposite contingency of a gain of work at the expense of heat is excluded in all cases. The conclusion is the same as before. While a wave of condensation may, perhaps, maintain a permanent regime as the result of dissipative agencies, a permanent wave of rarefaction is excluded.

It is remarked by Hugoniot that even when the ratio p_1/p_2 is indefinite, v_2/v_1 does not exceed $(\gamma+1)/(\gamma-1)$, which for common gases is equal to about 6. A similar remark is made by Duhem,[19] who discusses the whole question with great generality. With regard to perfect gases "lorsqu'une quasi-onde de choc se propage au sein d'un gas parfait, le fluide le plus condensé est toujours en amont de l'onde et le fluide le moins condensé en aval." But, so far as I see, neither of these authors proves that the propagation is possible in any case.

It is a question of great interest to inquire what is the influence of viscosity and especially whether alone, or in co-operation with heat-conduction, it allows a wave of condensation to acquire a permanent regime. We proceed to consider this question on the basis of the usual equations, although it must be admitted that their application to conditions which are somewhat extreme raises points of uncertainty.

Reverting to our original equations, we recognise that (45) is unaffected by the inclusion of viscosity, and that the change required in (46) is represented by writing[20]

$$\mathrm{X} = \frac{4}{3\rho}\frac{d}{dx}\left(\mu\frac{du}{dx}\right),$$

so that (46) takes the form

$$m\frac{du}{dx} + \frac{dp}{dx} - \frac{4}{3}\frac{d}{dx}\left(\mu\frac{du}{dx}\right) = 0,$$

whence ($v = 1/\rho$)

[19] 'Zeitschrift f. Physikal. Chem.,' 1909, vol. 69, p. 169.
[20] Lamb's 'Hydrodynamics,' §§ 314, 316.

$$p + m^2 v - \frac{4}{3} m\mu \frac{dv}{dx} = p_1 + m^2 v_1 = p_2 + m^2 v_2, \tag{87}$$

the terminal states (p_1, v_1), (p_2, v_2), being of uniformity, so that dv/dx there vanishes. From this it appears that (81), relating to the terminal states, holds good equally when viscosity is regarded.

A simple example under the head of viscosity is to suppose the temperature maintained uniform, as by a powerful radiation, so that the gas follows Boyle's law, making

$$pv = p_1 v_1 = p_2 v_2 = a^2.$$

From this and (87) we get

$$m^2\, v_1 v_2 = a^2,$$

and

$$p_1 + m^2 v_1 = a^2/v_1 + a^2/v_2.$$

Using these in (87), we find

$$\frac{3mdx}{4\mu} = -\frac{vdv}{(v_1 - v)(v - v_2)}, \tag{88}$$

as governing the distribution of v along the line of propagation. In a wave of condensation $v_1 > v > v_2$, so that the denominator on the right of (88) is positive. Thus when m is positive, dv/dx is negative, as should be the case. On integration (μ constant)

$$\frac{3mx}{4\mu} = \frac{1}{v_1 - v_2} \{v_1 \log(v_1 - v) - v_2 \log(v - v_2)\}, \tag{89}$$

the origin of x being chosen suitably.

The transition of volumes from v_1 to v_2 occupies, mathematically speaking, the whole range from $x = -\infty$ to $x = +\infty$, but practically it may be very sudden. Since in (88) dx/dv never changes sign, the condition of permanency for a condensational wave can always be satisfied, whatever may be the value of the ration v_1/v_2[21] or p_1/p_2, contrasting in this respect with the limitation found to be necessary on Rankine's conclusion relative to heat-conduction.

[21] But the limitation pointed out by Hugoniot still obtains; otherwise, one of the pressures would be negative.

As regards the velocity of wave propagation into the rarer medium, we have for its square

$$u_1^2 = m^2 v_1^2 = a^2 v_1/v_2. \tag{90}$$

Returning to the case where heat development and viscosity are both regarded, we see that in virtue of (81) Hugoniot's reasoning is still applicable without change, and it leads to the same final relation (85) as was found by Rankine when heat-conduction is alone considered.

In endeavouring to apply Rankine's method to the more general case where viscosity is retained, we shall find it more convenient to treat v, or $(1/\rho)$, rather than p, as independent variable. If, as before, dQ denotes the total quantity of heat received by unit mass of the gas, we have from (60), (62),

$$(\gamma - 1)\frac{dQ}{dx} = \gamma p\frac{dv}{dx} + v\frac{dp}{dx};$$

or, on elimination of p by means of (87),

$$(\gamma - 1)\frac{dQ}{dx} = \gamma\left\{p_1 + m^2 v_1 - m^2 v + \tfrac{4}{3} m\mu\frac{dv}{dx}\right\}\frac{dv}{dx}$$
$$+ v\left\{-m^2\frac{dv}{dx} + \tfrac{4}{3} m\frac{d}{dx}\left(\mu\frac{dv}{dx}\right)\right\}. \tag{91}$$

In (91) dQ consists of two parts, the first (dQ_1), with which alone Rankine dealt, the heat received by conduction, and the second (dQ_2) the heat developed internally under viscosity. As regards the latter, the heat developed in volume v and time dt is $\frac{4}{3} v\mu(du/dx)^2\, dt$,[22] in which we are to replace dt by dx/u, and u by mv, so that

$$\frac{dQ_2}{dx} = \tfrac{4}{3} m\mu\left(\frac{dv}{dx}\right)^2. \tag{92}$$

Multiplying this by $(\gamma - 1)$ and subtracting it from (91), we get

$$(\gamma - 1)\frac{dQ_1}{dx} = \gamma\left(p_1 + m^2 v_1\right)\frac{dv}{dx} - (\gamma + 1)m^2 v\frac{dv}{dx} + \tfrac{4}{3} m\frac{d}{dx}\left(\mu v\frac{dv}{dx}\right). \tag{93}$$

As in Rankine's investigation, the whole heat received by conduction in passing from one uniform state v_1 to another uniform state v_2 must vanish.

[22] Lamb's 'Hydrodynamics,' § 341.

Hence, on integrating between these limits, and dividing out the factor $(v_2 - v_1)$, we have

$$(\gamma+1)m^2(v_1+v_2)=2\gamma(p_1+m^2v_1)=2\gamma(p_2+m^2v_2)$$
$$=\gamma(p_1+p_2)+\gamma m^2(v_1+v_2),$$

or, as in (86),

$$m^2(v_1+v_2)=\gamma(p_1+p_2),$$

the same relation as was found by Rankine. Introducing it into (93), we get

$$(\gamma-1)\frac{dQ_1}{dx}=\tfrac{1}{2}(\gamma+1)m^2\left\{(v_1+v_2)\frac{dv}{dx}-\frac{dv^2}{dx}\right\}+\tfrac{4}{3}m\frac{d}{dx}\left(\mu v\frac{dv}{dx}\right). \tag{94}$$

A particular case arises when we suppose the conductivity to be zero, so that dQ_1/dx vanishes throughout. We have then

$$\tfrac{4}{3}\mu v\frac{dv}{dx}+\tfrac{1}{2}(\gamma+1)m(v_1-v)(v-v_2)=0, \tag{95}$$

differing from (88) only by the factor $\frac{1}{2}(\gamma+1)$. On the supposition that μ is constant, the solution is nearly the same as in (89), and, in fact, reduces to it when $\gamma = 1$, which represents Boyle's law. This case of no conduction is thus satisfactorily disposed of. Whatever be the ratio of pressures, a wave of condensation is always possible.

It should be remarked, however, that the supposition of constant μ does consist with the facts as known for actual gases when γ differs from unity. For such gases viscosity, though independent of *density*, varies with *temperature*, so that μ will not be constant in (95). But since μ is always positive, this complication merely affects the particular form of the integral and not the general conclusion as to the possibility of a permanent wave.

In general, from (94), if we reckon Q_1 from the terminal state v_1,

$$(\gamma-1)Q_1=\tfrac{1}{2}(\gamma+1)m^2(v_1-v)(v-v_2)+\tfrac{4}{3}m\mu v\frac{dv}{dx}. \tag{96}$$

The equation of conduction is the same (74) as before. And for θ, from (87),

$$\theta=\frac{pv}{R}=\frac{v}{R}\left\{\frac{\gamma+1}{2\gamma}m^2(v_1+v_2)-m^2v+\tfrac{4}{3}m\mu\frac{dv}{dx}\right\};$$

so that with regard to (62) the equation of conduction becomes

$$\frac{k}{mc}\left[m\frac{dv}{dx}\left\{\frac{\gamma+1}{2\gamma}(v_1+v_2)-2v\right\}+\tfrac{4}{3}\frac{d}{dx}\left(\mu v\frac{dv}{dx}\right)\right]$$
$$=\tfrac{1}{2}(\gamma+1)m(v_1-v)(v-v_2)+\tfrac{4}{3}\mu v\frac{dv}{dx}. \tag{97}$$

By omitting the terms containing μ we may of course fall back on Rankine's problem.

Equation (97), in its general form, is much more complicated than when either viscosity or heat-conduction is alone regarded, in consequence of the occurrence of the differential coefficient of the second order. In general, both k and μ are functions of temperature, and therefore of v; but, according to Maxwell's theory, which assumes a molecular repulsion inversely as the fifth power of the distance $c\mu/k$ is independent of temperature (as well as of density), and takes the value $\frac{2}{5}$. And it would seem that this independence of temperature and density is general, seeing that the ratio is of no dimensions, at least so long as the repulsive force can be represented by an inverse power of the distance.[23] We shall write h for the above ratio and assume that for a given gas it is an absolute constant. Thus, μ' being written for μ/m, (97) takes the form—

$$\mu'\frac{d}{dx}\left(\mu'\frac{dv^2}{dx}\right)+\mu'\frac{dv^2}{dx}\left\{\frac{3(\gamma+1)}{8\gamma}\frac{v_1+v_2}{v}-\tfrac{3}{2}-h\right\}=\tfrac{3}{4}h(\gamma+1)(v_1-v)(v-v_2); \tag{98}$$

in which v^2 may be regarded as the independent variable. For v^2 we shall write ξ, and if

$$\mathrm{U}=\mu' d\xi/dx, \tag{99}$$

our equation, since it contains x only through dx, may be reduced to one of the first order in U and ξ, *i.e.*,

$$\mathrm{U}\frac{d\mathrm{U}}{d\xi}+\mathrm{U}f(\xi)=\mathrm{F}(\xi), \tag{100}$$

where

$$f(\xi)=\frac{3(\gamma+1)}{8\gamma}\frac{\sqrt{\xi_1}+\sqrt{\xi_2}}{\sqrt{\xi}}-\tfrac{3}{2}-h, \tag{101}$$

and

$$F(\xi)=\tfrac{3}{4}h(\gamma+1)\left(\sqrt{\xi_1}-\sqrt{\xi}\right)\left(\sqrt{\xi}-\sqrt{\xi_2}\right). \tag{102}$$

---

[23] Compare 'Roy. Soc. Proc.,' 1900, vol. 66, p. 68; 'Scientific Papers,' vol. 4, p. 453.

If U can be found as a function of ξ from (100), x follows by simple integration of (99).

In considering equation (100) we may conveniently regard ξ as the linear co-ordinate of a material particle of unit mass moving in a straight line with velocity U. The first term, $\mathrm{U}d\mathrm{U}/d\xi$, then represents the acceleration of the particle; the second, $\mathrm{U}f(\xi)$, may be regarded as a *resistance*, proportional to the velocity, and at the same time variable with the position (ξ); and the third term on the right hand represents a force, which is also a function of position. If t be the time in this subsidiary problem, $\mathrm{U} = d\xi/dt$, and (100) may be written

$$\frac{d^2\xi}{dt^2} + f(\xi)\frac{d\xi}{dt} = \mathrm{F}(\xi), \tag{103}$$

while by (99)

$$dx = \mu' dt. \tag{104}$$

If μ' be constant, the substitution of $\mu' t$ for x in (98) is obvious.

If we take $h = 0{\cdot}4$, $\gamma = 1{\cdot}41$, (101), (102), become

$$f(\xi) = 0\cdot 641\frac{\sqrt{\xi_1}\sqrt{\xi_2}}{\sqrt{\xi}} - 1\cdot 900, \tag{103}$$

$$\mathrm{F}(\xi) = 0\cdot 723\left(\sqrt{\xi_1} - \sqrt{\xi}\right)\left(\sqrt{\xi} - \sqrt{\xi_2}\right). \tag{104}$$

It will be observed that, over the range from ξ_1 to ξ_2, $\mathrm{F}(\xi)$ is positive, but that the sign of $f(\xi)$ is doubtful. If ξ has the greater terminal value ξ_1, f is negative; but, when it has the smaller terminal value ξ_2, the sign depends upon the ratio ξ_1/ξ_2. If this ratio $<1{\cdot}21$, f is negative; otherwise it is positive.

I suppose that a complete analytical solution of our equation is not to be expected, and it is, indeed, hardly necessary for our purpose. What we most wish to know is whether a solution is possible which satisfies the prescribed conditions. Among these is the requirement that U in (100) vanish at both limits; and even then the manner of evanescence must be such as to secure that x, as determined by (99), shall be infinite at these limits. As the problem originally presents itself, we should have the representative particle travelling in the negative direction from ξ_1 to ξ_2, starting with no velocity and arriving with no velocity. It seems simpler to consider it in a modified form, *i.e.* with the motion reversed, so that it takes place in the direction of the force F. There is, then, no question of the particle stopping between the limits and returning upon its course. We may make this change, if in (105) we reverse the sign of f. We consider, then, the motion of the particle to be in the positive direction, from ξ_2 to ξ_1, with zero velocity at both limits, the motion between ξ_2 and ξ_1 being aided by the force F, which itself vanishes at these limits, and being also subject to a force of the nature of resistance, proportional to velocity. When ξ_1/ξ_2 does not exceed 1·21, the force is a resistance in the ordinary sense, *i.e.* it opposes the motion, and, in any case, it has this character near (and beyond) the arrival end ξ_1. But when ξ_1/ξ_2 exceeds 1·21, the force becomes what we may call a counter-resistance, and aids the motion near the initial end ξ_2. As regards F, in the neighbourhood of each limit it becomes a force proportional to distance therefrom, repulsive near ξ_1, and attractive near ξ_2. Thus, when ξ is nearly equal to ξ_2,

$$\mathrm{F}(\xi) = 0\cdot723\frac{\sqrt{\xi_1} - \sqrt{\xi_2}}{2\sqrt{\xi_2}}(\xi - \xi_2); \tag{107}$$

and when ξ is nearly equal to ξ_1,

$$\mathrm{F}(\xi) = 0\cdot723\frac{\sqrt{\xi_1} - \sqrt{\xi_2}}{2\sqrt{\xi_1}}(\xi_1 - \xi). \tag{108}$$

The particle, starting from ξ_2, is bound to go through to ξ_1. If it arrives at ξ_1 with zero velocity, we shall have, presumably, a solution of our problem. It is possible, however, that on first arrival at ξ_1, it may pass through, and only settle down after a number of oscillations. To this there does not appear to be any objection; but if on the return from ξ_1 it overshoots ξ_2, it can never again return to ξ_1, since on the left of ξ_2 the sign of f is negative; and then our problem has no solution. On the other hand, from the nature of F, it is not possible for the particle passing through ξ_1 in the positive direction to escape returning.

The character of the start from ξ_2 can be investigated with the aid of approximate equations. Thus, in (100) we may treat $f(\xi)$ as constant, say 2α, where α may be either positive or negative, and take, as in (107), $\mathrm{F}(\xi) = \beta(\xi - \xi_2)$, where β is positive, so that

$$\mathrm{U}\frac{d\mathrm{U}}{d\xi} + 2\alpha\mathrm{U} - \beta(\xi - \xi_2) = 0. \tag{109}$$

If in (109) we assume

$$\mathrm{U} = \lambda(\xi - \xi_2), \tag{110}$$

we find that the equation is satisfied provided that

$$\lambda = -\alpha \pm \sqrt{(\alpha^2 + \beta)}, \tag{111}$$

one value (λ_1) being positive and one (λ_2) negative. In the present case, where U must be positive when $\xi > \xi_2$, λ_1 is to be chosen.

The differential equation (109) can be made homogeneous, and its general solution,[24] when λ is real, can be put into the form

$$\frac{\{U-\lambda_1(\xi-\xi_2)\}^{\lambda_1}}{\{U-\lambda_2(\xi-\xi_2)\}^{\lambda_2}}=C, \tag{112}$$

when C is an arbitrary constant. This solution, of course, covers the cases where the particle starts from ξ_2 with a finite velocity (U_0), and it appears that $C=U_0^{\lambda_1-\lambda_2}$. We might conclude from this that when $U_0 = 0$, then $C = 0$, but the conclusion is not safe. If, however, U and $\xi - \xi_2$ are of the same order of magnitude, we may write $U = r(\xi - \xi_2)$, where r is not infinite. Substituting this in (112), we get

$$(r-\lambda_1)^{\lambda_1}\cdot(r-\lambda_2)^{-\lambda_1}\cdot(\xi-\xi_2)^{\lambda_1-\lambda_2}=C. \tag{113}$$

When $\xi - \xi_2$ vanishes, the third factor on the left is zero, and, since the first and second factors are not infinite, the conclusion follows that $C = 0$. This takes us back to (110), (111), the second solution (involving λ_2) relating to the case where U is negative, the motion being one of approach from the positive side to ξ_2.

These conclusions may be arrived at more easily from (103), of which the general solution in the present case is

$$\xi-\xi_2=Ae^{\lambda_1 t}+Be^{\lambda_2 t}, \tag{114}$$

giving

$$U=\lambda_1 Ae^{\lambda_1 t}+\lambda_2 Be^{\lambda_2 t}.$$

From these we may deduce

$$\begin{aligned} U-\lambda_1(\xi-\xi_2)&=(\lambda_2-\lambda_1)Be^{\lambda_2 t},\\ U-\lambda_2(\xi-\xi_2)&=(\lambda_1-\lambda_2)Be^{\lambda_1 t}; \end{aligned} \tag{115}$$

whence (112) follows by elimination of t. For our present purpose, $\xi - \xi_2$ is to vanish when $t = -\infty$, so that $B = 0$; and (110) follows with $\lambda = \lambda_1$. It will be remarked that (110) makes x, as determined by (99), infinite when $\xi = \xi_2$. The circumstances of the start from ξ_2 are thus definite and suitable. The question is as to the arrival at ξ_1.

[24] See, for example, Boole's 'Differential Equations,' p. 33.

In the neighbourhood of ξ_1 the approximate equation is

$$U\frac{dU}{d\xi}+2\alpha' U-\beta'(\xi-\xi_1)=0, \tag{116}$$

where α' is positive and, by (108), β' negative. If now

$$U=\lambda'\ (\xi-\xi_1), \tag{117}$$

the values of λ' are

$$\lambda' = -\alpha' \pm \sqrt{(\alpha'^2+\beta')}; \tag{118}$$

so that both values, if real, are negative. On the supposition of reality, (112) retains its form (with ξ_1 for ξ_2). If the velocity of arrival (U_0) be finite, $C = U_0^{\lambda_1'-\lambda_2'}$, as before; and it might be supposed that, if U_0 vanishes when $\xi - \xi_1 = 0$, C would have to vanish or become infinite. Such a conclusion would be incorrect. If in (113) we suppose λ'_1 to be numerically the smaller of the two values, the third factor indeed vanishes with $(\xi - \xi_1)$ as before; but the conclusion that $C = 0$ is evaded ultimately $r = \lambda'_1$.

The situation is most easily understood from the solution in terms of t as in (114), (115). Since λ'_1, λ'_2 are *both* negative, the condition that $\xi - \xi_1$ and U shall vanish together when $t = \infty$ is satisfied, whatever may be the values of A and B. There are now an infinite number of possible types of solution, instead of only one as in the former case. And it appears that the two simple types included under (117) are not at all upon an equal footing. Except in the particular case where $A = 0$, the solution always tends ultimately to the form $U = \lambda_1(\xi - \xi_1)$, and of course it may assume this form throughout. All these solutions satisfy the condition as to the infinitude of x when $U = 0$.

Whether the values λ' be real or not, the particle must ultimately settle down at ξ_1, unless it escape from the region to which the approximate equation applies. For in (118) the *real part* of is always negative.

Returning to (100) in its general form, let us consider the variation of U for a given ξ as dependent upon variations in f and F. We have

$$U\frac{d\delta U}{d\xi}+\delta U\frac{dU}{d\xi}+f.\delta U+U.\delta f-\delta F=0,$$

or

$$\frac{d\delta \mathrm{U}}{d\xi} + \mathrm{P}\delta\mathrm{U} - \mathrm{Q} = 0, \tag{119}$$

where P and Q are supposed to be known functions of ξ, viz.,

$$\mathrm{P} = \frac{f + d\,\mathrm{U}/d\xi}{\mathrm{U}}, \qquad \mathrm{Q} = \frac{\delta\mathrm{F} - \mathrm{U}.\delta f}{\mathrm{U}}. \tag{120}$$

The solution of the linear equation (119) is

$$\delta\mathrm{U} = e^{-\int \mathrm{P}d\xi}\left(\int e^{\int \mathrm{P}d\xi}\mathrm{Q}d\xi + c\right). \tag{121}$$

If Q = 0, δU has the same sign as c, so that an increment of velocity communicated at any point remains throughout of the same sign. Again, if Q be throughout of one sign, δU, as dependent upon it, has the same sign. For example, if U be positive over the range considered, δf positive, and δF negative, the δU is certainly negative. The increments δf, δF may be local, vanishing over any part of the range.

The application to the present problem is obvious. If the particle passing any point between ξ_2 and ξ_1, with velocity U, arrives at ξ_1 for the first time without velocity, it will still arrive at ξ_1 without velocity (it must in any case arrive, since F is positive), if U be diminished, or if f be increased, or if F be diminished, or if all these changes occur together. And in the limit, when ξ_1 is closely approached, the ratio of U to $(\xi_1 - \xi)$ is in general the same.

By use of this principle we may assure ourselves as to the possibility of a solution in certain cases where ξ_1/ξ_2 does not greatly exceed unity. We imagine a simplified problem which admits of analytical solution and is derived from the actual one by alterations which everywhere (over the range from ξ_2 to ξ_1) increase F and diminish f. If this modified problem admits of the solution required, *a fortiori* will the original problem do so.

If we consider the curve which according to (106) represents F as a function of ξ, we see that it is concave downwards, and that F will everywhere be increased if we substitute for the curve the two terminal tangents at ξ_2 and ξ_1, whose equations are given (107), (108). The abscissa of K, the point of intersection, is $\xi = \sqrt{(\xi_2\xi_1)}$. (Fig. 1.)

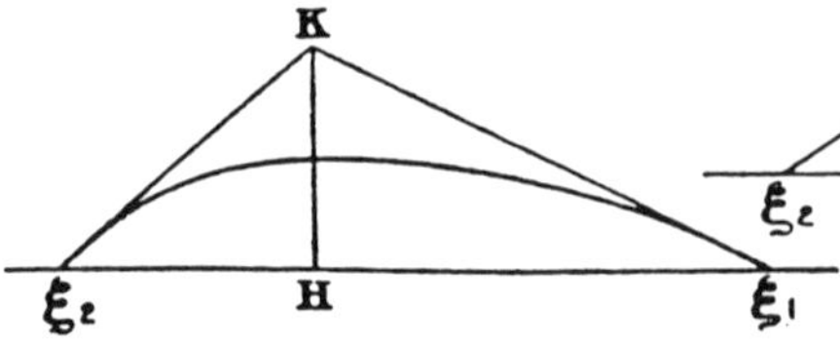

FIG. 1.

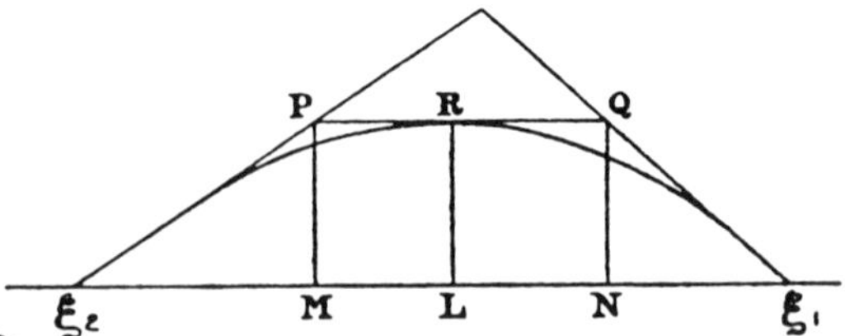

FIG. 2.

As regards f, its value is given by (105) with sign reversed, and is diminished when ξ is diminished. The changes will therefore be in the required direction if we represent f over ξ_2H by its value at ξ_2,

$$f_2 = 2\alpha = 1{\cdot}259 - 0{\cdot}641s \tag{122}$$

and from H to ξ_1 by its value at H, viz.,

$$f_1 = 2\alpha' = 1\cdot900 - 0\cdot641\left(s^{\frac{1}{2}} + s^{-\frac{1}{2}}\right), \tag{123}$$

if for brevity we write s for $\surd\ (\xi_1/\xi_2)$, so that s is the ration of terminal densities in the original problem.

As regards the first portion of the course, the solution already given (110), (111), determines the value of U on arrival at H. We have

$$\mathrm{U} = \lambda_1\xi_2(s-1), \quad \text{where} \quad \lambda_1 = \surd(\alpha^2 + \beta) - \alpha,$$

in which α is given by (122) while by (107) $\beta = 0{\cdot}361(s - 1)$. Using these, we get

$$\mathrm{U} = (s-1)\xi_2\ \surd\{(0{\cdot}630 - 0{\cdot}320s)^2 + 0{\cdot}361\ (s-1)\} - 0{\cdot}630 + 0{\cdot}320s]. \tag{124}$$

If $s = 1 + \sigma$, where σ is small, (124) becomes

$$\mathrm{U} = \sigma\xi_2 \times \frac{\beta}{2\alpha} = 0\cdot584\xi_2\sigma^2. \tag{125}$$

As regards the second portion of the course, the appropriate solution is provided by (117), (118), where λ' is restricted to be real. And in accordance with our suppositions α' is given by (123) while from (108) $\beta' = -0{\cdot}361(1 - s^{-1})$. In choosing between the values of λ' we are at liberty to take that which gives the largest value of U at H consistent with U = 0 at ξ_1, viz.,

$$\lambda'_2 = -\alpha' - \surd(\alpha'^2 + \beta').$$

Thus at H

$$U = (s^2 - s)\,\xi_2\,\{\alpha' + \sqrt{(\alpha'^2 + \beta')}\}, \tag{126}$$

or approximately, in terms of σ (supposed small),

$$U = \sigma\xi_2(0{\cdot}618 + 0{\cdot}033\sigma). \tag{127}$$

From (125), (127) we may infer that when σ is small, *i.e.* when s does not much exceed unity, the particle, starting from ξ_2, arrives at H with a velocity small enough to admit of its being stopped on arrival at ξ_1, even under the simplifying conditions that have been imposed, and therefore *a fortiori* under the actual conditions of the problem. Hence when ξ_1 does not too much exceed ξ_2, a wave of permanent regime is possible as the result of viscosity and heat-conduction.

But the range of ξ_1/ξ_2 thus proved admissible is rather severely limited. The postulated reality of λ' requires that $\alpha'^2 + \beta'$ be positive, and this again requires that $s<1{\cdot}34$. For values of s greater than this the motion in the simplified problem would become oscillatory. Calculation shows that, for $s = 1{\cdot}34$, (124) gives

$$U = (s-1)\xi_2 \times 0{\cdot}20,$$

while (126) gives

$$U = (s-1)\xi_2 \times 0{\cdot}43;$$

so that up to this limit the particle starting from ξ_2 in the simplified problem, and therefore also in the actual problem, would arrive at ξ_1 with zero velocity. Up to a ratio of densities equal to $1{\cdot}34$, the wave of permanent regime is certainly possible.

The next step in this line of procedure will be to replace the curve representing F by the broken line ξ_2PQξ_1 formed by three of its tangents, of which two are the same terminal tangents as before, while the third, PQ, may be taken to be the horizontal tangent parallel to $\xi_1\xi_2$. (Fig.2.) By (106) the point L where F is a maximum is determined by

$$\sqrt{\xi} = \tfrac{1}{2}\left(\sqrt{\xi_1} + \sqrt{\xi_2}\right) = \tfrac{1}{2}\sqrt{\xi_2}\,(s+1),$$

and the corresponding value of F is

$$0{\cdot}1875\xi_2(s-1)^2.$$

The abscissæ of the points of intersection of the of the horizontal tangent with the terminal tangents are given by (107), (108). For M

$$\xi = \tfrac{1}{2}\xi_2(s+1),$$

and for N

$$\xi = \tfrac{1}{2}\xi_2\left(s^2+s\right).$$

As for f we are to take along ξ_2M the value at ξ_2; along MN the value at M; and along Nξ_1 the value at N, as given by (105), with sign unchanged.

For the two terminal portions the solutions are of the same form as before, but for the middle portion a new form is required. Making F constant in (100) and writing $2\alpha''$ for f, we find on integration

$$-4\alpha''^2\xi = \mathrm{F}\log(\mathrm{F} - 2\alpha''\mathrm{U}) + 2\alpha''\mathrm{U} + \mathrm{C}, \tag{128}$$

where C is an arbitrary constant, to be determined so as to suit the velocity with which the particle arrives at M.

As before, the limiting value of s is determined by the consideration that, for our purpose, the arrival at ξ_1 must not be oscillatory. We get $s = 1{\cdot}633$ about, and with this value of s it is not difficult to show that the velocity of arrival at N is below the value prescribed by the solution for Nξ_1. The necessary conditions are thus fulfilled and we infer that the wave of uniform regime is possible so long as $s<1{\cdot}63$. Although this ratio is moderate, it exceeds that found admissible in Rankine's problem, which leaves viscosity out of account. We there found that the greatest admissible ratio of pressures was 1·61, which by (68) corresponds to 1·40 for the ratio of densities. Of course, in Rankine's problem the solution definitely fails at this point, while in the present problem all that we have so far proved is that the limit exceeds 1·63.

From the low limiting values of s found necessary in these two cases in order to secure the reality of the roots of (118), it might be inferred that the reality would fail for the limiting motion in the neighbourhood of ξ_1, when s had a considerable value, but such is not the case. If we use the value of f from (105) appropriate to the terminal point ξ_1 itself, we have

$$2\alpha' = 1{\cdot}900 - 0{\cdot}641\,(1 + s^{-1}) = 1{\cdot}259 - 0{\cdot}641 s^{-1};$$

and

$$\beta' = -0{\cdot}3615\,(1 - \mathrm{s}^{-1}).$$

Thus, even when $s = \infty$,

$$\alpha'^2 + \beta' = 0{\cdot}3969 - 0{\cdot}3615 = +0{\cdot}0048;$$

and the roots are always real. Hence, subsidence at ξ_1 is not ultimately oscillatory, but there is nothing in this argument to exclude a finite number of oscillations before subsidence into the region governed by the approximate equation.

The method of approximation already followed might be pushed further, but it seems preferable to use the general method of numerical calculation for the solution of differential equations as formulated by Runge.[25] The equation is that numbered (100), in which f (whose sign is to be reversed) and F are given by (105), (106). If we write $U' = U/\xi_2$, $\xi' = \xi/\xi_2$, our equation takes the form

$$\frac{dU'}{d\xi'} = \frac{0\cdot 723}{U'}\left(s - \sqrt{\xi'}\right)\left(\sqrt{\xi'} - 1\right) - 1\cdot 900 + \frac{0\cdot 642(1+s)}{\sqrt{\xi'}}. \tag{129}$$

The value of s being given, it is required to trace the connection between U' and ξ', simultaneous values being denoted respectively by a and b. If a receive the increment h, we have to calculate the corresponding increment k for b. If we call the function on the right of (129) ϕ (ξ', U'), Runge gives as a first approximation to k

$$k_1 = \phi\left(a + \tfrac{1}{2}h, b + \tfrac{1}{2}\phi_0.h\right)h, \tag{130}$$

where $\phi_0 = \phi(a, b)$. The next approximation is

$$k = k_1 + \tfrac{1}{3}\left(k_2 - k_1\right), \tag{131}$$

where

$$k_2 = \tfrac{1}{2}\left(k' + k'''\right) \quad \text{and} \quad k' = \phi_0.h,$$

$$k'' = \phi\left(a + h, b + k'\right)h, \qquad k''' = \phi\left(a + h, b + k''\right)h. \tag{132}$$

Having determined the new simultaneous values $a + h$, $b + k$, we make a fresh departure therefrom, and so trace out the function step by step.

In the present application the starting point is $a = 1$ (*i.e.* $\xi = \xi_2$), $b = 0$, and we have to trace the function until $\xi' = s^2$. The initial value of ϕ is to be found from (129) by putting $\xi' = 1$. Writing $U' = \phi_0(\xi' - 1)$, we get

$$\phi_0^2 + \{1\cdot 900 - 0\cdot 641(1+s)\}\phi_0 - \tfrac{1}{2} \times 0\cdot 723(s-1) = 0, \tag{133}$$

of which the positive root is to be chosen.

[25] See Forsyth's 'Differential Equations,' p. 51.

The extreme admissible value of s in our present problem is 6, and the first and rather elaborate calculation that I have made relates to this case. From (133) $\phi_0 = 3{\cdot}1591$, so that taking $a = 1$, $b = 0$, $h = 1$, we have $k' = 3{\cdot}1591$. Calculating from (13) we find $k_1 = 2{\cdot}255$, and from (132)

$$k'' = 1{\cdot}7076, \qquad k''' = 2{\cdot}0772,$$

making $k_2 = 2{\cdot}6182$. Hence the correction to k_1, viz., $\frac{1}{3}(k_2 - k_1)$, is equal to $0{\cdot}121$, and $k = 2{\cdot}376$. Thus, corresponding to $\xi' = 2$, we get $U' = 2{\cdot}376$. The following are the values of U′ obtained successively in this way:—

ξ'.	U′.	ξ'.	U′.	ξ'.	U′.
1	0·000	14	7·502	33	1·2942
2	2·376	18	6·807	34	0·8532
3	3·903	22	5·674	35	0·4160
4	4·986	26	4·241	35 1/2	0·2043
6	6·380	30	2·610	35 3/4	0·1012
10	7·526	32	1·736	36	0·0004

The correction to k_1 is everywhere subordinate. In the last step from $35\frac{3}{4}$ to 36 no correction to k_1 is applied.

There is little difficulty in tracing by this method and with full accuracy the final progress to zero when $\xi' = 36$. If any doubt be left, it may be removed by applying the former method to the course from 35 to 36, using the value of f appropriate to 35 and the terminal tangent as the representative of the curve for F. From this it appears that even if U′ at ξ' were as great as 0·7468, the moving particle could not pass $\xi' = 36$. The conclusion is that even in this extreme case of $s = 6$ the solution exists, and that a wave of permanent regime is possible. Further, from (99) we see that since U and μ' are both positive, $d\xi/dx$ is positive throughout, and the transition from the one density to the other takes place *without alternation*.

After what has been proved little doubt could remain but that a solution is possible when s has any value lower than 6. I have, however, thought it desirable to add a rough calculation (rough on account of the relative magnitude of the steps) for the case of $s = 3$:—

ξ'......	1	2	3	5	7	8	8 1/2	9
U′......	0·00	0·90	1·19	1·26	0·82	0·40	0·19	0·01

It is a question of some importance to consider what is the thickness of the transitional layer in the waves of uniform regime which have been proved to be possible. Mathematically speaking, the transition occupies an infinite space; but if we understand the expression to refer to a transition approximately complete, the thickness involved is finite, and indeed extremely small. Reference to (98) shows that x is of the order μ', or μ/m, for $\mu/\rho u$, where u is the velocity of the wave. For the present purpose we may take u as equal to the usual velocity of sound, *i.e.* 3×10^4 cm. per second. For air under ordinary conditions the value of μ/ρ in C.G.S. measure is 0·13; so that x is of the order $\frac{1}{3}\times 10^{-5}$ cm. That the transitional layer is in fact extremely thin is proved by such photographs as those of Boys, of the aerial wave of approximate discontinuity which advances in front of a modern rifle bullet; but that according to calculation this thickness should be well below the microscopic limit may well occasion surprise.

Resistance to Motion Through Air at High Velocities

According to the adiabatic law the pressures and velocities in a compressible fluid free from external force, see (47), are related by

$$\left(\frac{p_2}{p_1}\right)^{\frac{\gamma-1}{\gamma}} = 1 + \frac{\gamma-1}{2}\frac{u_1^2 - u_2^2}{a_1^2}, \tag{134}$$

in which p_1, ρ_1, u_1 denote the pressure, density, and velocity at one point of the path; p_2, ρ_2, u_2 the corresponding quantities at another point. Also $a_1{}^2 = \gamma p_1/\rho_1$, so that a_1 is the velocity of infinitesimal disturbances in the condition (1). In an early paper[26] I suggested the application of this formula to bodies moving through air at high velocities. Regarding the obstacle as stationary and the fluid in motion with velocity u_1 and pressure p_1, the pressure p_2 corresponding to the loss of this velocity is given by putting $u_2 = 0$ in (133), a_1 being the ordinary velocity of sound. This is the pressure which should obtain at the axial point on the nose of a symmetrical bullet, and although this value in strictness represents the maximum pressure, the analogy of an incompressible fluid suggests that the mean pressure on a flat surface would not be greatly inferior. But in a recent discussion,[27] Mr. Mallock has shown that this formula immensely over-estimates the resistance actually experienced by a bullet, and (so far as I am aware) the discrepancy remains unexplained.

If indeed the adiabatic law really prevailed throughout, there could be no escape from the conclusion formulated. A consideration of the photographs by Boys[28] will suggest the required explanation. At a short constant distance in front of the bullet there is an aerial bore, or place of approximate discontinuity. Along the axis, the fluid moving up to the bullet changes its density, and therefore

[26] 'Phil. Mag.,' 1876, vol. 2, p. 430; 'Scientific Papers,' vol. 1, p. 289.

[27] 'Roy. Soc. Proc.,' A, 1907, vol. 79, p. 266.

[28] 'Nature,' 1893, vol. 47, p. 440. The particular photograph reproduced by Mallock does not exhibit well the feature in question.

pressure and temperature, *suddenly*, so that there is here a special opportunity for viscosity and heat-conduction to take effect. The pressures and velocities on the two sides of the bore are related, not according to the adiabatic law, but according to Rankine's law already discussed. The changes which occur may be separated into two stages. The first is the sudden one in which the fluid passes from the atmospheric condition p_0, ρ_0, with velocity u_0 to the condition denoted by p_1, ρ_1, u_1. After passing the bore the fluid changes gradually according to the adiabatic law already stated until at the nose of the bullet the condition is represented by p_2, ρ_2, with $u_2 = 0$.

We are now in a position to calculate the final pressure p_2. For the first stage we have Rankine's formula (67), making

$$\rho_0 u_0^2 = \tfrac{1}{2}(\gamma - 1)p_0 + \tfrac{1}{2}(\gamma + 1)p_1,$$

or, if $a^2 = \gamma p_0/\rho_0$,

$$\frac{p_1}{p_0} = \frac{2\gamma}{\gamma + 1}\frac{u_0^2}{a^2} - \frac{\gamma - 1}{\gamma + 1}, \tag{135}$$

determining the pressure just inside the bore in terms of u_0 (the velocity of the bullet through quiescent air) and a, the ordinary velocity of sound. When $u_0 = a$, $p_1 = p_0$. For values of u_0 less than a, the first stage does not exist and we may suppose $p_1 = p_0$, $u_1 = u_0$.

In the second stage we use (134) with $u_2 = 0$. Thus

$$\left(\frac{p_2}{p_1}\right)^{\frac{\gamma-1}{\gamma}} = 1 + \frac{\gamma - 1}{2}\frac{\rho_1 u_1^2}{\gamma p_1}, \tag{136}$$

in which, by a formula analogous to (66),

$$\rho_1 u_1^2 = \tfrac{1}{2}(\gamma + 1)p_0 + \tfrac{1}{2}(\gamma - 1)p_1.$$

Hence

$$\left(\frac{p_2}{p_0}\right)^{\frac{\gamma-1}{\gamma}} = \frac{(\gamma + 1)^2}{4\gamma}\left(\frac{p_1}{p_0}\right)^{\frac{\gamma-1}{\gamma}}\left\{1 + \frac{\gamma - 1}{\gamma + 1}\frac{p_0}{p_1}\right\}. \tag{137}$$

When $u_0 > a$, p_1/p_0 is to be calculated from(135). When the resulting value is substituted in (137), p_2/p_0 is determined.

When $u_0 < a$, we have simply

$$\left(\frac{p_2}{p_0}\right)^{\frac{\gamma-1}{\gamma}} = 1 + \frac{\gamma-1}{2}\frac{u_0^2}{a^2}. \tag{138}$$

If u_0/a be *small*, (138) reduces to

$$\frac{p_2}{p_0} = 1 + \frac{\rho_0 u_0^2}{2p_0},$$

or

$$p_2 - p_0 = \tfrac{1}{2}\rho_0 u_0^2, \tag{139}$$

as for an incompressible fluid.

When $u_0 = a$, both systems give

$$\left(\frac{p_2}{p_0}\right)^{\frac{\gamma-1}{\gamma}} = \frac{\gamma+1}{2}. \tag{140}$$

When u_0/a is *large*, the second terms on the right of (135) and (137) may be neglected, and we obtain

$$\frac{p_2}{p_0} = \frac{\gamma+1}{2}\frac{u_0^2}{a^2}\left\{\frac{(\gamma+1)^2}{4\gamma}\right\}^{\frac{1}{\gamma-1}}; \tag{141}$$

or, when we put $\gamma = 1{\cdot}41$,

$$p_2/p_0 = 1\cdot 30\, u_0^2/a^2. \tag{142}$$

The following are some corresponding values of p_2/p_0 and u_0/a, calculated from (135), (137), with $\gamma = 1{\cdot}41$:—

u_0/a	1	2	3	4
p_2/p_0	1·90	4·49	11·7	20·7

From this point onwards the approximate formula (142) may suffice. The values found are in good agreement with Mr. Mallock's curve.

The question as to the linear interval between the bore and the nose of the bullet cannot be answered from the results of the present paper. In strictly one-dimensional motion, the bore and the plane wall constituting the obstacle could not move at the same speed.

Taylor
courtesy of University of Cambridge, Cavendish Laboratory

GEOFFREY INGRAM TAYLOR

G. I. Taylor was born on 7 March, 1886, into a family with a distinguished scientific background. His maternal grandfather was George Boole, the logician, and his mother's uncle was Sir George Everest, one of the founders of geodesy and the person after whom Mt. Everest was named. Taylor studied natural science at Trinity College, Cambridge. He received a scholarship for research at Trinity in 1908 and stayed at Cambridge for the rest of his life.

Taylor's original researches were in the field of optics. One of his first experiments (1909) was an investigation of the interference fringes formed by low-intensity light waves. Almost all of his later research was in the field of what we would now call applied mechanics, although he thought of himself as a classical physicist. The work that he did on the conditions for discontinuous motion in gases was communicated to the Royal Society in 1910, and was only the second paper he had published. In 1912 he spent six months aboard the HMS *Scotia*, a scientific research vessel in the North Atlantic, studying the mixing processes occurring in the lower layers of the atmosphere. Taylor's studies of turbulence are considered his most important contributions to the field of science.

He also did important work for the Royal Flying Corps during the First World War. With A. A. Griffith at the Royal Aircraft Factory, Farnborough, he investigated the stress distribution in cylindrical shafts under torsion. This led to his line of thinking concerning the physical processes which limit the strength of materials and his theory of dislocations in metals (1934). This also has important connections to field of shock compression in solids; plastic flow under the large shear stresses within the shock front and the unusual mechanical response upon release from the shocked state can be described in terms of elementary properties of dislocations. The interaction between Taylor and Griffith on the strength of materials points out very strongly the important role that military research can play in post-war scientific direction. The postulation of the Griffith crack and the dislocation helped form the foundations of modern materials science.

During World War II Taylor again served his country as a scientific consultant for war-related projects. He was involved with the Manhattan Project as an expert on blast and shock waves in the air. He was knighted in 1944.

Throughout his long scientific career, Taylor liked to work alone, with at most a technical assistant. He thought of himself as an "amateur" in science. By this he meant that he was not well-versed in modern mathematical techniques (Circa 1930), and always compared his work to that of his grandfather, George Boole, and others like Benjamin Franklin and the mathematician Ramanujan. Taylor died on June 27, 1975 following an extended period of poor health resulting from a stroke in 1972.

The Conditions Necessary for Discontinuous Motion in Gases

G. I. Taylor, B.A., Trinity College, Cambridge, *Proc. Roy. Soc. Lond.* A84, pp. 371-377 (1910).

The possibility of the propagation of a surface of discontinuity in a gas was first considered by Stokes* in his paper "On a Difficulty in the Theory of Sound." This paper begins with a physical interpretation of Poisson's integral of the equation of motion of a gas in one dimension. The integral in question is $w = f\{z-(a+w)t\}$; and it represents a disturbance of finite amplitude moving in a gas for which the velocity of propagation of an infinitesimal disturbance is a; w is the velocity of the gas in the direction of the axis z. It is shown that the parts of the waves in which the velocity of the gas is w travel forward with a velocity $a + w$, and that there is in consequence a tendency for the crests to catch up the troughs. After a certain time, and at a certain point in space, the value of $\partial w/\partial z$ will become negatively infinite; a discontinuity will then occur, and Poisson's integral will cease to apply.

Stokes then leaves the subject of oscillatory waves and proceeds to consider whether it is possible to maintain a sharp discontinuity in a gas which obeys Boyle's law $(p = a^2\rho)$. His argument, slightly modified by Lord Rayleigh, is as follows:

Suppose that a travelling discontinuity can exist. Give the whole gas such a motion that the discontinuity is brought to rest. Consider then a gas which is moving with uniform velocity u_1 up to a discontinuity. At this point the velocity suddenly changes to u_2; and the gas moves on uniformly at this speed. Let ρ_1 and ρ_2 be thc corresponding densities, p_1 and p_2 the corresponding pressures.

The equation of continuity of mass is

$$\rho_1 u_1 = \rho_2 u_2. \qquad (1)$$

* 'Phil. Mag.," 1848, vol. 33, p. 349; 'Collected Papers,' vol. 1.

The equation of conservation of momentum is

$$p_1 - p_2 = a^2\rho_1 - a^2\rho_2 = \rho_1 u_1 (u_1 - u_2). \tag{2}$$

If u_1 and ρ_1 be given, these two equations determine u_2 and ρ_2.

Against this theory, however, Lord Rayleigh* raised the objection that the equation of energy, $\frac{1}{2}u_1^2 - \frac{1}{2}u_2^2 = a^2 \log \rho_1 - a^2 \log \rho_2$, cannot, in general, be satisfied simultaneously with (1) and (2).

In a recent note† he adds a remark that it is possible that energy might be lost at the discontinuity, but it cannot be supposed that energy is gained. Lord Rayleigh further points out that the energy lost must be converted into heat, and that this complication must be taken into account. This has been done by C.V. Burton‡ and by Hugoniot,§ but their equations have the same disadvantage as those of Stokes, in that they contain no indication that the motion represented by them is irreversible.

In the case considered by Stokes it is evident that the motion is irreversible; in fact, it is only the front of a compression that can possibly travel unchanged. For, if for an instant the sharp discontinuity were to disappear, leaving a small transition layer in which the velocity might vary continuously from u_1 to u_2, then the back part of the layer would travel forward relatively to the front part with a velocity $u_1 - u_2$. Hence if u_1 exceeds u_2 any such transition layer will become obliterated owing to the greater velocity behind, and the discontinuity will thus be maintained. This is the case of the front of a wave of condensation. If, however, the wave is a wave of rarefaction, that is, if u_1 is less than u_2, then the layer of transition will get wider, and the sharp discontinuity will not be re-established.

The object of this paper is to discuss in detail what actually does occur at a discontinuity, and to determine, in the general case of a gas whose characteristics are known, whether a discontinuity obtained by the method of Stokes is a physically possible feature.

It is evident that a plane of absolutely sharp or mathematical discontinuity cannot occur in any real gas. When, owing to change of type, there is a sudden compression or rarefaction of the material in crossing any boundary, modified physical laws must come into operation whose effect is to prevent abrupt discontinuity from being formed. Some clue to the nature of the processes

---

* 'Theory of Sound,' vol. 2, p. 41.

† 'Roy. Soc. Proc.,' A, 1908, vol. 81. p, 449.

‡ 'Phil. Mag.,' 1893, vol. 35, p. 317.

§ See Lamb's 'Hydrodynamics,' note on p. 466, 3rd edition.

involved in this case is afforded by the kinetic theory of gases; for when the change in velocity is very sudden, the molecules which are moving faster will penetrate among those which are moving more slowly, and an irreversible redistribution of velocities will ensue. This suggests that heat conduction and viscosity are, in the case of a real gas, the causes of the production of dissipative heat; it will be shown that under certain conditions they are also sufficient to produce permanence of type in the layer of transition.

Consider a continuous disturbance of permanent type in a gas whose characteristic equations are known. Give the whole system such a velocity that the disturbance is brought to rest; the motion is then steady.

Let A and B be two planes which move with the gas, and let p', ρ', u', E′, and p, ρ, u, E, be the pressure, density, velocity, and internal energy of unit mass of the gas at A and B respectively. Since θ, the temperature, is a function of p and ρ, and E is a known function of p, ρ, and θ, therefore E may be regarded as a function of independent variable p and ρ.

The equation of continuity of matter is

$$\rho' u' = \rho u = \omega . \tag{3}$$

The rate of gain of momentum between A and B is $\omega\left(u-u'\right)$. The equation of momentum for the gas between A and B is therefore

$$\left(p+\mathrm{X}\right)=\omega u=\left(p'+\mathrm{X}'\right)=\omega u', \tag{4}$$

where X and X′ are the viscous normal forces which act over the planes B and A respectively.

The work done on the gas between A and B in a small interval of time dt is

$$\left(p+\mathrm{X}\right)+u'dt-\left(p+\mathrm{X}\right)udt .$$

The increase of its kinetic energy in time dt is

$$\tfrac{1}{2}\omega\left(u^2-u'^2\right)dt .$$

The increase in internal energy in the same time is $\left(\mathrm{E}-\mathrm{E}'\right)\omega\, dt$.

The amount of heat conducted away from the mass of gas between A and B in time dt is $\omega(\xi-\xi')dt$, where $\omega\xi$ and $\omega\xi'$ are the rates at which heat, measured in mechanical units, is conducted across the planes B and A.

The equation of energy for the gas between A and B is therefore

$$(p+X)u+\tfrac{1}{2}\omega u^2+\omega(E+\xi)=(p'+X')u'+\tfrac{1}{2}\omega u'^2+\omega(E'+\xi'). \tag{5}$$

Since E is a function of p and ρ and $u=\omega/\rho$, the state of the gas at any time may be completely represented by a point in a plane diagram such as is annexed; pressure is represented by the ordinates and density by the abscissæ. If C and D are the points which represent the state of the gas at the two ends of the transition layer in which the velocity changes from u_1 to u_2, then the state of the gas along that layer are represented by the points on some curved line L joining C and D. It will be possible by means of equations (4) and (5) to determine values of X and ξ so that any given line joining C and D may represent the state of the gas in the transition layer; but the motion so represented will not be thermodynamically possible unless the coefficients of conduction and viscosity are both positive. If x represents distance in the direction in which the gas is travelling, these conditions become

X and $\partial u/\partial x$ must have opposite signs,
ξ and $\partial\theta/\partial x$ must have opposite signs.

Hence at the front of a condensation X is positive and ξ is negative while for a rarefaction X is negative and ξ is positive.

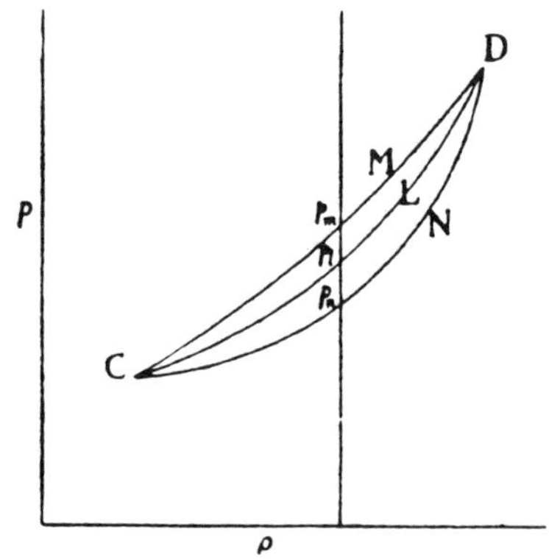

Construct curves M and N to represent the relations obtaining between p and ρ when $\mathrm{X} = 0$ and when $\xi = 0$ respectively. Let p_m, p, and p_n represent the pressures at points on M, L, and N corresponding to a particular value ρ of the density.

The equation to M is obtained by dropping X from (4), and is

$$p_m + \omega u = p_1 + \omega u_1,$$

and since (4) gives $(p + \mathrm{X}) + \omega u = p_1 + \omega u_1$, therefore $p_m = p + \mathrm{X}$. Similarly it can be shown that $\mathrm{E}_n = \mathrm{E} + \xi$.

In a condensation therefore $p_m > p$ and $\mathrm{E}_n < \mathrm{E}$, and in a rarefaction $p_m < p$ and $\mathrm{E}_n > \mathrm{E}$.

Also $\mathrm{E}_n - \mathrm{E} = (\theta_n - \theta)\mathrm{C}_v$ where C_v is the specific heat at constant volume; and $p_n - p$ must be of the same sign as $\theta_n - \theta$; hence in a condensation $p_m > p > p_n$, and in a rarefaction $p_m < p < p\omega$.

Now the equations to the lines M and N depend only on the relations which exist between pressure, density, temperature, and internal energy, that is on the characteristic equations of the gas, and not at all on its viscosity or its conductivity; for if either X or ξ is small the other can be eliminated.

Hence if a discontinuity is specified by the equations

$$\begin{aligned} u_1\rho_1 &= u_2\rho_2 = \omega, \\ p_1 + \omega u_1 &= p_2 + \omega u_2, \\ p_1 u_1 + \tfrac{1}{2}\omega u_1^2 + \omega \mathrm{E}_1 &= p_2 u_2 + \tfrac{1}{2}\omega u_2^2 + \omega \mathrm{E}_2, \end{aligned}$$

connecting the two uniform states between which it lies, and it is desired to find out whether it is thermodynamically possible, draw the lines M and N joining C and D, which are the points representing the states of the gas on the two sides of the discontinuity. If the line M lies above the line N (see diagram) so that p_m is greater than p_n then a condensation is possible. If the line M lies below the line N a rarefaction is possible. If the line M cuts the line N at any point between C and D, neither is possible.

The only special case of any importance is that of a perfect gas whose characteristic equations are

$$\frac{p}{\rho} = R\theta, \qquad E - E_1 = \frac{R}{\gamma - 1}(\theta - \theta_1),$$

where γ is the ratio of the specific heats.

The general criterion is as above; but if we also assume constant conductivity κ and viscosity μ, the circumstances can be followed out in detail. In this case it may be shown that

$$X = -\frac{4\mu}{3\omega}\frac{du}{dx} \quad \text{and} \quad \omega\xi = -Jk\frac{d\theta}{dx}.^*$$

Substituting these values in equations (4) and (5),

$$p - \frac{4\mu}{3\omega}\frac{du}{dx} + \omega u = p_1 + \omega u_1,$$

and

$$\left(p - \frac{4\mu}{3\omega}\frac{du}{dx}\right)u + \omega\frac{u^2}{2} + \omega E - Jk\frac{d\theta}{dx} = p_1 u_1 + \omega\frac{u_1^2}{2} + \omega E_1.$$

From these two equations, together with the equations

$$\frac{p}{\rho} = R\theta, \qquad u\rho = \omega, \qquad E - E_1 = \frac{R}{\gamma - 1}(\theta - \theta_1),$$

the quantities p, ρ, E and θ may be eliminated.

The resulting equation is

$$\omega Bu - \tfrac{1}{2}\omega u^2 + \frac{\omega}{\gamma - 1}\left(\frac{4\mu}{3\omega}u\frac{du}{dx} + Bu - u^2\right) - \frac{kJ}{R}\left\{\frac{4\mu}{3\omega}\frac{d}{du}\left(u\frac{du}{dx}\right) + (B - 2u)\frac{du}{dx}\right\} = \omega A,$$

where $\omega B = p_1 + \omega u_1$ and $\omega A = p_1 u_1 + \tfrac{1}{2}\omega u_1^2 + \omega E_1$.

It may be written in the simplified form

$$y\{P\,dy/du - Qu + S\} + Mu^2 - Nu + A = 0, \tag{6}$$

* See Rayleigh's 'Theory of Sound,' vol. 2, p. 315.

where
$$y = u\frac{du}{dx}, \qquad P = \frac{4\mu kJ}{3\omega^2 R}, \qquad Q = \frac{2kJ}{R\omega} + \frac{4\mu}{3\omega(\gamma-1)},$$
$$S = \frac{kBJ}{R\omega}, \qquad M = \frac{\gamma+1}{2(\gamma-1)}, \qquad N = \frac{B\gamma}{\gamma-1}.$$

If either $\kappa = 0$ or $\mu = 0$ (6) may be solved in the form

$$x = A \log (u_1 - u) + B \log (u - u_2),$$

where u_1 and u_2 are the roots of $Mu^2 - Nu + A = 0$.

If, however, neither μ nor κ vanish, (6) cannot be solved in finite terms, but if $u_1 - u_2$ be small compared with u_1 an approximate solution can be obtained.

If u_1 be the greater of the two roots of $Mu^2 - Nu + A = 0$, the solution is

$$x = \frac{Qu_1 - S}{M(u_1 - u_2)} \log \frac{u_1 - u}{u - u_2}. \tag{7}$$

By substituting for Q and S their values and remembering that $u_1 - u_2$ is small compared with u_1, it may be shown that

$$Qu_1 - S = \frac{kJ}{R\omega}\left(1 - \frac{1}{\gamma}\right) + \frac{4\mu u_1}{3\omega(\gamma-1)},$$

which is positive. Hence (7) represents a condensation; for when u approaches u_1, x approaches $-\infty$, and when u approaches u_2, x approaches $+\infty$. From (7) it is possible to calculate approximately the thickness of the transition layer. It is evident that the distance between the planes where the velocities are u_1 and u_2, is infinite; but to obtain some idea of the extent of the transition layer consider the thickness T of the layer in which the velocity changes from $\frac{9}{10}u_1 + \frac{1}{10}u_2$ to $\frac{1}{10}u_1 + \frac{9}{10}u_2$. Substituting these values for u in (7)

$$T = \frac{Qu_1 - S}{M(u_1 - u_2)} \cdot 2 \log 9,$$

and inserting the values of γ, ρ, μ, and κ for air,

$$\gamma = 1\cdot 4, \quad \rho = 1\cdot 3\times 10^{-3}, \quad \mu = 1\cdot 9\times 10^{-4}, \quad k = \frac{1\cdot 6\ \mu R}{J(\gamma - 1)},*$$

there is obtained $(Qu_1 - S)/M = 0\cdot 22$ and $T = (u_1 - u_2)^{-1}$ approximately.

In the case of waves of percussion it is known that the velocity differs appreciably from that sound. In that case $u_1 - u_2$ would be considerable and its reciprocal would be small, so that the motion would closely approximate to an abrupt discontinuous one. In the case of ordinary sounds, however, the relative velocities of air in different parts of a wave are small, so that T would be large compared with a wave-length, and nothing in the nature of a sharp discontinuity would ever be established.

Bethe
courtesy of AIP Emilio Segrè Visual Archives

HANS ALBRECHT BETHE

Hans Bethe was born in Strasbourg (then Germany) on July 2, 1906. He entered the University of Frankfurt in 1924 and went on to the University of Munich in 1926 where he earned his Ph.D. in physics under Sommerfeld. Bethe spent the years 1933-35 at Manchester and Bristol in England before going to Cornell University in the United States in 1935. He has been at Cornell University throughout his professional career except for brief periods during the Second World War. Bethe joined the Manhattan Project at Los Alamos in 1942 and served as director of the theoretical physics division during the war years. He received the Nobel Prize in physics in 1967 for work concerning the energy production in stars.

Professor Bethe still maintains connections with Los Alamos National Laboratory and one of the editors of this collection (JNJ) had the opportunity to discuss with him the circumstances that led to his work on the theory of shock waves. The following are notes from a discussion held on September 7, 1994, while Professor Bethe was visiting Los Alamos.

Professor Bethe says that much of the history of the circumstances and impetus that prompted work on the *"The Theory of Shock Waves for an Arbitrary Equation of State"* is somewhat obscure at the present time, although he is quite sure that the initial ideas grew out of discussions with John Von Neumann. Von Neumann asked the question "under what conditions can you have a shock wave in which the volume increases"? In other words, most materials with which we are familiar support shock waves in compression, just because of the nature of increasing acoustic wave speed with compression. Are there materials for which a shock wave might be generated as the specific volume increased? That is, in a rarefaction. Of course we know today that this can happen in certain materials; for example, a shock wave is formed in iron releasing through a pressure of 13 GPa. Also, the conditions that Bethe postulated for stability of the shock wave are violated for low-amplitude compression in fused quartz, and shocks are formed in release waves in this material.

It was recalled that while the work was published by the National Defense Research Committee of the Office of Scientific Research and Development, and perhaps distributed through the Ballistics Research Laboratories, Aberdeen Proving Ground, his involvement and interest was completely academic and the work was not under the direct financial support of OSRD. John Von Neumann was aware of the research of Bethe and Edward Teller on *"Deviations from Thermal Equilibrium in Shock Waves,"* previously distributed by BRL, and later by the University of Michigan. While the report with Teller has no date associated with it, it definitely precedes *"The Theory of Shock Waves for an Arbitrary Equation of State."*

Professor Bethe says that they (he and Von Neumann) knew from the outset that general theorems concerning conditions of shock-wave stability for arbitrary equations of state would be a valuable contribution to understanding shock compression. This was made clear to me when I suggested that the value of this work was apparent sometime *after* the appearance of the report. Professor Bethe replied "No! We were convinced from the beginning that this was important to the field of shock-wave research."

On the Theory of Shock Waves for an Arbitrary Equation of State

H. A. Bethe, Report No. 545 for the *Office of Scientific Research and Development*, Serial No. NDRC-B-237 (May 4, 1942)

Abstract

The fundamental equations of the hydrodynamic theory of one-dimensional shock waves — that is, the equations of conservation of mass, of momentum, and of energy — are developed. These are used to calculate the velocity, mass-velocity, temperature, and pressure rise in shock waves in air and in water. With one additional equation, they suffice to permit a calculation of detonation velocities in gaseous and in solid explosives. Predictions of detonation velocity as a function of loading density are thereby achieved, accurate to a few percent. Pressures, temperatures, and mass-velocities inside the explosive are also computed. The question of rarefaction waves following the detonation front in the explosive is investigated. The initial velocity, pressure, and so forth, of the shock wave produced at the end of a stick of explosive are calculated successfully. The dying away of shock waves, problems of reflection, and so forth, are also discussed briefly.

I. Introduction

The theory of shock waves thus far has been developed mainly for ideal gases. Even for these, the question of the stability of shock waves has received little attention. Recently the problem of shock waves in water has gained much practical importance. Therefore it seems worthwhile to investigate the properties of shock waves under conditions as general as possible.

1. Equation of state, notations, assumptions

We shall consider a material characterized by a certain equation of state

$$p = p(V, T) \tag{1}$$

where p is the pressure, V the specific volume, T the absolute temperature. We shall assume that p is a unique function of V and T. This will certainly be fulfilled if thermodynamic equilibrium is established behind the shock wave. Moreover, it will also be fulfilled if there are always the same deviations from equilibrium.

The most important example of such "regular" deviations from equilibrium is the non-occurrence of phase changes in shock waves in cases where they would be demanded by the phase diagram: For instance, Kirkwood has shown that water behind a violent shock wave (pressure about 10,000 atmospheres or higher) would transform to ice VII if thermodynamical equilibrium were established. However, as Kirkwood has pointed out, this will almost certainly *not* happen because the relaxation time for crystallization is probably many times longer than the duration of a shock condensation. In a case like this, we shall use the (metastable) continuation of the equation of state for liquid water to higher pressures.

For reasons to be explained below, we must *quite generally* exclude phase changes from our theory (cf. § 13, 14). The function $p(V, T)$ will, therefore, in general refer to that phase in which the material existed before the arrival of the shock wave, even if this phase becomes metastable at the density and temperature reached behind the shock wave. The justification for this procedure is the above-mentioned fact that the relaxation time of the phase changes will be long enough to preserve the metastable state behind the shock wave, No assumption about the relaxation time is required when the density and temperature behind the shock wave turn out to lie above their critical values. Then this state can be reached without a phase transition, both from the ordinary liquid and from the gaseous state.[1]

Besides p, V, and T, we shall use the internal energy E and the entropy S, both quantities per gram of the substance. If we agree in which phase the material is to be, its state can be described uniquely by E and V, or by S and V; for example, the pressure is a unique function of S and V. The variables S and V will turn out to be especially convenient.

Pressure and specific volume, on the other hand, are not a suitable pair of independent variables because they do not always define the state of the substance uniquely. Thus water at one atmosphere can occupy the same specific volume below and above 4°C. Ice I, at a few atmospheres and 0°C., has the same specific volume as liquid water at about 150°C and the same pressure. Both these facts are of course consequences of the negative expansion

---

[1] The transition of solids to states of extremely high temperatures and high densities will, in general, not involve any difficulties either, although there is probably no critical point for a solid (see Secs. 7 and 14).

coefficient of water. A negative expansion coefficient is not often found for substances other than water, and even for water the expansion coefficient becomes positive at higher pressures or temperatures. However, we wish to keep our considerations general and therefore admit both signs of the expansion coefficient--including, of course, the possibility of a positive coefficient for certain values of V and T, and a negative coefficient for other values.

We shall find in the following that certain assumptions must be made concerning the equation of state, in order to prove the existence and uniqueness of the solutions of Hugoniot's shock wave equations. The most important of these conditions (see Sec. 3) is

$$\left(\partial^2 p / \partial V^2\right)_S > 0 \tag{I}$$

Since $-(\partial p / \partial V)_S$ is the adiabatic compression modulus, condition (I) means that this modulus must increase with increasing compression. Condition I is very plausible; its validity will be investigated in detail in Sec. 12. It will be shown that for *all single-phase systems the condition is very well fulfilled.* Only in extreme cases, as for a gas at a pressure of 10^{-44} atmospheres (1), the condition is violated for certain temperatures .

On the other hand, condition I is violated for most phase changes (Sec. 13). Only for evaporation and condensation does condition I remain true. But for the theory of compressional waves, which forms the main problem of this paper, only phase transformations between condensed phases, that is, from liquid to solid, or from one solid phase to another, would be of interest. For these transitions condition I is in general violated, and this is the reason why we had to exclude phase transitions from our theory. If phase changes could take place sufficiently rapidly to occur in shock waves, the violation of condition I at the phase boundary would have serious consequences for the structure and stability of the shock waves (Sec. 14).

The second requirement for our theory is

$$\Gamma = V\left(\frac{\partial p}{\partial E}\right)_V > -2 \quad . \tag{II}$$

This is only a sufficient, not a necessary, condition in the general proof that compressional waves correspond to an increase of entropy (Sec. 5). Condition (II) is of course fulfilled whenever the substance expands with increasing temperature (at constant pressure). But it is also fulfilled for a small negative expansion coefficient; for example, for liquid water at 0°C, the left-hand side of condition (II) is -0. 015. For melting ice I, $V(\partial p / \partial E)_V$ seems to go down to just about the critical value, namely, -2.1. Calculations for water and ice will be given in Sec. 15. We believe that condition (II) is valid for all substances in practically all states.

Conditions (I) and (II) are sufficient for the general proof of the existence and uniqueness of the solutions of the shock equation. However, in proving the stability of shock waves against breaking up in any way (Sec. 11), we also use the inequality

$$(\partial p / \partial V)_E < 0 \tag{III}$$

which again is only a sufficient, not a necessary, condition of stability. Condition (III) is obviously fulfilled for ideal gases for which constant energy is equivalent to constant temperature. We have also found it to be valid for all one-phase systems we have investigated (Sec. 16). However, it can be shown that condition (III) breaks down again for some phase transformations, namely, if the energy and entropy change in opposite directions.

2. *The shock equations*

We shall denote by the subscript 1 the quantities referring to the "undisturbed" substance in front of the shock wave by 2 those referring to the material behind the shock wave. The velocity of the material relative to the shock wave will be u. V, E, S, p are volume, energy, entropy and pressure, the first three quantities per gram. By ΔV, Δp, etc., we denote the change of volume, pressure, etc, namely,

$$\Delta p = p_2 - p_1, \qquad \Delta V = V_2 - V_1 \qquad \text{and so forth.} \tag{2}$$

The conservation laws for mass, momentum and energy are, respectively

$$\frac{u_1}{V_1} = \frac{u_2}{V_2} \quad , \tag{3}$$

$$p_1 + \frac{u_1^2}{V_1} = p_2 + \frac{u_2^2}{V_2} \quad , \tag{4}$$

$$E_1 + p_1 V_1 + \frac{1}{2} u_1^2 = E_2 + p_2 V_2 + \frac{1}{2} u_2^2 \quad . \tag{5}$$

It is convenient to transform Eq. (4) by use of Eq. (3). We obtain

$$\Delta p = p_2 - p_1 = \frac{u_1{}^2}{V_1} - \frac{u_2{}^2}{V_2} = \frac{u_1}{V_1}(u_1 - u_2)$$
$$= \frac{u_1{}^2}{V_1}\left(1 - \frac{V_2}{V_1}\right) = -\frac{u_1{}^2}{V_1{}^2}\Delta V \quad , \tag{4a}$$

or

$$-\frac{\Delta p}{\Delta V} = \frac{u_1{}^2}{V_1{}^2} = \frac{u_2{}^2}{V_2{}^2} \quad . \tag{6}$$

Similarly, Eq. (5) may be transformed:

$$\Delta E = E_2 - E_1 = p_1 V_1 - p_2 V_2 + \frac{1}{2}\left(u_1{}^2 - u_2{}^2\right) \quad . \tag{5a}$$

Now we have from Eq. (4a)

$$u_1{}^2 - u_2{}^2 = (u_1 + u_2)\Delta p \frac{V_1}{u_1} \quad ,$$

and, using Eq. (3):

$$u_1{}^2 - u_2{}^2 = \Delta p(V_1 + V_2) \quad . \tag{7}$$

Inserting into Eq. (5a):

$$-\frac{\Delta E}{\Delta V} = \frac{1}{2}(p_1 + p_2) \quad . \tag{8}$$

This is the famous Hugoniot equation. It will be noted that the equation no longer contains the velocities, u_1 or u_2 but that it is an equation between thermodynamic quantities only. It is, therefore, most suitable for a determination of the possible "final" states of the material behind the shock wave when the "initial" state (state 1), in front of the shock wave, is given. For a given V_1, E_1, p_1, and given V_2, Eq. (8) gives a linear relation between E_2 and p_2. Another relation between the same quantities is provided by the equation of state. At the intersection of these two relations (if such an intersection exists), we find the possible "final" state of the material for the given value of V_2. This makes it easy to determine all possible final states for a given initial state. Having determined p_2 and V_2, we can immediately find the corresponding velocities u_1 and u_2 from Eq. (6).

II. General Theory for Ordinary Substances[2]

3. Small shock waves

Eqs. (6) and (8) are obviously generalizations of the equations for infinitesimal (sound) waves. Eq. (8) goes over, for infinitesimal changes, into

$$\frac{\partial E}{\partial V} = -p \quad , \tag{9}$$

which is the condition for an adiabatic change of state. Eq. (6) then takes the form

$$u_1^2 = u_2^2 = -V^2\left(\frac{\partial p}{\partial V}\right)_S = \left(\frac{\partial p}{\partial \rho}\right)_S \quad , \tag{10}$$

where ρ is the density. The right-hand side is the well-known expression for the square of the velocity of sound; that is, the material moves with respect to the "infinitesimal shock wave," with the velocity of sound, and this is, of course, also the velocity of the wave relative to the material.

We shall now consider shock waves of finite, but still very small, amplitude. We choose volume and entropy as independent variables, and expand ΔE and p_2, in Eq. (8), in powers of ΔV and ΔS. Retaining only those powers which will turn out to be relevant, we find

$$\Delta E = \left(\frac{\partial E}{\partial V}\right)_S \Delta V + \frac{1}{2}\left(\frac{\partial^2 E}{\partial V^2}\right)_S \Delta V^2 + \frac{1}{6}\left(\frac{\partial^3 E}{\partial V^3}\right)_S (\Delta V)^3 + \cdots$$
$$+\left(\frac{\partial E}{\partial S}\right)_V \Delta S + \cdots \tag{11a}$$

All derivatives are to be taken at the original volume and entropy, V_1 and S_1. Remembering the thermodynamic relations

$$\left(\frac{\partial E}{\partial V}\right)_S = -p \tag{11}$$

and

[2] That is, substances fulfilling the conditions I to III of Sec. 1. (See Secs. 12 to 16.)

$$\left(\frac{\partial E}{\partial S}\right)_V = T \quad , \tag{12}$$

we find

$$-\frac{\Delta E}{\Delta V} = p + \frac{1}{2}\left(\frac{\partial p}{\partial V}\right)_S \Delta V + \frac{1}{6}\left(\frac{\partial^2 p}{\partial V^2}\right)_S (\Delta V)^2 + \cdots$$

$$-T\frac{\Delta S}{\Delta V} + \cdots \quad . \tag{12a}$$

Likewise, since $p = p_1$,

$$\frac{1}{2}(p_1 + p_2) = p + \frac{1}{2}\left(\frac{\partial p}{\partial V}\right)_S \Delta V + \frac{1}{4}\left(\frac{\partial^2 p}{\partial V^2}\right)_S (\Delta V)^2 + \cdots$$

$$+\frac{1}{2}\left(\frac{\partial p}{\partial S}\right)_V \Delta S + \cdots \quad . \tag{12b}$$

The last term in Eq. (12b) is negligible compared with the last term in Eq. (12a) since ΔV is assumed to be small. Comparison of Eqs. (12a) and (12b), which must be equal according to Eq. (8), shows that the two first terms are identical, and therefore

$$\Delta S = S_2 - S_1 = -\left(\frac{\partial^2 p}{\partial V^2}\right)_S \frac{(\Delta V)^3}{12T} \quad , \tag{13}$$

neglecting higher powers of ΔV.

We have assumed that the shock wave moves into material 1, which is thereby converted into material 2. If the shock wave is to be thermodynamically stable, the entropy must increase in this process, that is, ΔS must be positive. For a compressive wave, $\Delta V = V_2 - V_1$ is negative; if at the same time ΔS is to be positive, we must have [see Eq. (13)]

$$\left(\frac{\partial^2 p}{\partial V^2}\right)_S > 0 \quad . \tag{I}$$

This condition seems to be fulfilled for all substances as long as only one phase is present (Sec. 12). At phase boundaries condition (I) is usually violated (Sec. 13). Therefore, if a phase change has time to occur, rarefaction waves might be stable and compression waves unstable, a case discussed in detail in Sec. 14.

As was already mentioned in Sec. 1, phase changes usually cannot occur because of the long time required for them. We can therefore consider condition (I) quite generally as valid. In any case, we shall assume it to be valid in the main part of this paper (Secs. 3 to 11).

If condition (I) is valid, then Eq. (13) shows that compressive shock waves of small amplitude ΔV are thermodynamically stable since the entropy is greater behind the wave than in front of it. On the other hand, rarefaction waves (positive ΔV) cannot have a finite amplitude because the entropy would decrease in that case. Therefore, in any substances obeying condition (I) rarefaction waves will dissolve into trains of infinitesimal waves, a behavior well known from ideal gases. All these statements have been proven only for small ΔV -- small enough to make the higher powers of ΔV negligible in Eqs. (12a, b). The general proof will be given in the next section (Sec. 4).

We have also shown in Eq. (13) that the entropy change is proportional to the third power of the volume change. This is understandable since we have seen that in first approximation the entropy is unchanged [see Eq. (9)]. Moreover, ΔS must be proportional to an odd power of ΔV because, when the states 1 and 2 are interchanged, the sign of both ΔS and ΔV must change. The third power relation is thus the simplest possible.

We shall now calculate the velocity of the shock wave relative to the two media, u_1 and u_2. For this purpose, we evaluate the left-hand side of Eq. (6):

$$-\frac{\Delta p}{\Delta V}=-\left(\frac{\partial p}{\partial V}\right)_{S,1}-\frac{1}{2}\left(\frac{\partial^2 p}{\partial V^2}\right)_{S,1}\Delta V+\cdots \quad , \tag{14}$$

where we have neglected terms of order $(\Delta V)^2$ and of order $\Delta S/\Delta V$ which is also proportional to $(\Delta V)^2$ [see Eq. (13)]. The subscript 1 means that the derivatives are to be taken at $V=V_1$, $S=S_1$. Now

$$-\left(\frac{\partial p}{\partial V}\right)_{S,1}=\frac{a_1{}^2}{V_1{}^2} \quad , \tag{14a}$$

where a_1 is the velocity of sound in the medium in front of the shock wave, and $(\partial^2 p/\partial V^2)_S$ is positive (condition I). Using Eq. (6), we find then

$$u_1{}^2=a_1{}^2+\frac{1}{2}V_1{}^2\left(\frac{\partial^2 p}{\partial V^2}\right)_{S,1}(V_1-V_2)+\cdots \quad . \tag{15}$$

Therefore, if state 2 is denser than state 1, $V_1 > V_2$ and

$$u_1 > a_1 \quad . \tag{15a}$$

In other words, the shock wave velocity in the medium of smaller density (medium 1) is greater than the velocity of sound in that medium. Conversely, relative to the denser medium 2, the shock wave moves more slowly than sound. The general proof of these theorems, for arbitrarily large pressure change $p_2 - p_1$, will be given in Sec. 9.

4. General proof of the increase of entropy for compression waves

We now admit arbitrarily large changes of the variables of state V, p, E and S. We shall keep the "initial state" V_1, S_1 fixed and consider all possible "final" states V_2, S_2 which satisfy the shock equation, Eq. (8), and the equation of state.

We shall prove in this section that in any compression wave the entropy must increase, whatever the amount of the compression, whereas in the last section this theorem was proved only for small shock waves. Let $V_1 > V_2$ and let the states 1 and 2 be connected by Eq. (8), namely

$$E_2 - E_1 = \frac{1}{2}(p_2 + p_1)(V_1 - V_2) \quad . \tag{16}$$

Now let us consider the state 1' which has the same specific volume as state 1, but the same entropy as state 2. Its energy is related to that of state 2 by the thermodynamic relation, Eq. (11), which gives upon integration

$$E_2 - E_1' = \int_{V_2}^{V_1} p\, dV \quad , \tag{17}$$

the integral to be taken along the adiabatic passing through the states 2 and 1'. Now the equation of state has already been assumed to fulfill the condition

$$\left(\partial^2 p / \partial V^2\right)_S > 0 \quad . \tag{I}$$

Therefore, if V and p are used as coordinates, the adiabatic from 2 to 1' will lie below the straight line joining 2 and 1', explicitly:

$$p(V, S_2) < p_2 + (p_1' - p_2)\frac{V - V_2}{V_1 - V_2} \qquad \text{if} \qquad V_2 < V < V_1 \quad . \tag{18}$$

Integration of Eq. (17) gives

$$E_2 - E_1' < \frac{1}{2}(p_1' + p_2)(V_1 - V_2) \quad . \tag{18a}$$

Subtracting Eq. (16) from Eq. (18a) yields

$$E_1 - E_1' < \frac{1}{2}(p_1' - p_1)(V_1 - V_2) \quad . \tag{19}$$

We are now going to prove that E_1 *must be smaller than* E_1'. For this purpose let us assume that the reverse is the case, that is, $E_1 - E_1' > 0$. Then, since $V_1 > V_2$ by assumption, Eq. (19) requires that $p_1' > p_1$. This means that, for fixed volume $V = V_1$ the pressure must decrease with increasing energy (and temperature). This requires a negative expansion coefficient (see Sec. 1) which in itself is quite possible.

However, Eq. (19) requires also that the decrease of pressure with energy should exceed a certain amount. It is most convenient to divide both sides of Eq. (19) by $\frac{1}{2}(E_1 - E_1')$ which has been assumed to be positive; then we get

$$-\frac{p_1 - p_1'}{E_1 - E_1'}(V_1 - V_2) > 2 \quad , \tag{19a}$$

or

$$= -\overline{(\partial p / \partial E)_V}(V_1 - V_2) > 2 \quad , \tag{20}$$

where $\overline{(\partial p / \partial E)_V}$ denotes the average of the derivative between E_1' and E_1. If $V_1 - V_2$ is small, Eq. (20) can certainly not be fulfilled so that our original assumption, $E_1 > E_1'$, is proved to be wrong for this case, in agreement with our results in Sec. 3. However, for larger volume changes, we can assert that Eq. (20) is false *only* if $(\partial p / \partial E)_V$ is limited in absolute value. A certainly sufficient condition is

$$V\left(\frac{\partial p}{\partial E}\right)_V > -2 \tag{II}$$

everywhere, because if condition II is fulfilled, then Eq. (20) is certainly false since $V_1 - V_2$ smaller than V_1. It will be proved in Sec. 15 that condition II is very well fulfilled even for water. The only exception we have found (Sec. 15) is melting ice I, but in this case also condition I is violated. We believe that condition II is probably fulfilled for all substances for which condition I is satisfied, and at least for all one-phase systems. If condition II is valid, we have shown that E_1 must be smaller than E_1' Since, at constant volume, the entropy increases together with the energy, S_1' must be greater than S_1. But S_1' is equal to S_2 by definition. Therefore we have proved that

$$S_2 > S_1 \quad \text{if} \quad V_2 < V_1 \tag{21}$$

The denser state bounding a shock wave has a greater entropy than the state of smaller density. Therefore *compressional shock waves are always accompanied by an increase in entropy and are therefore thermodynamically stable. Rarefaction waves of finite amplitude are always unstable.*

These results are valid for any substance with any arbitrary equation of state, and for any change of the specific volume, and of the pressure in the shock wave, provided only conditions I and II are fulfilled .

5. *Existence of solutions*

Let again $V_2 < V_1$. Then, if there is a solution of the shock equations for a given value of V_2 we have shown in the preceding section that $S_2 > S_1$. We ask now whether or not there actually exists a solution for a given V_2 and given "initial" state V_1, E_1.

For fixed V_2, V_1 and E_1 the shock equation, Eq. (8)

$$E_2 - E_1 = \frac{1}{2}(p_2 + p_1)(V_1 - V_2) \quad , \tag{16}$$

is represented by a straight line in a p_2, E_2 diagram. The equation of state again for the same V_2, is represented by another curve in this diagram, in general not a straight line. The solution (or solutions) of the shock equation, if any, is given by the intersection of the two curves. We shall examine the existence of such intersections by investigating, at two points p_2 which of the two curves lies higher.

(a). Take the state of volume V_2 which has the same entropy as state 1, let us call it $p_2'E_2'$. Then, in analogy to Eqs. (17), (18a) we have

$$E_2' - E_1 = \int_{V_2}^{V_1} p dV < \frac{1}{2}(p_1 + p_2')(V_1 - V_2) \tag{22}$$

Denoting by $\tilde{E}_2'$ the value of the energy obtained from the shock equation (16) for $p_2 = p_2'$ we see that

$$E_2' < \tilde{E}_2' \quad . \tag{22a}$$

In words: For the pressure p_2' the equation of state gives a lower energy than the shock equation.

(b). Take a state of extremely high energy (temperature) and of specific volume V_2. The high temperature will cause all molecules to dissociate completely into nuclei and electrons, and the kinetic energy of their thermal motion will be large compared with their Coulomb interaction. Therefore, the substance, whatever it may be, will behave as an ideal monatomic gas (more accurately, as a mixture of such gases -- namely, the gas of nuclei and the electron gas). For any monatomic gas, we have the relation

$$E = \frac{3}{2} pV \quad , \tag{23}$$

which remains valid even when the gas is partly or wholly degenerate. Therefore, if p_2'' denotes some very high pressure and E_2'' and $\tilde{E}_2''$ are the energies corresponding to p_2'' according to the equation of state and the shock equation respectively, we have

$$E_2'' = \frac{3}{2} p_2'' V_2 \quad , \tag{23a}$$

$$\tilde{E}_2'' = \frac{1}{2} p_2''(V_1 - V_2) \quad . \tag{23b}$$

Therefore

$$E_2'' > \tilde{E}_2'' \text{ if } V_2 > \frac{1}{4} V_1 \qquad \text{and} \qquad E_2'' < \tilde{E}_2'' \text{ if } V_2 < \frac{1}{4} V_1 \quad . \tag{24}$$

Since $E_2' < \tilde{E}_2'$ [Eq. (22a)], we find that for $V_2 > \frac{1}{4} V_1$ there must be at least one intersection of the curves representing the equation of state and the shock equation. If there is more than one intersection, the number of intersections must be odd. Therefore *for any specific volume greater than one quarter of the original volume, there must be at least one solution, and always an odd number of solutions, of the shock equations.*

For any small value of the specific volume, $V_2 < \frac{1}{4} V_1$, there need not be any solution, or, if there are solutions, their number must be <u>*even*</u>. We are now going to show that for real substances there will be at least two solutions for volumes

V_2 between $\frac{1}{4}V_1$ and a certain smaller value $V_{2\,\min}$. The simplest example is an ideal gas which is not monotonic: If we use the abbreviation

$$\beta = \frac{E}{pV} \quad , \tag{25}$$

it is shown in the ordinary theory of shock waves in ideal gases that the gas may be compressed down to a specific volume

$$V_{2\,\min} = \frac{V_1}{2\beta + 1} \quad . \tag{25a}$$

The constant β has, at low temperatures, the value 2.5 for diatomic gases and higher values for polyatomic gases. Inserting the value of 2.5 into Eq. (25a) we find the well-known result that diatomic gases may be compressed in shock waves to one sixth of their original volume.

However, much greater compressions are found to be possible when the vibration of the molecule and, especially, its dissociation and ionization are considered. When these processes take place, the energy content increases tremendously, and β may easily reach values of 6 or 7.[3]

According to Eq. (25a), the density of the gas behind the shock wave may thus be as much at 15 times the original density. Even higher values may be expected for polyatomic gases. If the temperature behind the shock wave is very much higher than that in front, it can easily be shown that for a given β, V_2 has very nearly the value of Eq. (25a).

The temperature region in which β has high values is likely to be quite extensive because when the dissociation is completed, ionization will follow and will affect first the outer, then successively the inner electron shells. However, at still higher temperatures, when ionization is almost complete, β must decrease again and must finally reach the value 1. 5 for monatomic gases, because then the nuclei and electrons may be considered as free particles with small interaction.

We thus find that any gas will have a certain maximum β which will be reached at some temperature in the dissociation or ionization region, probably

---

[3] (See a paper by H. A. Bethe and E. Teller, published by the Ballistic Laboratory of the Aberdeen Proving Ground in 1940, which gives the energy content of air up to 5000°, and new calculations by H. A. Bethe and J. F. Whitney in which the temperature range is extended to $25{,}000^{\circ}$).

when both processes are already well along; that is, at temperatures of about 10,000°. Correspondingly, for a given initial state, V_2 will have a minimum,

$$V_{2\,\min} \approx \frac{V_1}{2\beta_{\max}+1}\quad , \tag{25b}$$

which is certainly less than $\frac{1}{4}V_1$, thus, there will be at least two solutions for every value of V_2 between $V_{2\min}$ and $\frac{1}{4}V_1$.

Our considerations are not restricted to di- and poly-atomic gases. Monatomic gases will also be subject to ionization, and are only distinguished by having the same value of β at low and at very high temperatures. Further, if we start from a condensed system, the compression will not become very large as long as the system is a true fluid or solid. But, as the temperature behind the shock wave increases, the substance will become quasi-gaseous (the transformation takes place above the critical density and therefore does not lead to a real gas) and the transition to this state will give rise to similar phenomena as dissociation and ionization. The latter processes themselves will also occur, at higher temperatures .

It can, of course, not be predicted in general whether or not there will be only two solutions for a given V_2. It may happen that β, as a function of temperature, has several maxima. This will occur, for example, if two dissociation or ionization processes occur at widely different temperatures. Ordinary air is an example of this behavior, having one maximum due to dissociation near 8000°, and another, due to ionization, near $20{,}000^{\circ}$. Then V_2 will have two (or even more) minima, and one (or more) maxima in between. In general the maxima of V_2 will be smaller than $\frac{1}{4}V_1$ because β must be at least 1. 5 at any temperature, and V_2 is close to the value of Eq. (25a). Therefore there is, in general, only one solution for every V_2 between $\frac{1}{4}V_1$ and V_1 In Fig. 2 we shall give some examples of possible shock curves (see Sec. 8).

The most important result, however, is *that there is at least one solution for* $V_2 > \frac{1}{4}V_1$, *and at least two between* $\frac{1}{4}V_1$ *and some smaller value,* $V_{2\min}$. *In general, the minimum of* V_2 *occurs at very high temperature.*

6. The shock curve

It is convenient to consider all the solutions of the shock equation, Eq. (8), for a given "initial" state V_1, E_1, including both those with $V_2 < V_1$ and $V_2 > V_1$.

The former group (compressed states) will occur behind shock waves propagating into a medium which is in state l; the latter group (expanded states $V_2 > V_1$), will occur in front of shock waves behind which the state of the medium is given by V_1, E_1.

We are going to show now that all the solutions of the shock equation must lie on a continuous curve in the *V*-*E* diagram which we shall denote as the "shock curve." For this purpose, we consider two solutions of the shock equation whose specific volume differs by the small amount dV. Then we have, by differentiation of Eq. (8) with respect to the quantities defining state 2:

$$2dE = -(p_1 + p_2)dV - (V_2 - V_1)dp \quad . \tag{26}$$

Now the equation of state gives the pressure as a unique function of V and E; therefore we may write,

$$dp = \left(\frac{\partial p}{\partial E}\right)_V dE + \left(\frac{\partial p}{\partial V}\right)_E dV \quad , \tag{27}$$

where the derivatives are again unique functions of V and E, determined by the equation of state of the substance. Inserting into Eq. (26), we get

$$\frac{dE}{dV} = -\frac{p_1 + p_2 + (V_2 - V_1)(\partial p / \partial V)_E}{2 + (V_2 - V_1)(\partial p / \partial E)_V} = f(V_2, E_2) \ . \tag{28}$$

The derivatives $\partial p / \partial E$ and $\partial p / \partial V$ are to be taken at V_2, E_2.

Equation (28) is a first order differential equation for the shock curve. If the equation of state is known, the entire shock curve can be obtained by integrating Eq. (28) using any standard method. The integration can be started at the point V_1, E_1; at this point Eq. (28) becomes

$$\frac{dE}{dV} = -p_1 \quad , \tag{28a}$$

a relation already given in Eq. (9). Since Eq. (28) is a first-order differential equation, the shock curve cannot end or start anywhere in the *V*-*E* plane. It must therefore either go to infinity or leave the region of definition of the equation of state or be closed. The last alternative will be proved to be impossible in the next section and it will be shown that the shock curve goes to infinite energy E_2 at $V_2 = \frac{1}{4}V_1$ (see Sec. 5), whereas on the side $V_2 > V_1$ it goes either to zero temperature or to a phase boundary .

If the numerator or the denominator of Eq. (28) becomes zero separately, no difficulty arises. A simultaneous vanishing of both numerator and denominator would give a singular point but it will be shown in Sec. 7 that the shock curve cannot start or end in singular points.

For many purposes it is more convenient to use the entropy rather than the energy as an independent variable. According to Sec. 4, we have $S_2 > S_1$ for the compressed states $(V_2 < V_1)$ and $S_2 < S_1$ for the expanded states $(V_2 > V_1)$. Therefore, in a V-S diagram with V_1, S_1 as the origin, the part of the shock curve containing the compressed states will lie in the second quadrant, that containing the expanded states in the fourth quadrant. The shock curve in the V-S diagram will be tangent to the V axis near the origin, according to Eq. (13).

The entropy can be introduced by means of the thermodynamic relation (applied to state 2);

$$T_2 dS = dE + p_2 dV \quad . \tag{29}$$

Inserting this into Eq. (26) gives

$$2T_2 dS = \Delta p dV - \Delta V dp \quad , \tag{30}$$

a most convenient relation. Expressing p as a function of V and S, we may write

$$dp = \left(\frac{\partial p}{\partial V}\right)_S dV + \left(\frac{\partial p}{\partial S}\right)_V dS \tag{30a}$$

and obtain

$$\frac{dS}{dV} = \frac{\Delta p - \Delta V(\partial p / \partial V)_S}{2T_2 + \Delta V(\partial p / \partial S)_V} \quad . \tag{31}$$

The denominator of (31) is the same as that of (28), except for the factor T_2, because

$$(\partial E / \partial S)_V = T . \tag{31a}$$

For positive ΔV (expanded states), the denominator can never become zero if condition II is fulfilled by the equation of state. Therefore, *on the side of expanded states, the shock curve must go monotonically to greater specific volume.* For compressed states $(\Delta V < 0)$, the denominator may easily become zero; this corresponds to a minimum (or maximum) of the specific volume. We have shown at the end of the last section that for all real substances there exists at least one minimum of V_2, for a given V_1, and that this minimum occurs at rather high temperatures (about $10,000^{\circ}$).

The numerator of Eq. (31) may be written

$$\Delta V\left[\frac{\Delta p}{\Delta V}-\left(\frac{\partial p}{\partial V}\right)_S\right]. \tag{31b}$$

Therefore, if *dS*/*dV* were zero anywhere on the shock curve (which as we are going to show in the next section, does not occur except for $\Delta V = 0$), we should have $\Delta p / \Delta V = (\partial p / \partial V)_S$. Physically, this would mean [See Eq.(6)] that the shock wave velocity relative to the medium behind the shock wave,

$$u_2 = V_2\sqrt{-\frac{\Delta p}{\Delta V}} \quad , \tag{31c}$$

becomes equal to the sound velocity in that medium,

$$a_2 = V_2\sqrt{-\left(\frac{\partial p}{\partial V}\right)_S} \quad . \tag{31d}$$

It should be remarked that, while *dS*/*dV* is well defined by Eq. (31) for any value of V and S, it need not be continuous. For a mixture of two phases, $(\partial p / \partial V)_S$ will have a different (greater) value than for the pure phases (Sec. 13); hence *dS*/*dV* will in general have a discontinuity and $S(V)$ a kink where phase transitions begin or are completed. However, as was pointed out repeatedly, most phase changes must be excluded from our theory because condition I is violated.

7. *Proof of the monotonic behavior of the entropy*

In this section we shall prove the central theorem of the theory, namely: *If the state in front of the shock wave* (V_1, S_1) *is given, there is one and only one solution of the shock equations for any given value of the entropy,* S_2, *behind the shock wave,* $(S_1 < S_2 < \infty)$.

If the state behind the shock wave is given (V_1, S_1) and if phase changes are excluded, there is one and only one solution of the shock equations for any given value of the entropy, S_2, *in front of the wave which is larger than a certain* S_B where the state S_B lies on a phase boundary and S_B is a function of (V_1, S_1).

The main problem is to show that there is *only* one solution then from the considerations of Sec. 6 it can be proved easily that there is actually one. To prove the main statement, we shall show that the assumption of two solutions with the same entropy leads to a contradiction.

Suppose we have two solutions, denoted by subscripts 2 and 3, which have the same entropy

$$S_2 = S_3 \quad . \tag{32}$$

Without loss of generality, we can assume that

$$V_3 > V_2 \quad . \tag{32a}$$

Since both solutions fulfill the shock equation, Eq. (8), for the same initial state V_1, p_1, E_1, we have

$$2(E_2 - E_1) = (p_2 + p_1)(V_1 - V_2) \quad , \tag{32b}$$

$$2(E_3 - E_1) = (p_3 + p_1)(V_1 - V_3) \quad . \tag{32c}$$

Therefore

$$2(E_2 - E_3) = (p_2 - p_3)(V_1 - V_2) + (p_3 + p_1)(V_3 - V_2) \quad . \tag{33}$$

Since $S_2 = S_3$, we have in analogy to Eq. (18a)

$$2(E_2 - E_3) = 2\int_{V_2}^{V_3} p dV < (p_2 + p_3)(V_3 - V_2) \quad . \tag{33a}$$

Subtracting Eq. (33) from Eq. (33a) we obtain

$$(p_2 - p_1)(V_3 - V_2) + (p_2 - p_3)(V_2 - V_1) > 0 \quad . \tag{34}$$

Since $V_3 > V_2$, we may divide this inequality by $V_3 - V_2$ and obtain

$$p_1 - p_2 < \frac{p_2 - p_3}{V_3 - V_2}(V_2 - V_1) \quad , \tag{34a}$$

which holds irrespective of the sign of $p_1 - p_2$. Adding $p_2 - p_3$ on both sides, we obtain the corresponding equation

$$p_1 - p_3 < \frac{p_2 - p_3}{V_3 - V_2}(V_3 - V_1) \quad . \tag{34b}$$

We can now deduce inequalities for the energy by inserting relations (34a, b) into Eqs. (32b, c). To obtain results independent of the sign of $V_1 - V_2$ or $V_1 - V_3$, we divide Eqs. (32b, c) by these quantities, then we get

$$2\frac{E_1 - E_2}{V_2 - V_1} = p_1 + p_2 < 2p_2 + \frac{p_2 - p_3}{V_3 - V_2}(V_2 - V_1) \quad , \tag{35}$$

$$2\frac{E_3 - E_1}{V_1 - V_3} = p_1 + p_3 < 2p_3 + \frac{p_2 - p_3}{V_3 - V_2}(V_3 - V_1) \quad . \tag{35a}$$

From these inequalities, we shall deduce an inequality for the *entropy* of state 1. We have proved in Sec. 4 that

$$S_1 > S_2 = S_3 \qquad \text{if} \qquad V_1 < V_2(< V_3) \quad , \tag{36}$$

$$S_1 < S_2 = S_3 \qquad \text{if} \qquad V_1 > V_3(> V_2) \quad . \tag{36a}$$

Now let us consider the state p_1', E_1' which has the volume V_1 but the entropy $S_2 = S_3$. The adiabatic which goes through the three states (V_2, p_2), (V_3, p_3) and (V_1, p_1') is convex in the V,p diagram according to condition I $\left[\left(\partial^2 p / \partial V^2\right)_S > 0\right]$. Therefore, everywhere *outside* the range from V_2 to V_3, the adiabatic will lie above the straight line joining the points V_2, p_2 and V_3, p_3 (Fig.1). Expressed in formulas,

$$p(V, S_2) > p_2 + \frac{p_2 - p_3}{V_3 - V_2}(V_2 - V) = p_3 + \frac{p_2 - p_3}{V_3 - V_2}(V_3 - V) \quad \text{if}$$

$$V < V_2 \qquad \text{or} \qquad V > V_3 \quad . \tag{37}$$

Now it is clear that V_1 must lie outside the range from V_2 to V_3 because the states 2 and 3 must either both be compressed states, or both expanded states [See Eqs. (36, 36a)]. If $V_1 < V_2$, we have, using Eqs. (17) and (37)

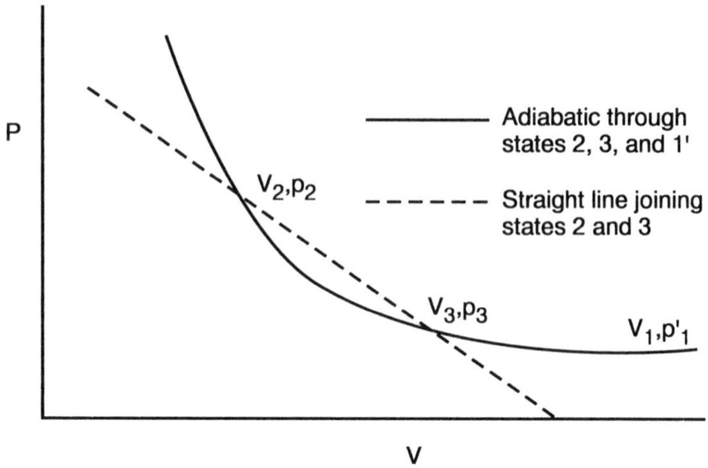

Fig. 1. Specific volume-pressure diagram used in the proof of the central theorem. The adiabatic lies below the straight line between states 2 and 3 (Sec. 4), above it everywhere else (Sec. 7).

$$2(E_1' - E_2) = 2\int_{V_1}^{V_2} p dV > 2\int_{V_1}^{V_2}\left[p_2 + \frac{p_2 - p_3}{V_3 - V_2}(V_2 - V)\right]dV$$

$$= \left[2p_2 + \frac{p_2 - p_3}{V_3 - V_2}(V_2 - V_1)\right](V_2 - V_1) \quad . \tag{38}$$

Since $V_2 - V_1$ is positive, it follows by comparison with Eq. (35) that

$$E_1' > E_1 \quad , \tag{39}$$

or, since E_1' and E_1 are states of the same specific volume, and since the entropy of state E_1' is S_2,

$$S_2 > S_1 \quad , \tag{40}$$

in contradiction to Eq. (36).

Similarly, if $V_1 > V_3$, we calculate the energy E_1' by integration over the part of the adiabatic between V_3 and V_1, rather than V_2 and V_1. We have then from Eq. (37)

$$2(E_3 - E_1') = 2\int_{V_3}^{V_1} pdV > 2\int_{V_3}^{V_1}\left[p_3 + \frac{p_2 - p_3}{V_3 - V_2}(V_3 - V)\right]dV$$

$$= \left[2p_3 + \frac{p_2 - p_3}{V_3 - V_2}(V_3 - V_1)\right](V_1 - V_3) \quad . \tag{38a}$$

Since now $V_1 - V_3$ is positive, it follows by comparison with Eqs. (35a) that

$$E_1' < E_1 \tag{39a}$$

and therefore

$$S_2 < S_1 \quad , \tag{40a}$$

in contradiction to Eq. (36a) which is applicable to our case.

We have thus obtained a contradiction, both for $V_1 < V_2$ and for $V_1 > V_3$. Therefore our original assumption must be false and we find: *For a given initial state* V_1, S_1, *there are never two solutions of the shock equation belonging to the same "final" entropy* S_2.

The shock curve (Sec. 6) will therefore go monotonically from low S *to high* S. *It cannot have any maximum or minimum* because this would imply that the same value of S is taken on at least twice, namely once on each side of the maximum or minimum. This means also that *there cannot be any closed shock curve* because such a curve would necessarily have a maximum and a minimum.

Likewise, there cannot be any singularity on the shock curve which is approached in a spiral. If there is any singularity at all, it can at most cause a kink or a cusp in the shock curve, without interrupting the monotonic increase of the entropy.

We have shown in Sec. 6 that the shock curve cannot end anywhere in the V,S -plane but must either be closed or leave the region of definition of V and S, or go to infinity. We have just proved that the curve cannot be closed. The region of definition would extend from 0 to ∞ for both V and S if we considered all phases of the substance. Actually, we have agreed in Sec. 1 to exclude phase changes, firstly because they probably do not occur because of their long relaxation time, and secondly in order to insure the validity of condition I. If we consider only *one* phase, for example, the liquid, certain large values of the volume and certain small values of the entropy cannot be reached; we have, therefore, a definite boundary of the phase towards large V and small S. On the other hand, states of *high entropy* corresponds to temperatures above the critical one. From the solid, states of arbitrarily high S can also be reached by raising the temperature without phase change, it being only questionable whether they are identical with the states obtained by heating the liquid or gas; but metastable phases have explicitly been permitted in our consideration. [4] For the same volumes may be reached by a liquid, and also by a gas above the critical temperature. Summarizing, we find that for compressed states V is limited only by zero and S by infinity, whereas for expanded states the limits are usually given by phase boundaries.

We shall now examine the behavior of the shock curve in the various quadrants. On the side of small $V_2(< V_1)$, we know that the entropy is high $S_2 > S_1$. We know further that there are states of very high energy and entropy, and of volume $V_2 = \frac{1}{4}V_1$, which satisfy the shock equation, Eqs. (23a, b). It follows that *on the high density side, the shock curve goes to* $S_2 = \infty$ *rather than to* $V_2 = 0$.

That the curve does not go to $V_2 = 0$ can be demonstrated in various ways of which the following may be the simplest. At very small specific volume, the atoms of the substance will be crushed by pressure ionization even at low temperatures since the mean distance between nuclei can be made smaller than the radius of the K-shell. Then the electrons and nuclei form again a perfect though highly degenerate gas. The Coulomb energy becomes negligible in the limit of very small specific volume, being proportional to $V^{-1/3}$ while the kinetic energy of the degenerate electrons is proportional to $V^{-2/3}$. As has been mentioned above (Sec. 5) the relation

$$E = \frac{3}{2} pV \tag{23}$$

is valid for a degenerate gas just as for a non-degenerate one, provided only the interactions are negligible. But if Eq. (23) is valid then there cannot be any solutions of the shock equation for $V_2 < \frac{1}{4}V_1$ [see Eqs. (25a, b)]. Therefore there is certainly no solution for very small V_2.

---

[4] The high temperature--high density state of a solid may consist of an ordered arrangement of the nuclei, with the electrons moving practically freely. Under equilibrium conditions, the "lattice of nuclei" will melt at a certain high temperature. In view of the high temperature, this phase change will probably occur rather rapidly. However, as will be shown in Sec. 14, such a phase transition at very high temperature will almost certainly leave our general theory valid.

On the side of expanded states $(V_2 > V_1)$, the entropy S_2 is less than S_1. Now it can easily be seen that there is no solution for very large V_2. For it follows from the fundamental equation, Eq. (8), that for $V_2 >> V_1$

$$E_2 - E_1 \approx -\frac{1}{2}(p_1 + p_2)V_2 < -\frac{1}{2}p_1 V_2 \quad , \tag{41}$$

which goes to $-\infty$ as V_2 increases. Since there is a lower bound to the energy, no solutions exist for V_2 very large. Therefore the shock curve must leave the range of definition either at $S_2 = 0$, or at a phase boundary. In the latter case, we denote the value of S_2 reached at the phase boundary, by S_B.

We have thus proved:

The shock curve is a continuous curve, beginning at $S_2 = 0$ *(absolute zero temperature) or at phase boundary, and at finite* $V_2 > V_1$, *and going to* $S_2 = \infty$ *and* $V_2 = (1/4)V_1$. *The entropy has no maxima and minima. Every solution of the shock equation (for the given phase) lies on this shock curve.*

These statements contain the theorem enounced at the beginning of this section -- namely, that there is one and only one solution for any value of S_2 greater than S_1. It is possible to arrange all solutions for given V_1, S_1 in a single sequence according to the value of S_2.

8. *Behavior of volume, energy and pressure on the shock curve*

The specific volume V does not show such monotonic behavior as the entropy S. Only for expanded states $V_2 > V_1$, which are of relatively little interest, can we show that there is at most one solution for any given V_2. This has been done in Sec. 6, by means of Eq. (31a), [see the remarks after Eq. (31a)].

Thus, on the side of expanded states, the volume increases monotonically with decreasing entropy. As we have shown in Sec. 7, the shock curve ends in general on a phase boundary or possibly on the line $S_2 = 0$, but in any case with a finite volume V_2 which we may denote by V_B. In other words, the possible states in front of shock wave, behind which the material is in state V_1, S_1, can be ordered in a single sequence of monotonically increasing volume, ranging from V_1 to V_B.

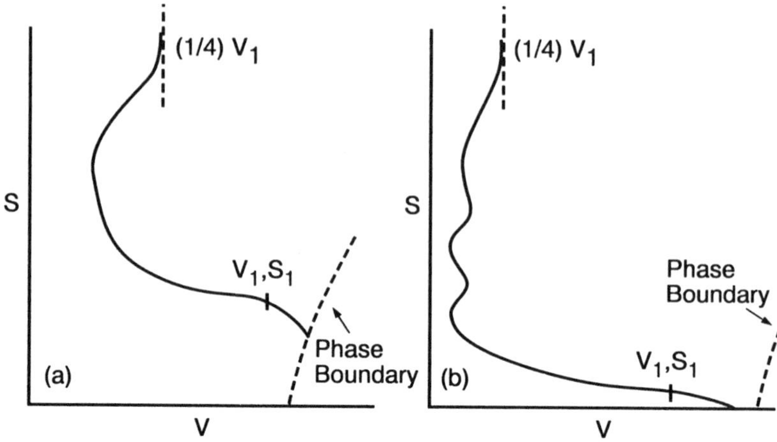

Fig. 2. Examples of shock curves in a specific volume - entropy diagram. Both curves approach V = $V_1/4$ asymptotically for large S. (a) Only one minimum for V; curve ends for large V on a phase boundary. (b) Three minima of V; curve goes to S = 0.

For compressed states $(V_2 < V_1)$, there is no restriction on the number of solutions for a given V_2. In fact, we have shown in Sec. 5 that for any real substance there will be at least two solutions for any V_2 between $V_{2\,\min}$ and $V_1 / 4$ Typical shock curves in the *V-S* diagram are shown in Fig. 2. The (first) minimum of V_2 will ordinarily occur at a rather high temperature at which dissociation of the molecules (or ionization) is rather far advanced for air, it lies at about 8,000°, and has a value of about $V_1 / 14$. For higher temperature T_2 (or entropy S_2), V_1 will increase again. For some substances, V_2 will have further maxima and minima (Fig. 2b) while for others it may increase monotonically from $V_{2\,\min}$ to the asymptotic value, $V_1 / 4$ (Fig. 2a).

The part of the shock curve between V_1 and $V_{2\,\min}$ is ordinarily the part of greatest practical importance and may therefore be called the "main section."

We shall now consider the *energy* E. We have the thermodynamic relation

$$dE = T_2 dS - p_2 dV \quad . \tag{29}$$

where, as usual, *dE*, *dS*, *dV* denote small changes of E_2, S_2, V_2, with E_1, S_1, V_1 kept constant. Over the "main section" of the shock curve (see preceding paragraph), *dV* is negative for positive *dS*. Therefore *dE* is positive, that is, the energy increases with increasing S_2.

When V_2 has passed its minimum, and S_2 increases further, the second term in Eq. (29) becomes negitive, but, directly behind the minimum of V_2, dV will be small so that the energy will still increase with S. The same will again be true at extremely high entropy where (Sec. 5) the volume remains almost constant, $V_2 = V_1 / 4$. However, in between there may, in principle, be a maximum of E_2 (followed by a minimum at a higher value of S_2). In order to derive the condition for such a maximum, we use the fundamental equation

$$2(E_2 - E_1) = (p_2 + p_1)(V_1 - V_2) \tag{8}$$

and consider p_2 as a function of E_2 and V_2. Differentiating, we obtain

$$2dE = (\partial p / \partial E)_V (V_1 - V_2)dE + (V_1 - V_2)(\partial p / \partial V)_E dV - (p_1 + p_2)dV \quad . \tag{42}$$

Therefore, if dE / dV is to be zero, we must have:

$$\left(\frac{\partial p}{\partial V}\right)_E = \frac{p_1 + p_2}{V_1 - V_2} \quad . \tag{43}$$

The right hand side is positive for compressed states. Therefore, a necessary, though by no means sufficient, condition for the occurrence of a maximum of E_2 is:

$$\left(\frac{\partial p}{\partial V}\right)_E > 0 \quad . \tag{44}$$

Therefore, we can exclude the occurrence of any energy maximum on the shock curve if we require

$$\left(\frac{\partial p}{\partial V}\right)_E < 0 \quad . \tag{III}$$

Condition III can easily be shown to be satisfied for ideal gases and for dissociating gases. We have not found any exception from it for any one-phase system (Sec. 6). In fact, condition III is even fulfilled for most phase transitions for which condition I is violated. We believe therefore that condition III is generally valid when condition I is valid, i.e., when our entire theory is justified.

If condition III is satisfied, the energy of "compressed states" increases monotonically with the entropy, from E_1 to ∞. For "expanded states," we have shown in the beginning of this section that the volume V_2 increases monotonically with decreasing S_2. Therefore both terms in Eq. (29) are negative, i.e., the energy decreases monotonically with decreasing entropy. Therefore we find:

If condition III is fulfilled, the energy E_2 *increases monotonically with the entropy, along the entire shock curve.* (For the monotonic increases of S, condition III need not be fulfilled.) In Sec. 17 we shall discuss what happens if, for some substance, condition III should be violated while conditions I and II are satisfied.

For the *pressure,* we shall prove first that $\Delta p = p_2 - p_1$ cannot be zero anywhere, except at the "origin," $V_2 = V_1$, $S_2 = S_1$. For compressed states, $V_2 < V_1$, this follows most easily from a relation proved in the next section, Eq. (50), which states that $u_1{}^2 > a_1{}^2$ if $V_2 < V_1$. Using this result we find from Eq. (6)

$$p_2 - p_1 = (V_1 - V_2)\frac{u_1{}^2}{V_1{}^2} > 0 \quad \text{if } V_2 < V_1 \quad . \tag{45}$$

Since the shock equations are symmetrical in states 1 and 2, it follows from Eq. (45) that also

$$p_2 < p_1 \quad \text{if} \quad V_2 > V_1 \quad . \tag{45a}$$

Summarizing, we have proved that *the material behind a shock wave must have greater density, entropy, energy and pressure than the material in front of the wave.*

Next, we can show that p_2 increases monotonically as the entropy increases from S_1 to ∞ provided the energy also increases monotonically, that is, provided condition III is fulfilled. It can be seen immediately that p_2 certainly increases S_2 as long as V_2 decreases. This follows from Eq. (30) which gives, for negative ΔV,

$$|\Delta V| dp = 2T_2 dS - \Delta p dV \quad . \tag{46}$$

Since Δp is positive (see previous paragraph), it follows that dp must be positive if dS is positive and dV negative. On the other hand, if the volume *increases* with increasing entropy, we may use Eq. (8), viz

$$p_2 + p_1 = \frac{2(E_2 - E_1)}{V_1 - V_2} \quad . \tag{47}$$

Then, as long as E_2 increases with the entropy, the numerator will increase, the denominator decrease; therefore p_2 will still increase with increasing S_2.

For expanded states, very little can be said about p_2, beyond the statement $p_2 < p_1$ [Eq. (45a)]. Equation (30) gives, for positive ΔV and negative Δp:

$$\Delta V dp = -2T_2 dS - |\Delta p| dV \quad . \tag{48}$$

The first term is positive for decreasing entropy, the second negative, and we do not have much information about their relative magnitude. Therefore p_2 may easily have maxima and minima, for $p_2 < p_1$.

Thus we find that energy and entropy increase monotonically together, being greater than E_1, S_1 for compressed, smaller than E_1, S_1 for expanded states. The pressure increases monotonically with the entropy for compressed states; for expanded states, it need not behave monotonically but will always remain below p_1. The volume increases monotonically with decreasing entropy for expanded states; for compressed states, it will in general not decrease monotonically but will always remain smaller than V_1.

9. *Relation between velocity and entropy*

In Eq. (30) we have shown that

$$2T_2 dS = \Delta p dV - \Delta V dp \quad , \tag{30}$$

if S_2, V_2, p_2 and $S_2 + dS, V_2 + dV, p_2 + dp$ are both solutions of the shock equation for the same initial state S_1, V_1, p_1. These two neighboring solutions will correspond to different velocities, u_1 and $u_1 + du_1$, of the shock wave relative to medium 1. Differentiating Eq. (6), logarithmically, we get

$$\frac{2du_1}{u_1} = \frac{dp}{\Delta p} - \frac{dV}{\Delta V} \quad . \tag{49}$$

Comparing this with Eq. (30), we find immediately

$$\frac{du_1}{u_1} = -\frac{T_2}{\Delta p \Delta V} dS \quad , \tag{49a}$$

and multiplying again by Eq. (6),

$$u_1 du_1 = \frac{V_1^2}{(\Delta V)^2} T_2 dS \quad . \tag{50}$$

By Eq. (50), the velocity change du_1 is uniquely related to dS, the coefficient being positive definite. Therefore, since the entropy increases monotonically on the shock curve from S_B to infinity, the velocity u_1, must also increase monotonically. Moreover, there can be *only one* solution of the shock equations for any given value of u_1, but we have not yet shown whether or not a solution will exist for a given u_1.

For $S_2 = S_1$, we have shown already in Sec. 3, Eq. (10), that u_1 is the sound velocity, a_1. For $S_2 > S_1$. (compression waves), Eq. (50) shows that u_1 must be greater than a_1. As S_2 increases, V_2 approaches the value $V_1 / 4$ [see Eq. (23)] while p_2 can increase indefinitely; therefore, according to Eq. (6), u_1 will also increase indefinitely. The same conclusion can also be deduced from Eq. (50). Therefore:

For any initial state of the material, V_1, S_1, there exists one and only one solution of the shock equations for any shock wave velocity, u_1, greater than the velocity of sound in the material, a_1.

For all these solutions, the specific volume V_2 behind the shock wave will be less than that in front, V_1, while pressure, entropy and energy are higher behind the wave than in front of it. If conditions I, II and III are satisfied, pressure, energy and entropy behind the shock wave are monotonically increasing functions of the shock wave velocity u_1. In the case of the entropy S_2, this statement is valid even if condition III should be violated.

For $S_2 < S_1$ (expanded states; for their physical meaning, see beginning of Sec. 6), u_1 must be smaller than the velocity of sound, a_1. As S_2 decreases, u_1 will also decrease. However, when S_2 becomes equal to S_B (phase boundary) or even equal to zero, u_1 will certainly not vanish because V_2 is finite, as shown in the last section, and p_2 must be smaller than p_1 [see Eq. (45a)]; therefore Eq. (6) gives

$$\frac{u_1^2}{V_1^2} = \frac{p_1 - p_2}{V_2 - V_1} > 0 \quad \text{for } S_2 = S_B \quad . \tag{51}$$

Let us denote the value of u_1 for $S_2 = S_B$ (or 0), by u_{1B}.

Then we find: For given V_1, S_1, there is one and only one solution for every u_1 between u_{1B} and a_1. These solutions correspond to $V_2 > V_1, p_2 < p_1, E_2 < E_1$, and $S_2 < S_1$, and represent those states which may exist in front of a shock wave when the material behind the wave is in state V_1, S_1. For $u_1 < u_{1B}$, there is no solution of the shock equations.

The result which will be most important for the stability considerations (Secs. 10, 11) is this:

The velocity of a shock wave with respect to the material in front of it (less dense material) is always greater than the sound velocity in that material; the velocity relative to the material behind the wave (denser material) is always less than the corresponding sound velocity.

10. Stability of shock waves against splitting into waves moving in the same direction

Consider a shock wave which is preceded or followed by infinitesimal waves, either compression or rarefaction waves. Since infinitesimal waves move with the velocity of sound of the medium, the shock wave will, according to the last theorem of Sec. 9, move faster than the infinitesimal waves in front of it, and slower than those behind it. Therefore the shock wave will catch up with the sound waves preceding it, and will be overtaken by the sound waves following it.

Now consider two shock waves moving in the same direction. With respect to the material between the two waves, the "front" shock wave will move more slowly than sound, the "rear" shock wave faster than sound. Therefore the rear shock wave will overtake the front shock wave.

The same will be true for any arrangement of waves moving in the same direction, however many shock waves and infinitesimal waves it may contain. Each shock wave in the system will move faster than the wave preceding it and slower than the wave following it, and there is therefore always a tendency towards combination of waves. We shall now prove the stability of a single shock wave against splitting into several waves moving in the same direction. Let the shock wave be at $x = 0$ at time $t = 0$. If the wave splits at this instant, all the partial waves must start from the same point, $x = 0$. But, according to our discussion, the "preceding" waves move more slowly than those "following" them, which is obviously impossible if they all start from the same point.

No shock wave can split into "partial" waves traveling in the same direction, whether these partial waves be shock waves or infinitesimal ones.

Quite generally, it is impossible that a shock wave and any other wave start from the same point at the same time in the same direction. There can only be either a shock wave, or a train of infinitesimal rarefaction waves.

11. Stability against any splitting

The result of the last section still leaves the possibility open that a shock wave may split spontaneously into two shock waves moving in opposite directions (instead of the shock wave moving opposite to the direction of the original shock wave, we may substitute a train of infinitesimal rarefactions waves). In addition to the two shock waves, as von Neumann has pointed out, there will in general exist a discontinuity of the density (and entropy) which is stationary with respect to the material and remains at the material point where the splitting of the original shock wave has occurred. The pressure is continuous at this point.

Let 0 be the original shock wave, A that shock wave after the split which moves in the same direction as 0 which we call "to the right," C the wave which moves "to the left," and B the stationary discontinuity of density. Further, let 1 be the material to the right of wave 0 or A which is as yet unaffected by the shock waves, 2 the material between A and B, 3 that between B and C, and 4 that behind C or, originally, behind 0.

Then materials 4 and 1 must be connected by the shock equations since they are originally separated by shock wave 0, thus

$$2(E_4 - E_1) = (p_4 + p_1)(V_1 - V_4) \tag{52}$$

and

$$\frac{u_{01}^2}{V_1^2} = \frac{u_{04}^2}{V_4^2} = \frac{p_4 - p_1}{V_1 - V_4} \quad . \tag{53}$$

where u_{01} and u_{04} denote the velocity of the shock wave 0 relative to the media 1 and 4, respectively. Medium 4, then, moves to the right relative to 1 with the velocity

$$v = u_{01} - u_{04} \quad . \tag{54}$$

This velocity must, of course, remain unchanged after the split of the shock wave.

Materials 2 and 1 must also be connected by the shock equations because wave A must obviously be a shock wave rather than a rarefaction wave. Consequently,

$$2(E_2 - E_1) = (p_2 + p_1)(V_1 - V_2) \quad , \tag{55}$$

$$\frac{u_{A1}^2}{V_1^2} = \frac{u_{A2}^2}{V_2^2} = \frac{p_2 - p_1}{V_1 - V_2} \quad , \tag{55a}$$

and the velocity of 2 with respect to 1 will be

$$\upsilon_2 = u_{A1} - u_{A2} \tag{55b}$$

towards the right.

Materials 3 and 2 have the same pressure and velocity,

$$\begin{aligned} p_3 &= p_2 \quad , \\ \upsilon_3 &= \upsilon_2 \quad , \end{aligned} \tag{55c}$$

while there is no condition on $V_3 - V_2$ or on $E_3 - E_2$. Wave C may either be a shock wave or a rarefaction wave. We shall not use the detailed theory of that wave but only the following simple consideration. If u_{C3} and u_{C4} are the velocities of wave C relative to the media 3 and 4, then, since wave C moves to the left relative to the material, the velocity to the *right* of medium 4 relative to 3 will be

$$\upsilon_4 - \upsilon_3 = u_{C4} - u_{C3} \quad . \tag{56}$$

Using Eqs. (55b,c), we get for the velocity to the right of medium 4 relative to medium 1:

$$\upsilon_4 = u_{A1} - u_{A2} + u_{C4} - u_{C3} \quad . \tag{56a}$$

This quantity must be equal to the velocity v given by Eq. (54), i.e.,

$$u_{01} - u_{04} = u_{A1} - u_{A2} + u_{C4} - u_{C3} \quad . \tag{57}$$

This will be the fundamental equation of the theory.

Now we have from Eq. (3)

$$\frac{u_{C4}}{u_{C3}} = \frac{V_4}{V_3} \quad . \tag{58}$$

Therefore

a) if wave C is a compression wave,

$$p_2 = p_3 > p_4 \quad . \tag{59}$$

Therefore, according to Sec. 8

$$V_4 > V_3 \quad ;$$

and according to Eq. (58)

$$u_{C4} > u_{C3} \quad .$$

Inserting this into Eq. (57) gives

$$u_{01} - u_{04} > u_{A1} - u_{A2} \quad . \tag{59a}$$

b) if wave C is a rarefaction wave,

$$p_2 < p_4 \quad , \tag{60}$$

we obtain by a similar reasoning

$$u_{01} - u_{04} < u_{A1} - u_{A2} \quad . \tag{60a}$$

Equations (59a) and (60a) are in a convenient form since they require the comparison of two solutions of the shock equation belonging to the same initial state, 1. From Eq. (58) we find

$$(u_{01} - u_{04})^2 = (p_4 - p_1)(V_1 - V_4) \quad , \tag{61}$$

and similarly

$$(u_{A1} - u_{A2})^2 = (p_2 - p_1)(V_1 - V_2) \quad . \tag{61a}$$

Obviously, Eqs. (59) to (61) are completely symmetrical with respect to states 2 and 4; therefore we can assume without loss of generality that $p_4 > p_2$, that is, Eq. (60). Then Eq. (60a) must hold, and therefore we find from Eqs. (61) and (61a) the condition for split:

$$(p_4 - p_1)(V_1 - V_4) < (p_2 - p_1)(V_1 - V_2) \quad . \tag{62}$$

Since $p_4 > p_2$, Eq. (62) is obviously impossible if

$$V_4 < V_2 \quad , \tag{62a}$$

However, Eq. (62a) follows automatically from $p_4 > p_2$ as long as both states, 2 and 4, lie on a part of the shock curve on which the volume decreases with increasing entropy, for example, on the "main section" (Sec. 8). For most substances, this main section probably covers all temperatures up to about 10,000° (Sec. 8).

However, at very high temperatures which may still be experimentally accessible, V_2 increases with increasing entropy (Sec. 8) while p_2 continues to increase. Then Eq. (62a) does not hold but is reversed, that is, $V_4 < V_2$. In this case, we have certainly

$$2p_1(V_1 - V_4) < 2p_1(V_1 - V_2) \quad . \tag{62b}$$

Adding this inequality to Eq. (62) and using the fundamental shock equations (52) and (55), we get the condition

$$E_4 < E_2 \quad , \tag{63}$$

which is necessary but not sufficient for the split of the original shock wave. (It is almost sufficient since $p_1 << p_2$ beyond the minimum of the volume.) Now, as we have shown in Sec. 8, the energy will increase monotonically with the pressure if the material fulfills condition III. Then Eq. (63) can certainly not be fulfilled for $p_4 > p_2$. Thus we find:

A shock wave can never split in a material whose equation of state fulfills the three conditions I, II, III.

As these conditions appear to be valid for practically all materials, as long as there are no phase transitions, *we have proved the complete stability of one-dimensional shock waves in all ordinary materials.*

III. Investigation of the Three Conditions

In this part of the paper, we shall investigate the validity of our three postulates

$$(\partial^2 p / \partial V^2)_S > 0 \quad , \qquad \text{I}$$

$$V(\partial p / \partial E)_V > -2 \quad , \qquad \text{II}$$

$$(\partial p / \partial V)_E < 0 \quad , \qquad \text{III}$$

which we have used throughout in our theory of shock waves (Part I). Of these conditions only the first two are needed in the general proof (up to Sec. 10); the last one is only required to prove the stability of shock waves against any kind of splitting (Sec. 11).

Analysis shows that none of the three conditions is required by any general thermodynamic or statistical argument because it can be shown that for each one of the three conditions there exist some substances for which the condition is violated at certain temperatures and densities. Therefore, we can only ascertain the range of validity of the conditions by investigating a sufficient number of different physical states. For all single-phase systems which we have investigated, all three conditions have been found valid, by a wide margin. Therefore we believe that they are valid for all single phase systems of any practical importance.

The cases in which one or more of the conditions are violated all refer to phase transitions. Therefore we had to exclude phase transitions in Part I of this paper. In Sec. 14 and Sec. 17 we shall discuss briefly some of the phenomena which might occur if phase transitions could take place in shock waves.

12. The condition $\left(\partial^2 p / \partial V^2\right)_S > 0$ *for single-phase systems*

The simplest equation of state is that of a perfect gas with constant specific heat. Then the adiabatics are given by

$$p = \text{constant} \cdot V^{-\gamma} \quad , \tag{64}$$

where γ is the ratio of the specific heats, C_p / C_V, and the constant depends only on the entropy. Therefore

$$\left(\frac{\partial^2 p}{\partial V}\right)_S = \gamma(\gamma+1)\frac{p}{V^2} > 0 \quad , \tag{64a}$$

which is certainly positive.

For most cases it is convenient to express the adiabatic derivative in terms of isothermal derivatives. For any function f (V, T), we have

$$\left(\frac{\partial f}{\partial V}\right)_S = \left(\frac{\partial f}{\partial V}\right)_T - \frac{(\partial S / \partial V)_T}{(\partial S / \partial T)_V}\left(\frac{\partial f}{\partial T}\right)_V \quad . \tag{65}$$

Here we may use the thermodynamic relations

$$\left(\frac{\partial S}{\partial V}\right)_T = \left(\frac{\partial p}{\partial T}\right)_V \quad , \tag{66}$$

$$\left(\frac{\partial S}{\partial T}\right)_V = \frac{C_V}{T} \quad , \tag{66a}$$

where C_V is the specific heat at constant volume, a positive definite quantity. Applying Eq. (65) to f = p, we find

$$\left(\frac{\partial p}{\partial V}\right)_S = \left(\frac{\partial p}{\partial V}\right)_T - \frac{T}{C_V}\left(\frac{\partial p}{\partial T}\right)_V^2 \quad . \tag{67}$$

Since $(\partial p / \partial V)_T$ is negative, we find that the adiabatic modulus of compression $-(\partial p / \partial V)_S$, always greater than the isothermal one. Another differentiation gives

$$\left(\frac{\partial^2 p}{\partial V^2}\right)_S = \frac{\partial^2 p}{\partial V^2} - \frac{2T}{C_V}\frac{\partial p}{\partial T}\frac{\partial^2 p}{\partial V \partial T} + \frac{T}{C_V{}^2}\left(\frac{\partial p}{\partial T}\right)^2 \frac{\partial C_V}{\partial V}$$

$$- \frac{T}{C_V}\frac{\partial p}{\partial T}\left[\frac{\partial^2 p}{\partial V \partial T} - \frac{1}{C_V}\left(\frac{\partial p}{\partial T}\right)^2 + \frac{T}{C_V{}^2}\left(\frac{\partial p}{\partial T}\right)^2 \frac{\partial C_V}{\partial T} - \frac{2T}{C_V}\frac{\partial p}{\partial T}\frac{\partial^2 p}{\partial T^2}\right] . \tag{67a}$$

On the right-hand side, p and C_V are considered as functions of V and T, so that $\partial / \partial T$ implies that V is kept constant, and vice versa. Equation (67a) may be slightly simplified by means of the thermodynamic relation.

$$T\frac{\partial^2 p}{\partial T^2} = \frac{\partial C_V}{\partial V} \quad . \tag{67b}$$

Then Eq. (67a) becomes

$$\left(\frac{\partial^2 p}{\partial V^2}\right)_S = \left(\frac{\partial^2 p}{\partial V^2}\right)_T - \frac{3T}{C_V}\frac{\partial p}{\partial T}\frac{\partial^2 p}{\partial V \partial T} + \frac{3T}{C_V{}^2}\left(\frac{\partial p}{\partial T}\right)_V^2 \frac{\partial C_V}{\partial V}$$

$$+ \frac{T}{C_V{}^2}\left(\frac{\partial p}{\partial T}\right)^3 \left(1 - \frac{T}{C_V}\frac{\partial C_V}{\partial T}\right) \quad . \tag{68}$$

From Eq. (68) we can easily get an idea about the terms which might theoretically cause $\left(\partial^2 p / \partial V^2\right)_S$ to be negative. Beginning with the last term of Eq. (68), we have the following possibilities:

(a). The specific heat may increase rapidly with the temperature. Then, if $\partial p / \partial T > 0$ which is the normal behavior, the last term of Eq. (68) is negative. It will be large in absolute value if, simultaneously with a large $\partial C_V / \partial T$, we have a small specific heat C_V. This points to low temperatures as the place where $\left(\partial^2 p / \partial V^2\right)_S$ might be most likely to become negative. The following cases of rapidly increasing specific heat will be discussed below:

(i) Ideal gases with internal degrees of freedom, such as vibration or electronic excitation.
(ii) Dissociating, but otherwise ideal, gases (ionization is a special case of dissociation.
(iii) Solids at very low temperature.

In cases (i) and (ii), the first term in Eq. (68) can be shown to be numerically larger than the last one, because C_V is never very small (at least $3k/2$ for monatomic, $5k/2$ for diatomic gases, etc.). In case (iii), the term with $\partial C_V / \partial V$ is positive and numerically greater than the last term.

(b). The specific heat may decrease with increasing volume, that is, $\partial C_V / \partial V < 0$, This case is realized for a number of liquids. However, from the empirical data it can easily be shown that the third term in Eq. (68) is usually less than one per cent of the first.

(c). The pressure may decrease with the temperature $\left(\partial p / \partial T < 0\right)$ while the modulus of compression, $-\partial p / \partial V$ increases with T $\left(\text{i.e., } \partial^2 p / \partial V \partial T < 0\right)$. Then the second term in Eq. (68) is negative. Water below 4° is an example, but the second term is again less than one per cent of the first.

(d). The derivative at constant temperature, $\partial^2 p / \partial V^2$, may be negative. This happens at and near the critical temperature, for volumes greater than the critical volume. In this case, the (positive) second term of Eq. (68) more than outweighs the first.

In all cases mentioned, the resulting value of $\left(\partial^2 p / \partial V^2\right)_S$ is positive. We shall now discuss the various cases in order.

(a) *Specific heat increasing rapidly with temperature.*

(i) *ideal gas with variable specific heat.* For an ideal gas, we have

$$pV = RT \quad , \tag{69}$$

$$\left(\frac{\partial p}{\partial T}\right)_V = \frac{R}{V} > 0 \quad , \tag{69a}$$

$$\frac{\partial C_V}{\partial V} = T\frac{\partial^2 p}{\partial T^2} = 0 \quad , \tag{69b}$$

[See Eq. (67b)], Eq. (68) becomes

$$\left(\frac{\partial^2 p}{\partial V^2}\right)_S = \frac{p}{V^2}\left[\left(2+\frac{R}{C_V}\right)\left(1+\frac{R}{C_V{}^2}\right) - \frac{R^2}{C_V{}^2}\frac{T}{C_V}\frac{dC_V}{DT}\right] \quad . \tag{70}$$

The critical term is obviously the last one. To get an estimate of its value, we consider a vibration of the molecules of frequency $\nu = k\theta / h$. Then the specific heat is

$$\frac{C_V}{R} = \beta + \left(\frac{\theta}{T}\right)^2 \frac{e^{\theta / T}}{(e^{\theta/T} - 1)^2} \quad , \tag{70a}$$

where βR is the specific heat without the vibration (β = 2.5 for linear, β = 3 for non-linear molecules). The highest value of the last term in Eq. (70) is obtained for $T = \theta / 4$; at this temperature, the contribution of the oscillation to C_V is about 0.3R while the derivative, TdC_V / dT, is about 0.65R. Then, if β = 2.5, the last term in Eq. (70) becomes -0.030, as against 3.2 for the first term. The negative term is thus only about one per cent of the positive one.

It might be expected that the negative term in Eq. (70) will become greater if the molecule has several modes of vibration, because each mode will contribute to TdC_V / dT. However, this effect will be largely offset by the increase of C_V itself. For example, for a molecule having as many as 100 different modes of vibration, all of the same frequency, and having β = 3, the maximum of the negative term occurs at T ≈ 0.120 and has a value of about 0.11, as against 2.7 for the positive term. Only for a molecule with more than 10^{66}(!) vibrational modes all of the same frequency, would the derivative $\left(\partial^2 p / \partial V^2\right)_S$ become negative at certain (low) temperatures. From this we see that, *while it is in principle possible that postulate I is violated*, this will never occur due to the excitation of vibrations for any real gas. It can easily be seen that the same holds for the excitation of the higher electronic states.

(ii) *Dissociation of molecules.* In this case, the exact calculation would become exceedingly complicated. We can, however, get a rather good approximation by remembering that the significant term in Eq. (68) is the last one, and that this term is greatest when the specific heat rises steeply but is not yet itself very large. This will occur when the degree of dissociation, α, is still very low; in fact, from our calculation we shall find that a value of α less than one per cent is most favorable. Then we can neglect α compared with 1, but we must, of course, not neglect $T\partial\alpha / \partial T$.

When a molecule dissociates into n+1 atoms, the degree of dissociation is given by

$$\frac{\alpha^{n+1}}{(1-\alpha)V^n} = K(T) = A^n e^{-Q/RT} \quad , \tag{71}$$

where K is the dissociation constant, Q the dissociation energy per gram, R the gas constant per gram of the molecular gas. A^n is the ratio of the "a priori probabilities" of dissociated and molecular states and depends only slightly on temperature. The numerical value of AV is of the order 10^4 to 10^8 for ordinary densities and temperatures.

With our assumption, $\alpha \ll 1$, we have

$$\alpha = (AV)^{\frac{n}{n+1}} e^{-q} \quad , \tag{71a}$$

with

$$q = \frac{Q}{(n+1)RT} \quad . \tag{71b}$$

We shall need the derivatives

$$\frac{\partial\alpha}{\partial V} = \frac{n}{n+1}\frac{\alpha}{V} \quad , \tag{72}$$

$$\frac{\partial\alpha}{\partial T} = q\frac{\alpha}{T} \quad . \tag{72a}$$

Because of the large value of the "a priori probability" AV, we get appreciable dissociation already for quite large values of q. If we require $(n+1)\alpha \approx 0.001$ to 0.01 (see below) and have AV = 10^4 to 10^8 then

$$q = 10 \text{ to } 30 \quad . \tag{72b}$$

The pressure is given by

$$p = \frac{RT}{V}(1 + n\alpha) \quad . \tag{73}$$

The derivatives required in Eq. (68) are, neglecting $n\alpha$ but keeping $qn\alpha$:

$$\frac{\partial p}{\partial T} = \frac{R}{V}\left(1 + n\alpha + nT\frac{\partial \alpha}{\partial T}\right) = \frac{R}{V}(1 + qn\alpha) \quad , \tag{73a}$$

$$\frac{\partial^2 p}{\partial V \partial T} = -\frac{R}{V^2}\left(1 + qn\alpha - qnV\frac{\partial \alpha}{\partial V}\right) = -\frac{R}{V^2}\left(1 + q\frac{n}{n+1}\alpha\right) \quad , \tag{73b}$$

$$\frac{\partial^2 p}{\partial V^2} = \frac{2RT}{V^3} \quad . \tag{73c}$$

The specific heat may be written

$$C_V = C_{V,0} + Q\frac{\partial \alpha}{\partial T} \quad , \tag{74}$$

where $C_{V,0}$ is the value of the specific heat if the degree of dissociation does not change, i.e., essentially the C_V of the molecule. Putting

$$C_{V,0} = \beta R \quad , \tag{74a}$$

where β is a slowly changing function of the temperature, and using Eq. (71b), we find

$$C_V = R\left[\beta + (n+1)q^2\alpha\right] \quad . \tag{75}$$

Therefore

$$\frac{\partial C_V}{\partial V} = \frac{R}{V}nq^2\alpha \quad , \tag{75a}$$

$$\frac{\partial C_V}{\partial T} = R\left[\frac{\partial \beta}{\partial T} + (n+1)\frac{\alpha}{T}q^2(q-2)\right] \quad . \tag{75b}$$

Inserting the results of the last paragraph into Eq. (68), we obtain, term by term

$$\left(\frac{\partial^2 p}{\partial V^2}\right)_S = \frac{p}{V^2}\left\{2+3\frac{(1+nq\alpha)\left(1+\frac{n}{n+1}q\alpha\right)}{\beta+(n+1)q^2\alpha}+3\frac{nq^2\alpha(1+nq\alpha)^2}{\left[3+(n+1)q^2\alpha\right]^2}\right.$$

$$\left.+\frac{(1+nq\,\alpha)^3}{\left[\beta+(n+1)q^2\alpha\right]^2}-\frac{(1+nq\,\alpha)^3}{\left[\beta+(n+1)q^2\,\alpha\right]^3}\left[T\frac{d\beta}{dT}+(n+1)q^2(q-2)\alpha\right]\right\} \quad . \tag{76}$$

Only the last term is negative, as expected. We know already that the term $Td\beta / dT$ is harmless [see case (i)]. The other negative term is largest for large q and relatively small α. It can easily be shown that the maximum of this term, for fixed q (>> 1) and variable is α obtained for

$$(n+1)q^2\alpha \approx \frac{1}{2}\beta \quad . \tag{76a}$$

Then $nq\alpha < \beta / 2q$, and the last term of Eq. (76) becomes approximately

$$\frac{\frac{1}{2}\beta\left(q+\frac{3}{2}\beta-2\right)}{\left(\frac{3}{2}\beta\right)^3} = \frac{4}{27}\frac{q+\frac{3}{2}\beta-2}{\beta^2} \quad . \tag{76b}$$

The highest value which q can take [see Eq. (72b)] is about 30, and this can only occur for large n in which case β is at least 3, and in most cases very much higher. For q = 30, β = 3, we obtain for the last term of Eq. (76),

$$\text{Last term} = \frac{4}{27}\frac{32.5}{9} = 0.54 \quad , \tag{76c}$$

while the positive terms become 2.94. The *negative term* is thus *less than one fifth of the positive ones*, although we have made conditions most favorable for a large negative term. The value of $(n+1)\alpha$ becomes, with our assumptions, about 0.0017, justifying the neglections made $(n\alpha \ll 1)$ and also the value of α used in computing Eq. (72b).

The case of the ionization of monatomic gases may seem of interest because for these, β is only 1.5. However, at the same time the a priori weight A is reduced because of the small mass of the electron. Values of AV between 1 and 10^4 are usual, which, with $(n+1)\alpha = 0.01$, gives q between 5 and 14. In the most favorable case, this gives about 1 for the negative term in Eq. (76) against about 4 for the positive terms.

As in the case of an ideal gas with variable β, there exists the possibility of negative $\left(\partial^2 p / \partial V^2\right)_S$ but only for extreme dilution of the gas. If we consider the ionization of a monatomic gas which is most favorable for negative $\left(\partial^2 p / \partial V^2\right)_S$, and if AV=$10^4$ for one atmosphere pressure, we expect AV=10^{48} for a pressure of 10^{44} atmospheres (!). For this value of AV, and for $(n+1)\alpha$ =10^{-4}, we get q =120, which would make the negative term in Eq. (76) just greater than the positive ones. Thus we see again that $\left(\partial^2 p / \partial V^2\right)_S > 0$ is not required on statistical grounds but is very well fulfilled for all experimentally obtainable pressures.

(iii) *Solids at Very Low Temperatures*. The specific heat is given in good approximation by Debye's relation

$$C_V = \mathrm{a}(T / \theta)^3 \quad , \tag{77}$$

where θ is the Debye temperature and a is a constant. We have thus a rapid increase of C_V with temperature, and at the same time we can make C_V itself as small as we like, in contrast to the two previous cases where C_V was at least equal to the specific heat of translation and rotation. It seems therefore that the negative term with $\partial C_V / \partial T$, can be made as large as we like compared with the first term in Eq. (68). However, as we shall see, the term with $\partial C_V / \partial V$ saves the inequality $\left(\partial^2 p / \partial V^2\right)_S > 0$.

We have from Eq. (77)

$$\frac{T}{C_V}\frac{\partial C_V}{\partial T} = 3 \quad ; \tag{77a}$$

hence the three last terms in Eq. (68) become

$$\frac{T}{C_V{}^2}\left(\frac{\partial p}{\partial T}\right)^2\left(3\frac{\partial C_V}{\partial V} - 2\frac{\partial S}{\partial V}\right) \quad , \tag{77b}$$

where we have used the thermodynamic relation Eq. (66). At zero temperature, the entropy is zero for any V; further we have from Eq. (77) for a given V,

$$S = \int_0^T \frac{C_V}{T} dT = \frac{1}{3} C_V \quad . \tag{77c}$$

Inserting into Eq. (77b), the term in parentheses becomes

$$\left(3 - \frac{2}{3}\right) \frac{\partial C_V}{\partial V} = -7C_V \frac{\partial \log\theta}{\partial V} \quad , \tag{78}$$

which is certainly positive because the Debye temperature depends on the strength of the elastic forces and therefore increases upon compression.

(b) *Specific heat decreasing with increasing volume, liquids and solids at ordinary temperature* -- Here we may use the Tait equation of state

$$p = B(T)\left[e^{(V_o - V)/\kappa} - e^{(V_o - V_T)/\kappa}\right] \quad , \tag{79}$$

where κ is a certain constant of the dimension of a volume, usually about one-tenth of the volume of the substance, V_T is the volume at temperature T and zero pressure, and V_o a suitably chosen standard volume (constant). The function B, in all cases the author is aware of, increases with T; for water it has a value of about 3000 bars (1 bar $\approx$ 1 kg/cm^2 $\approx$ 1 atm).

We have then

$$\frac{\partial^2 p}{\partial V^2} = \frac{B}{\kappa^2} e^{(V_o - V)/\kappa} > 0 \quad , \tag{79a}$$

$$\frac{\partial^2 p}{\partial V \partial T} = -\frac{1}{\kappa} \frac{dB}{dT} e^{(V_o - V)/\kappa} < 0 \quad . \tag{79b}$$

For all substances with positive expansion coefficient, $\partial P / \partial T$ is positive, therefore the first two terms in Eq. (68) are positive. Usually $\partial C_V / \partial T$ is very small so that the last parenthesis in Eq. (68) is also positive. There remains the term with $\partial C_V / \partial V$. This is "generally assumed to be zero"[5] but may actually be slightly negative, as for[5] CCl_4 and[6] C_6H_6. For CCl_4 at 45°C and 1 atmosphere pressure, we have

---

[5] R. E. Gibson, and D. H. Loeffler, Journ. Am. Chem. Soc., *63*, 898, 1941.

[6] R. E. Gibson, and D. H. Loeffler, Journ. Am. Chem. Soc., *61*, 2515, 1939.

$$\partial C_V / \partial V = -9.5 \ bar/degree, \qquad \mathrm{B} = 740 \ bar,$$

$$dB / dT = 2.2 \ bar/degree, \qquad \kappa = 0.0600 \ cm^3 / g,$$

$$T \partial p / \partial T = p + \partial E / \partial V = 3250 \ bars,$$

and

$$C_V = 0.21 \ cal / g \ degree = 8.9 \ bar \ cm^3 / g \ degree \quad .$$

Therefore

$$\frac{\partial^2 p}{\partial V^2} = 206{,}000 \quad bar \, (g / cm^3)^2 \quad ,$$

$$-\frac{3T}{C_V}\frac{\partial p}{\partial T}\frac{\partial^2 p}{\partial V \partial T} = 40{,}000 \quad bar \, (g / cm^3)^2 \quad ,$$

$$\frac{3T}{C_V^2}\left(\frac{\partial p}{\partial T}\right)^2 \frac{\partial C_V}{\partial V} = -12{,}000 \quad bar \, (g / cm^3)^2 \quad .$$

The negative term is thus seen to be only about one-seventeenth of the leading (first) term. It is also clear from the nature of the quantity $\partial C_V / \partial V$ that it cannot be very large because otherwise C_V would reach exceedingly high values for high compression.

(c) *Pressure decreasing with temperature, water below 4°C.* -- Below 4°C, water has a negative $\partial p / \partial T$. The expansion coefficient at 0°C, is[7]

$$\left(\partial V / \partial T\right)_p = -3.1 \times 10^{-5} \, cm^3 / g \ degree \quad . \tag{79c}$$

The characteristic pressure B, extrapolated from reference 5, is about 2750 bars, while $\kappa = 0.137 cm^3 / g$. Therefore

$$\left(\frac{\partial p}{\partial T}\right)_V = \frac{B}{\kappa}\left(\frac{\partial V}{\partial T}\right)_p = -0.62 \ \frac{bars}{degree} \quad . \tag{79d}$$

Further, $dB / dT \approx 10$ bar/deg (likewise extrapolated), so that [see Eq. (79b)]

[7] Dorsey, Properties of Ordinary Water Substance, Reinhold, 1940 (p. 231).

$$\partial^2 p / \partial T \partial V \approx -70 \ \ bar\ g / cm^3 degree \quad .$$

The second term in Eq. (68) becomes then, with $C_V = 42 \ \ bar\ cm^3 / g\ degree$,

$$-\frac{3T}{C_V}\frac{\partial p}{\partial T}\frac{\partial^2 p}{\partial V \partial T} = -850 \text{ bar} \left(\frac{g}{cm^3}\right)^2 \quad ,$$

while the first term

$$\frac{\partial^2 p}{\partial V^2} = \frac{B}{\kappa^2} = 145{,}000 \text{ bar}\left(\frac{g}{cm^3}\right)^2 \quad .$$

(d) *Derivative at constant temperature, $(\partial^2 p / \partial V^2)$, negative; neighborhood of the critical point.* [8]

On the critical isotherm we have $(\partial p / \partial V)_T = 0$ at the critical volume and $(\partial p / \partial V)_T < 0$ for larger volumes, so that $\left(\partial^2 p / \partial V^2\right)_T$ is negative at volumes somewhat above the critical one. This contribution may be compensated by the second term of Eq. (68): $(\partial p / \partial T)_V$ is positive for a gas, and $\partial^2 p / \partial V \partial T$ is negative, that is, the modulus of compression, $-\partial p / \partial V$, increases with T.

The calculation is simplest if we use reduced temperatures, volumes and pressures, namely

$$\upsilon = \frac{V}{V_C}, \quad \tau = \frac{T}{T_c}, \quad \pi = \frac{p}{p_c} \quad , \tag{80}$$

where V_C, T, and p_c are the critical volume, temperature, and pressure. According to Van der Waals' equation,

$$p_c V_C = \frac{3}{8} R T_c \quad , \tag{80a}$$

and the equation itself has the form

$$\pi = \frac{8}{3}\frac{\tau}{\upsilon - \frac{1}{3}} - \frac{3}{\upsilon^2} \quad . \tag{80b}$$

---

[8] I am indebted to Dr. G. Placzek for the calculations reported in this section.

The derivatives are:

$$\left(\frac{\partial^2 \pi}{\partial \upsilon^2}\right)_\tau = \frac{16}{3}\frac{\tau}{\left(\upsilon - \frac{1}{3}\right)^3} - \frac{18}{\upsilon^4} \quad , \tag{80c}$$

$$\left(\frac{\partial \pi}{\partial \tau}\right)_\upsilon = \frac{8}{3}\frac{1}{\upsilon - \frac{1}{3}} \quad , \tag{80d}$$

$$\left(\frac{\partial^2 \pi}{\partial \tau^2}\right)_\upsilon = 0, \qquad \text{therefore [see Eq. (67b)]} \quad \frac{\partial C_V}{\partial V} = 0 \quad , \tag{80e}$$

$$\left(\frac{\partial^2 \pi}{\partial \upsilon \partial \tau}\right) = -\frac{8}{3}\frac{1}{\left(\upsilon - \frac{1}{3}\right)^2} \quad . \tag{80f}$$

Inserting into Eq. (68), we obtain after simplification,

$$\left(\frac{\partial^2 p}{\partial V^2}\right)_S = \frac{16}{3}\frac{p_c}{V_c^2}\frac{1}{\left(\upsilon - \frac{1}{3}\right)^2}\left[\tau - \frac{27}{8}\frac{\left(\upsilon - \frac{1}{3}\right)^3}{\upsilon^4} + \frac{3\tau}{2\beta} + \frac{\tau}{2\beta^2}\right] \tag{81}$$

The first two terms in the square bracket arise from $\left(\partial^2 p / \partial V^2\right)_T$, the third term in Eq. (81) represents the second term in Eq. (68), and the last one is due to the fourth term in Eq. (68). The third term in Eq. (68) is zero because of the special form of the Van der Waals equation [see Eq. (80e)]; the last term in Eq. (68) has been assumed to be zero, i.e., C_V is assumed to be independent of temperature, and to be given by

$$C_V = \beta R \quad . \tag{81a}$$

The isothermal derivative, $\left(\partial^2 p / \partial V^2\right)_T$, i.e., the first two terms in Eq. (81), is negative if

$$\tau < \tau_\ell = \frac{27}{8}\frac{\left(\upsilon - \frac{1}{3}\right)^3}{\upsilon^4} \quad . \tag{81b}$$

The maximum of τ_ℓ is obtained for $\upsilon = 4/3$ and has the value

$$\tau_{\ell\,\max} = \frac{2187}{2048} = 1.0678 \quad . \tag{81c}$$

For a given v, the substance is a gas only if τ is greater than a certain τ_g, while for lower temperature, we would obtain an unstable state. If the Van der Waals equation is used, the minimum temperature for v = 4/3 is[9]

$$\tau_g(4/3) = 0.9838 \quad . \tag{81d}$$

For actual gases, τ_g is even higher than the value given by Van der Waals' equation.

According to Eqs. (81c, d), the isothermal derivative $\left(\partial^2 p/\partial V^2\right)_T$ is negative for v = 4/3 in the narrow temperature region from τ_g to $\tau_{\ell\,\max}$ and the region would be even narrower for other values of v. However, according to Eq. (81), the *adiabatic* derivative $\left(\partial^2 p/\partial V^2\right)_S$ can be negative only for

$$\tau < \tau_\ell - \left(\frac{3}{2\beta} + \frac{1}{2\beta^2}\right)\tau \quad . \tag{81e}$$

If this is to be greater than τ_g, we must have

$$\frac{3}{2\beta} + \frac{1}{2\beta^2} < \left(\tau_\ell - \tau_g\right)/\,\tau_g \quad .$$

With the values of Eqs. (81b, d), this gives

$$\beta > 17.5 \quad .$$

Such a large value of β is practically impossible for any gas at a temperature as low as the critical temperature where values of β between 2.5 and 4 are the rule. We find therefore that $\left(\partial^2 p/\partial V^2\right)_S$ remains positive near the critical point for all real gases obeying the Van der Waals equation.

13. The condition $(\partial^2 p/\partial V^2)_S > 0$ *for phase transitions*

We consider the transition between two phases, 1 and 2, and represent the state of the substance in a p,V-diagram (Fig. 3). Let phase 1 be that of smaller volume; then the pure phase 1 will lie on the left side of the diagram, the pure

---

[9] From Kuenen, Zustandsgleichung, 1907, p. 94 (According to a table calculated by J. P. Dalton)

phase 2 on the right side, and the mixture of the two phases in the middle region (Fig. 3). Isotherms will be horizontal in the two-phase region, and slope downwards for both the pure phases. Of course $\left(\partial^2 p / \partial V^2\right)_T$ is infinite at the boundaries between single-phase and two-phase regions; it is positive at the boundary of the denser phase (1), negative at that of the less dense phase (2).

Of course $\left(\partial^2 p / \partial V^2\right)_S$ will also have discontinuities at the two boundaries, and we wish to find out the sign of these discontinuities. In the pure phase 1, we have from Eq. (67)

$$\left(\frac{\partial p}{\partial V}\right)_{S,1} = \left(\frac{\partial p}{\partial V}\right)_{T,1} - \frac{T}{C_{V,1}}\left(\frac{\partial p}{\partial T}\right)_{V,1}^2 \quad . \tag{82}$$

In the two-phase region, p is a function of T only and is independent of V. Therefore

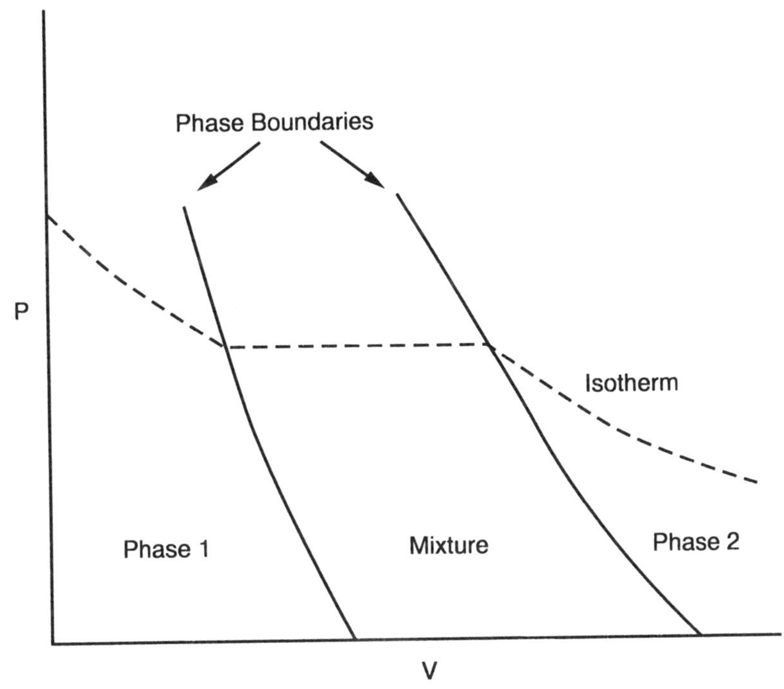

Fig. 3. Diagram of a phase transition.

$$\left(\frac{\partial p}{\partial V}\right)_{SM} = -\frac{T}{C_{VM}}\left(\frac{dp}{dT}\right)^2 \quad , \tag{82a}$$

where the subscript M refers to the mixture of the two phases, and total derivatives, like dp/dT, refer to the phase boundary. The Clapeyron equation gives

$$\frac{dp}{dT} = \frac{\Delta S}{\Delta V} = \frac{S_2 - S_1}{V_2 - V_1} \quad , \tag{82b}$$

where ΔS and ΔV are the differences of entropy and volume between the two phases, at the temperature T.

The quantity most difficult to calculate is C_{VM}. If the volume of the mixture is to be kept constant, while the temperature changes, the concentration of the two phases must change. If x is the concentration of phase 2, the volume of the mixture is

$$V = xV_2 + (1-x)V_1 \quad , \tag{83}$$

where V_2 and V_1 are the volumes of the pure phases. Therefore, if the volume is to remain constant with change of temperature, we have

$$\frac{dV_1}{dT} + x\frac{d\Delta V}{dT} + \frac{dx}{dT}\Delta V = 0 \quad . \tag{83a}$$

Since we want to consider a state near the phase boundary, x is negligible. Then the specific heat of the mixture becomes

$$C_{VM} = \left(\frac{\partial E}{\partial T}\right)_V = \frac{dE_1}{dT} + \frac{dx}{dT}\Delta E = \frac{dE_1}{dT} - \frac{\Delta E}{\Delta V}\frac{dV_1}{dT} \quad . \tag{84}$$

The derivative dE_1 / dT can be expressed as follows:

$$\frac{dE_1}{dT} = \left(\frac{\partial E}{\partial T}\right)_{V,1} + \left(\frac{\partial E}{\partial V}\right)_{T,1}\frac{dV_1}{dT} \quad . \tag{84a}$$

Using the thermodynamic relations

$$\left(\frac{\partial E}{\partial V}\right)_T = T\left(\frac{\partial p}{\partial T}\right)_V - p \tag{84b}$$

and

$$T\Delta S = \Delta E + p\Delta V \quad , \tag{84c}$$

and remembering (82b), Eq. (84) becomes

$$C_{VM} = C_{V,1} + T\frac{dV_1}{dT}\left[\left(\frac{\partial p}{\partial T}\right)_{V,1} - \frac{dp}{dT}\right] \quad . \tag{84d}$$

Now, in analogy to Eq. (84a), we have

$$\frac{dp}{dT} = \left(\frac{\partial p}{\partial T}\right)_{V,1} + \left(\frac{\partial p}{\partial V}\right)_{T,1}\frac{dV_1}{dT} \quad ; \tag{84e}$$

therefore Eq. (84d) reduces to

$$C_{VM} = C_{V,1} - T\left(\frac{\partial p}{\partial V}\right)_{T,1}\left(\frac{dV_1}{dT}\right)^2 \quad . \tag{85}$$

Since $(\partial p/\partial V)_T$ is always negative, Eq. (85) shows that the specific heat of the mixture, near the boundary, is always greater than that of the adjacent pure phase.

Now let us calculate the difference between the values of $(\partial p/\partial V)_S$ for pure phase and mixture, or rather this difference multiplied by C_{VM}:

$$C_{VM}\left[\left(\frac{\partial p}{\partial V}\right)_{SM} - \left(\frac{\partial p}{\partial V}\right)_{S,1}\right] = -T\left(\frac{dp}{dT}\right)^2 - \left[\left(\frac{\partial p}{\partial V}\right)_{T,1} - \frac{T}{C_{V,1}}\left(\frac{\partial p}{\partial T}\right)_{V,1}^2\right]\times$$

$$\left[C_{V,1} - T\left(\frac{\partial p}{\partial V}\right)_{T,1}\left(\frac{dV_1}{dT}\right)^2\right] = -C_{V,1}\left(\frac{\partial p}{\partial V}\right)_{T,1} + T\left[\left(\frac{\partial p}{\partial T}\right)_{V,1}^2 - \left(\frac{dp}{dT}\right)^2\right] +$$

$$T\left(\frac{\partial p}{\partial V}\right)_{T,1}^2\left(\frac{dV_1}{dT}\right)^2 - \frac{T^2}{C_{V,1}}\left(\frac{\partial p}{\partial V}\right)_{T,1}\left(\frac{\partial p}{\partial T}\right)_{V,1}^2\left(\frac{dV_1}{dT}\right)^2 \quad . \tag{85a}$$

The quantity in the square bracket can be transformed, using Eq. (84e)

$$\left(\frac{\partial p}{\partial T}\right)_{V,1}^2 - \left(\frac{dp}{dT}\right)^2 = -\left(\frac{\partial p}{\partial V}\right)_{T,1}\frac{dV_1}{dT}\left\{2\left(\frac{\partial p}{\partial T}\right)_{V,1} + \left(\frac{\partial p}{\partial V}\right)_{T,1}\frac{dV_1}{dT}\right\} \quad . \tag{85b}$$

The second term in Eq. (85b) cancels the second last term in Eq. (85a). The remaining terms in Eqs. (85a,b) give a full square, and we obtain

$$\left(\frac{\partial p}{\partial V}\right)_{SM} - \left(\frac{\partial p}{\partial V}\right)_{S,1} = -\left(\frac{\partial p}{\partial V}\right)_{T,1} \frac{C_{V,1}}{C_{VM}} \left[1 + \frac{T}{C_{V,1}} \left(\frac{\partial p}{\partial T}\right)_{V,1} \frac{dV_1}{dT}\right]^2 . \quad (86)$$

Since $(\partial p / \partial V)_T$ is always negative, and C_{V1} and C_{VM} always positive, the right-hand side of Eq. (86) is always positive; therefore

$$\left(\frac{\partial p}{\partial V}\right)_{SM} > \left(\frac{\partial p}{\partial V}\right)_{S,1} . \quad (86a)$$

We have derived the result [Eq. (86a)] without any assumption regarding the relative magnitudes of V_1 and V_2, or S_1 and S_2. The result will therefore be valid for *both* boundaries of the two-phase region. We find therefore:

At the boundary between a two-phase region (in the p,V-diagram) and a single phase, the adiabatic compression modulus, $-(\partial p / \partial V)_S$, *will always be greater for the single phase than for the mixture of the two phases.*

We are interested in the sign of the second derivative, $(\partial^2 p / \partial V^2)_S$, or, more correctly, in the sign of the discontinuity of $(\partial p / \partial V)_S$ when we follow the adiabatic in the direction of increasing volume V. This sign will depend on the direction in which the adiabatic crosses the boundary between two-phase and one-phase region: If adiabatic *expansion* will lead to the phase transition, then $(\partial p / \partial V)_S$ will *increase* discontinuously as the adiabatic enters the two-phase region. Then $(\partial^2 p / \partial V^2)_S$ is positive (infinite) at the boundary, and postulate I remains true. If, however, *adiabatic compression* leads from the pure phase to the mixture of two phases, $(\partial p / \partial V)_S$ will *decrease* discontinuously at the boundary, if we proceed in the direction of increasing volume. Then $(\partial^2 p / \partial V^2)_S$ is negative at the boundary, and *condition I is violated.*

The transition between a condensed phase and the vapor has generally the property that adiabatic expansion leads to the phase transition, both if we start from the condensed phase and if we start from the vapor. The former is rather obvious since adiabatic expansion of a liquid or solid at low pressure is almost identical with isothermal expansion, and will therefore ultimately lead to evaporation. The other part of the statement is a well-known experimental fact: Adiabatic expansion of nearly saturated vapor leads to condensation (principle of cloud chamber). Therefore, *condition I is generally satisfied for evaporation and condensation.*

On the other hand, for transitions between two condensed phases --liquid and solid, or two solid modifications--the adiabatics usually run similar to the isothermals, i.e., proceeding in the direction of increasing volume, the adiabatics start in the denser phase region, and finally into the less dense phase. *At the boundary of the less dense phase,* $(\partial p / \partial V)_S$ will therefore decrease discontinuously and *condition I will be violated.*

Generally, the direction of the crossing of the boundary can be deduced from thermodynamic quantities. Let us consider the boundary of the phase of *smaller* density. Under which conditions is postulate I still satisfied at this boundary, i.e., when do the adiabatics go with increasing volume from the dilute phase into the two-phase region? The condition for this is (see Fig. 4)

$$\left(\frac{\partial V}{\partial p}\right)_{S,1} > \frac{dV_1}{dp} \quad , \tag{87}$$

the total derivative referring, as usual, to the equation of the phase boundary. We have

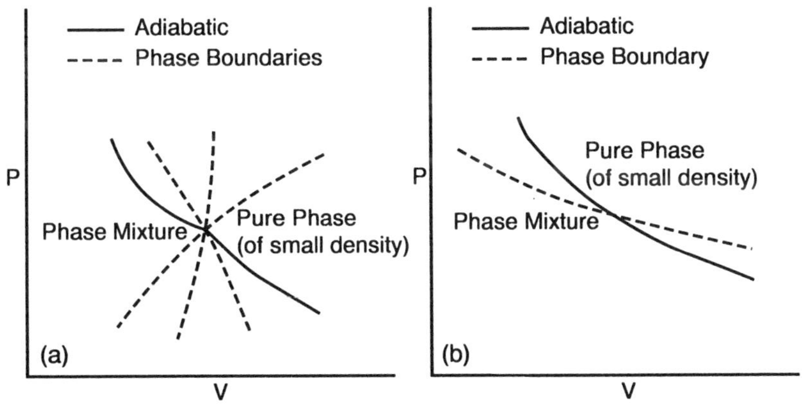

Fig. 4. Crossing of phase boundary by adiabatic. To the right of the dotted lines is the pure phase (small density); to the left is the phase mixture. (a) The adiabatic goes with increasing volume from the mixture to the pure phase. By turning the diagram through 90° it can be seen that the slope $-\partial V / \partial p$ is greater for the adiabatic than for the phase boundaries. (b) The adiabatic goes with increasing volume from the pure phase to the mixture. The slope $-\partial V / \partial p$ is smaller for the adiabatic than for the phase boundary.

$$\left(\frac{\partial V}{\partial p}\right)_{S,1} = \left(\frac{\partial V}{\partial p}\right)_{T,1} + \left(\frac{\partial V}{\partial T}\right)_{p,1}\left(\frac{\partial T}{\partial p}\right)_{S,1} \quad , \tag{87a}$$

$$\frac{dV_1}{dp} = \left(\frac{\partial V}{\partial p}\right)_{T,1} + \left(\frac{\partial V}{\partial T}\right)_{p,1}\frac{dT}{dp} \quad . \tag{87b}$$

If the expansion coefficient of the dilute phase, $(\partial V / \partial T)_{p,1}$ is positive--and we do not know any exception from this--Eq. (87) is equivalent to

$$\left(\frac{\partial T}{\partial p}\right)_{S,1} > \frac{dT}{dp} \quad ; \tag{88}$$

that is, the temperature must rise more rapidly for adiabatic compression than for compression along the phase boundary. For dT/dp we have the Clapeyron equation (82b) whereas

$$\left(\frac{\partial T}{\partial p}\right)_{S,1} = -\frac{(\partial S / \partial p)_{T,1}}{(\partial S / \partial T)_{p,1}} = \frac{T}{C_{p,1}}\left(\frac{\partial V}{\partial T}\right)_{p,1} \quad , \tag{88a}$$

using a well known thermodynamic relation. Then Eq. (88) becomes

$$T\left(\frac{\partial V}{\partial T}\right)_{p,1} > C_{p,1}\frac{\Delta V}{\Delta S} \quad , \tag{89}$$

which is the desired condition.

If phase 1 obeys the ideal gas equation, we have

$$T\left(\frac{\partial V}{\partial T}\right)_{p,1} = V \approx \Delta V \quad , \tag{89a}$$

so that Eq. (89) reduces to

$$\Delta S > C_{p,1} \quad . \tag{89b}$$

This is ordinarily fulfilled with a wide margin, e. g., for water at 100°C, ΔS = 6.05 joule/g deg. [Dorsey,[7] p. 616] while $C_p \approx 2$ joule/g deg. [Dorsey,[7] p. 101). For solids and liquids, on the other hand, $(\partial V / \partial T)_p$ is usually quite small and

Eq. (89) will, in general, not be fulfilled. For example, for the transition from water to ice VI we have at 30°C (data from Dorsey, reference 7)

p =10,250 atm [Dorsey, p. 476]

$$\left(\frac{\partial V}{\partial T}\right)_p = 3.3\times 10^{-4} \quad cm^3 / g\ deg$$ [Dorsey, p. 216]

$$C_V = 4.072 \times (1 - 0.210) = 3.22\ joule / g\ deg$$

at 40° and 10,250 atm , therefore about 3.1 at 30°

[Dorsey, p. 261]

$$\approx 3.1\ joule / g\ deg \quad at\ 30^o \quad .$$

$$C_p - C_V = -T\left(\frac{\partial p}{\partial V}\right)_T \left(\frac{\partial V}{\partial T}\right)^2 \quad ,$$

$$-\frac{\partial p}{\partial V} = 1.0\times 10^5 \ \text{bar g / cm}^3 \quad .$$ [Dorsey, p. 216]

Therefore

$$C_p - C_V = 3.3\ \frac{\text{bar cm}^3}{\text{g deg}} = 0.33\ \frac{\text{joule}}{\text{g deg}} \quad ,$$

$$C_p = 3.4\frac{\text{joule}}{\text{g deg}} \quad ,$$

$$\Delta S = 1.09\ \frac{\text{joule}}{\text{g deg}} \quad ,$$ [Dorsey, p. 617]

$$\Delta V = 0.0663\ \text{cm}^3 / g \quad .$$ [Dorsey, p. 613]

Inserting into Eq. (89), we get

$$T\,\partial V / \partial T = 0.100 \quad cm^3 / g \quad ,$$

$$C_p \frac{\Delta V}{\Delta S} = \frac{3.4\times 0.0663}{1.09} = 0.21 \ cm^3 / g \quad .$$

Condition I is therefore *not* fulfilled at the boundary between water and water + ice VI.

On the boundary of the *denser* phase, the condition for the validity of I is just reversed, as can easily be seen. We have then instead of Eq. (89):

$$T\left(\frac{\partial V}{\partial T}\right)_{p,\,dense} < C_{p,\,dense}\frac{\Delta V}{\Delta S} \tag{90}$$

For evaporation, this is practically always fulfilled because ΔV is almost equal to the volume of the vapor and therefore very large compared with the volume of the liquid or solid which in turn is large compared with $T(\partial V / \partial T)$. On the other hand, for transitions between condensed phases Eq. (90) is not always valid; in particular, if $\Delta V / \Delta S$ happens to be small or negative for a transformation, so that Eq. (89) is fulfilled, Eq. (90) will in general not be fulfilled. This is the case, for example, for the transition from ice I to water for which $\Delta V / \Delta S$ is negative: Then condition I breaks down at the boundary of the *denser* phase, that is, of the water.

With possibly a few exceptions, we can therefore state: *Condition I will break down for transitions between two condensed phases at one of the two boundaries between pure phase and mixture. Condition I will remain valid for evaporation and condensation.*

14. Consequences of the breakdown of condition I at phase boundaries

If we compress a liquid adiabatically, we finally come to the phase boundary with the solid. As we have shown in the last section, the derivative $\left(\partial^2 p / \partial V^2\right)_S$ will in general be negative (infinite) at one of the boundaries between pure phase and phase mixture. For water, this occurs at the boundary between liquid water and the mixture of water and ice VI. Therefore, if the phase transition is not forbidden by its long relaxation time, all proofs given in part A will break down.

It is easy to show that the negative discontinuity of $(\partial p / \partial V)_S$ will actually have serious consequences for the theory of shock waves. Consider an initial state V_1, S_1 close to the phase boundary, but still on the side of liquid water. Then we may neglect the change of $(\partial p / \partial V)_S$ from V_1, S_1 to the phase boundary; moreover, the shock curve will cross the phase boundary at an entropy S_C very close to S_1, and at a certain volume $V_C < V_1$. For the state V_C, S_C, we have very nearly

$$\frac{\Delta p}{\Delta V} = \frac{p_c - p_1}{V_c - V_1} = \left(\frac{\partial p}{\partial V}\right)_{SL} , \tag{91}$$

where the subscript L refers to the liquid. If we then go a small distance into the two-phase region, we have from Eq. (31)

$$\left(\frac{dS}{dV}\right)_M = \frac{\Delta p - \Delta V(\partial p / \partial V)_{SM}}{2T_2 + \Delta V(\partial p / \partial S)_{VM}} , \tag{91a}$$

where the index M refers to the mixture. Neglecting ΔV in the denominator, and inserting Eq. (91), we find

$$\left(\frac{dS}{dV}\right)_M = \frac{\Delta p}{2T_2}\left(1 - \frac{(\partial p / \partial V)_{SM}}{(\partial p / \partial V)_{SL}}\right) . \tag{92}$$

Now we know from Eq. (86a) that

$$\frac{(\partial p / \partial V)_{SM}}{(\partial p / \partial V)_{SL}} < 1 \tag{92a}$$

since $(\partial p / \partial V)_S$ is negative. Therefore Eq. (92) is positive, i.e., *the entropy will decrease with further compression*, is contrast to our theorems in Secs. 3, 4 and 7. Since we have assumed that S_C is very close to S_1, the entropy will soon fall below S_1. Therefore the *compressional shock waves* will become *unstable.*

The solution of this difficulty seems to be that there will be *two shock waves proceeding in the same direction.* In the space between the two shock waves the state of the material is given by V_C, S_C, i.e., the state lies just on the phase boundary. Behind the second shock wave, the material is a mixture of liquid and solid. (There may, of course, be some question whether such a mixture can occur behind a shock wave.) The second shock wave moves approximately with the velocity of sound characteristic for the mixture, and will therefore remain behind the first wave which moves nearly with the velocity of sound of the pure liquid. This result is in striking contrast to that found in Sec. 10, for a material obeying condition I: We proved that for such a material there can always be only one shock wave starting from a given point in a given direction.

As the difference between the specific volumes in front of the first wave (V_1) and behind the second one (V_2) increases, the velocity u' of the second wave will increase, following the ordinary laws[10] developed in Part A, with the state V_C, S_C playing the role of "initial state." The velocity u of the first shock wave remains constant. Ultimately, u' will become equal to u. Suppose this happens for $V_2 = V_D$, $p_2 = p_D$, then V_D, p_D are given by:

$$V_C^2 u'^2 = \frac{p_D - p_C}{V_C - V_D} = \frac{p_C - p_1}{V_1 - V_C} = V_C^2 u^2 \tag{93}$$

(u' and u are velocities relative to the material between the shock waves). *If the volume V_2 is decreased below V_D, we shall again get one shock wave*, with state 1 in front and state 2 behind the shock front. The state p_D, V_D satisfying Eq. (93) may occur either in the two-phase region or may be already in the pure dense phase (in our case, ice VI).

For the two separate shock waves, the stability considerations of Part A will all be valid because in the liquid and in the mixture *separately*, condition I is fulfilled. Likewise, the proofs of stability will hold for the single wave occurring at pressures greater than p_D; in this case, the theorems may be proved by starting from state D and continuing along the shock curve to higher pressures and entropies.

Thus far we have considered an initial state V_1, S_1 very close to the phase boundary. On the other hand, the shock curve may cross a phase boundary at a density and entropy far above those of the initial state. Then it may easily happen that the *single* shock wave will remain stable even if the state behind the shock wave lies beyond the phase boundary. If u_{C2} is the shock wave velocity at the point C at which the shock curve belonging to state 1 crosses the phase boundary, then the condition for the single shock wave to remain stable is that u_{C2} be smaller than the sound velocity in the phase mixture, i.e.,

$$\frac{p_C - p_1}{V_1 - V_C} < \left| \left(\frac{\partial p}{\partial V} \right)_{SM} \right| \tag{94}$$

[see also Eqs. (31c, d)]. Equation (94) may be fulfilled for large shock waves because their velocity, with respect to the medium C behind the wave is known to be considerably smaller (Sec. 9) than the sound velocity in that medium, i.e., [see Eqs. (6) and (10)]

$$\frac{p_C - p_1}{V_1 - V_C} << \left| \left(\frac{\partial p}{\partial V} \right)_{SL} \right| \quad . \tag{94a}$$

---

[10] We have not proved that condition I is fulfilled *in* the two-phase region but we believe this to be true.

Whether or not Eq. (94) is true for a given initial state and a given phase boundary, must be investigated in each particular case. If the initial state is water at 1 atmosphere, and the phase boundary to ice VI is crossed at 30°C, we have p = 10,250; further [Dorsey,[7] p. 467] $V_C = 0.8055$, $V_1 = 1.00$, $p_1 = 0$; therefore

$$\frac{p_C - p_1}{V_1 - V_C} = 52{,}600\frac{atm\ cm^3}{g} = 53{,}400\frac{bar\ cm^3}{g} \quad . \tag{94b}$$

The data for the calculation if $(\partial p / \partial V)_{SM}$ are mostly given previously in this report; we have [see Eqs. (82a,b), (84e) and (85)]

$$\frac{dp}{dT} = \frac{\Delta S}{\Delta V} = 164\ \frac{bar}{deg} \quad ,$$

$$\frac{dV_1}{dT} = 1.2 \times 10^{-3}\ \frac{cm^3}{g\ deg} \quad , \qquad \text{[Dorsey, p. 467]}$$

$$C_{V,1} = 31\ bar\ cm^3 / g\ deg \quad . \qquad \text{[This report]}$$

$$C_{VM} = 31 + 303 \times (1.2 \times 10^{-3})^2 \times 10^5 = 75\ bar\ cm^3 / g\ deg \quad . \qquad \text{[This report]}$$

$$-\left(\frac{\partial p}{\partial V}\right)_{SM} = \frac{303}{75} \times 164^2 = 108{,}000\frac{bar\ cm^3}{g} \quad . \qquad \text{[Eq. (82a)]} \tag{94c}$$

This is greater than Eq. (94b) so that Eq. (94) is fulfilled. Therefore, even if the transformation of water to ice VI could take place in shock waves, the theorems of Part A would remain valid. (This holds if the initial state is at 1 atmosphere pressure; if we started from a high pressure, let us say 8,000 atmospheres, we should obtain two shock waves as described above.)

We thus get the following picture: If the initial state V_1, p_1 lies fairly close to a phase boundary, there will be a region of "final" pressures p_2 between p_C and p_D [see Eq. (93)], for which two shock waves exist behind one another. In this whole region, the velocity of the first wave remains constant, equal to that for $p_2 = p_c$. For greater final pressure, $p_2 > p_D$, there will be a single shock wave whose velocity increases with increasing p_2. If we move now the initial state farther away from the phase boundary, the region from p_C to p_D will become smaller and will finally disappear.

This result may be applicable to extremely large shock waves in solids. As we have pointed out, in footnote 4, Sec. 7, a solid under compression and extremely high temperature may not go over smoothly into the quasi-gaseous state; in other words, there may be no critical point for a solid. However, the phase transition, if any, will occur at such a high pressure that the single shock wave will almost certainly remain stable.

The double shock wave will never occur for *gases*. For we have shown in Sec. 13 that a vapor just on the verge of condensation will be removed from the phase boundary by adiabatic compression. Therefore a gas, under any amount of adiabatic compression, will always remain a gas, and this will be *a fortiori* true under shock compression which leads to an even higher temperature. Therefore, for gases, whether ideal or not, the results of Part I are valid.

We shall now turn to rarefaction waves. Consider as initial state V_1, S_1 a state on the high density side of a phase boundary at which condition I breaks down, so that adiabatic expansion leads from V_1, S_1 across the boundary. (This can occur only when the initial state is a mixture of two condensed phases.) If we follow the shock curve to larger volumes, we can see from an argument exactly similar to Eqs. (91) to (92a), that the entropy will first decrease to the phase boundary and then increase. We thus obtain the possibility of stable *rarefaction waves of finite amplitude*. In fact, if we tried to work with infinitesimal rarefaction waves, those corresponding to the further expansion of the pure phase would travel with the greater sound velocity of the pure phase, which leads to a contradiction. Thus we not only can but actually must have a rarefaction shock wave.

With increasing expansion, we shall come to a point V_E, p_E at which the velocity of sound in the pure phase has dropped to a value equal to that of the rarefaction shock wave. For any expansion beyond this point, we shall get a rarefaction shock wave in which the substance expands to the volume V_E, followed by a train of infinitesimal waves of further rarefaction.

As the initial state V_1, p_1 is removed from the phase boundary, the point V_E, p_E will move also; we have not investigated in which direction. If the initial state is in the pure dense phase, there must still be a rarefaction shock wave since the rarefaction still cannot be accomplished by infinitesimal waves, because the sound velocity in the pure dilute phase is greater than in the phase mixture. It does not seem obvious which part of the rarefaction is accomplished by a shock and which part by infinitesimal waves. It seems reasonably certain that there will be infinitesimal waves in which the material is expanded down to the boundary of the dense phase; then possibly a shock wave will follow which carries the material over the mixed region and into the dilute phase, and this may be followed by another train of infinitesimal waves. But it may also be that the shock wave "starts" with a state in the region of mixed phases. In any case, these rarefaction shock waves can only start from initial states of very high pressure and can only occur at all if the phase transformation is rapid enough to happen in the shock wave.

The strange phenomenon of rarefaction shock waves will not occur for gases, nor for liquids or solids initially at atmospheric or other low pressures. By adiabatic expansion, we obtain in each case a mixture of gas and condensed phase, and we have shown in Sec. 13 that condition I is valid for the transition into this mixed phase, both from the condensed and from the gaseous state.

15. Condition II: $V(\partial p / \partial E)_V > -2$

This condition is automatically fulfilled for all substances with a positive expansion coefficient which incudes gases as well as practically all liquids and solids. We need only investigate the case of negative expansion coefficient, of which water is the most notable example.

(a) Liquid Water

The greatest negative expansion coefficient is reached for $0^o C$ and 1 atmosphere pressure. (At lower temperatures, water is stable only under higher pressures, for which the expansion coefficient becomes positive.) The value of $\partial p / \partial T$ for $0^o C$ was given in Eq. (79d) and is -0.62 bar/degree. With $C_V = 42\ bar\ cm^3 / g\ degree$, and $V = 1\ cm^3 / g$, we get

$$V\left(\frac{\partial p}{\partial E}\right)_V = -\frac{0.62}{42} = -\ 0.015 \quad , \tag{95}$$

as compared with the limit of -2 set by condition II.

(b) Ice at Extremely Low Temperatures

A relatively large negative expansion coefficient was found by Jakob and Erk for ice at extremely low temperatures [Dorsey,[7] p. 473]. At -250°C the *linear* expansion coefficient is -6.1×10^{-6}. If we assume the cubic expansion coefficient to be three times as large, we have

$$\beta = \frac{1}{V_o}\left(\frac{\partial V}{\partial T}\right)_p = -18.3\times10^{-6} \quad , \tag{95a}$$

where V_o [=1.09] is the specific volume at $0^o C$. The compressibility has not been measured at these low temperatures. Near the melting point, the experimental results differ widely [Dorsey,[7] p. 471]: Bridgman finds $\gamma = -(1 / V_0)(\partial V / \partial p)_T$ to decrease from 33×10^{-6} at 0°C to 19×10^{-6} at -10°C and 18×10^{-6} at -15°C while Richards and Speyers find γ as low as 12×10^{-6} at -7°C. If we assume the last-mentioned figure for -250°C, we get

$$\frac{\partial p}{\partial T} = -\frac{18.3\times10^{-6}}{12\times10^{-6}} = -1.5\ bar\,/\,deg \quad . \tag{95b}$$

The specific heat, according to Pollitzer [see Dorsey,[7] p. 479] is 1.28 *bar* cm^3/g *degree*, therefore

$$V\left(\frac{\partial p}{\partial E}\right)_V = -1.07\times\frac{1.5}{1.28} = -1.25 \quad , \tag{95c}$$

which is of the same order of magnitude, but smaller in absolute value, than the value permitted by condition II.

(c) Melting Ice

The largest contraction with increasing energy is found for melting ice I. From the Clapeyron equation we have

$$\frac{dp}{dT} = \frac{\Delta S}{\Delta V} \quad , \tag{96}$$

where p is the melting pressure at temperature T. The entropy change from ice I to water is [Dorsey, p. 617]

$$\Delta S = 12.2\frac{bar\,cm^3}{g\,deg} \text{ at } 0^\circ C, \qquad 9.55 \text{ at - } 20^o \quad ,$$

The change of volume is

$$\Delta V = -0.0900\frac{cm^3}{g} \text{ at } 0^\circ C, \qquad -0.1313 \text{ at - } 20^o \quad .$$

Accordingly,

$$\frac{dp}{dT} = -136\frac{bar}{deg}\ at\ 0^o C, \qquad or\ -73\ \ at\ -20^0 \quad . \tag{96a}$$

These values are very much greater than $\partial p / \partial T$ for water at $0^o C$., which we found to be -0.62 *bar/deg* [Eq. (84d)]. In fact, if the specific heat C_V of the mixture of water and ice were between the specific heat C_V of the mixture of water and ice were between the specific heats of water (42 $bar\ cm^3\ /\ deg$) and of ice (20 $bar\ cm^3\ /\ deg$), condition II would be violated by a large amount.

Actually, we know from Eq. (85) that C_V for the mixture is larger than for the pure phases. It can easily be shown that C_{VM} is smallest when the mixture contains mostly ice, firstly because the C_V of pure ice is smaller than that of pure water, and second because the volume change $dV_1\ /\ dT$, is also smaller. We have

$$C_{V\,ice} = 21.15\ bar\ cm^3\ /\ deg \text{ at } 0^o C, \quad and\ 19.59\ at\ \text{-}\ 20^o \quad ,$$

$$dV_1\ /\ dT \approx 3.5\times 10^{-3}\frac{cm^3}{g\ deg}\ at\ 0^o C, \quad and\ \ 0.8\times 10^{-3}\ at\ \text{-}\ 20^o C \quad ,$$

$$-\partial p\ /\ \partial V \approx 4\times 10^4\frac{bar\ g}{cm^3}\ at\ 0^o C, \quad and\ \approx 10.5\times 10^4\ \frac{bar\ g}{cm^3}\ at\ \text{-}\ 20^o \quad .$$

Therefore from Eq. (85), at 0°C,

$$C_{VM} = 21.15 + 273\times 4\times 10^4\times\left(3.5\times 10^{-3}\right)^2 = 156\frac{bar\ cm^3}{deg} \quad , \tag{96b}$$

and at -20°C,

$$C_{VM} = 19.59 + 253\times 10.5\times 10^4\times\left(0.8\times 10^{-3}\right)^2 = 36.5\ bar\ cm^3 deg \quad . \tag{96c}$$

Therefore

$$V\left(\frac{\partial p}{\partial E}\right)_V = \frac{V}{C_{VM}}\left(\frac{dp}{dT}\right)_V = -\frac{1.09\times 136}{156} = -0.95 \text{ at } 0^o C \quad , \tag{96d}$$

and

$$V\left(\frac{\partial p}{\partial E}\right)_V = -\frac{1.06\times 73}{36.5} = -2.12\ \ at\ \ -20^\circ\ C \quad . \tag{96e}$$

The last mentioned value is just slightly below the limit set by condition II, the difference being within the probable error of measurement of $dV_1\ /\ dT$. It seems therefore that melting ice forms an exception to condition II, but that this condition is fulfilled for pure water as well as pure ice, and probably for most other substances.

16. Condition III: $(\partial p / \partial V)_E < 0$

This condition is obviously fulfilled for ideal gases because for these constant energy is equivalent with constant temperature, and the isothermal value of $\partial p / \partial V$ must always be negative.

The condition is fulfilled *a fortiori* for practically all substances for which the internal energy increases with isothermal expansion. We have

$$\left(\frac{\partial p}{\partial V}\right)_E = \left(\frac{\partial p}{\partial V}\right)_T - \left(\frac{\partial p}{\partial T}\right)_V \frac{(\partial E / \partial V)_T}{C_V} \quad . \tag{97}$$

If $(\partial p / \partial T)_V > 0$, which is true for almost all substances (except water below 4°C), the last term is negative for positive $(\partial E / \partial V)_T$.

Therefore

$$\left(\frac{\partial p}{\partial V}\right)_E < \left(\frac{\partial p}{\partial V}\right)_T < 0 \tag{97a}$$

if

$$\left(\frac{\partial E}{\partial V}\right)_T > 0 \quad . \tag{97b}$$

A great many substances fulfill the condition $(\partial E / \partial V)_T > 0$. Some of the most important are:

(a) *Dissociating gases.* The dissociation increases with the volume, at constant temperature, and the dissociation increases the internal energy. This case is the most important of all because condition III is required (Sec. 8) to prove that the energy has no extremes in the part of the shock curve beyond the minimum of the volume V_2, i.e., at very high temperatures at which dissociation progresses rapidly.

(b) *Any fairly dilute, imperfect gas.* The attractive forces (Van der Waals forces) must be taken into account while the repulsive forces are unimportant. Since the attractive forces decrease with increasing volume, the energy must increase with V. An example is provided by a gas obeying Van der Waals' equation, namely

$$\left(p + \frac{a}{V^2}\right)(V - b) = RT \quad . \tag{98}$$

From the thermodynamic relation

$$\left(\frac{\partial E}{\partial V}\right)_T = T\left(\frac{\partial p}{\partial T}\right)_V - p \tag{98a}$$

we find

$$\left(\frac{\partial E}{\partial V}\right)_T = \frac{a}{V^2} > 0 \quad . \tag{98b}$$

(c) *Most solids and liquids at ordinary temperatures.* From measurements such as those of Gibson and Loeffler,[5,6] $\partial E / \partial V$ turns out to be positive for practically all liquids at ordinary temperatures. The only notable exception is water below $4^o C$ (see below). The argument given in the literature for the "normal " behavior, is the same as that given under Sec. 16(b) for gases, namely that the most important volume dependent contribution to E arises from attractive forces, which give a positive $\partial E / \partial V$. This argument may be expected to hold for solids as well as for liquids.

Other substances still fulfill condition III although $(\partial E / \partial V)_T$ is negative. Examples follow:

(d) *Water below 4°C.* Here $(\partial p / \partial T)_V$ at zero pressure is negative so that Eq. (98a) yields a negative $(\partial E / \partial V)_T$. However, since at the same time $\partial p / \partial T$ is negative, the last term of Eq. (97) remains negative. It is true that there is a region of pressures for which $\partial E / \partial V$ is still negative while $\partial p / \partial T$ is already positive [because of the term -p in Eq. (98a)], but in this region the last term in Eq. (97) is entirely negligible compared with the first, viz. about 1 part in 10,000.

(e) *Solids at low temperature.* The last term in Eq. (97) may be expected to be large and positive (that is, unfavorable for relation III) if C_V is small, $\partial p / \partial T$ positive, and p large [see Eq. (98a)]. All conditions are fulfilled for solids at low temperature. C_V is given by Eq. (77), namely

$$C_V = a(T / \theta)^3 \quad , \tag{77}$$

and goes to zero for zero temperature. Fortunately, $\partial p / \partial T$ also goes to zero; we have from Eqs. (77c) and (77)

$$\left(\frac{\partial p}{\partial T}\right)_V = \left(\frac{\partial S}{\partial V}\right)_T = \frac{1}{3}\frac{\partial C_V}{\partial V} = -\frac{1}{3}C_V\frac{\partial \log \theta}{\partial V} \quad . \tag{99}$$

This expression is positive because the Debye temperature decreases with increasing volume. Since θ is proportional to the frequency of the vibrations of the crystal lattice, therefore approximately

$$\theta = bV^{-1/3}\,\mathrm{a} \quad , \tag{99a}$$

where b is a constant depending on the crystal structure but not on V, while a is the velocity of sound, namely

$$\mathrm{a}^2 = -V^2(\partial p / \partial V)_T \quad . \tag{99b}$$

Therefore

$$\frac{\partial \log\theta}{\partial V} = \frac{2}{3V} + \frac{(\partial^2 p / \partial V^2)_T}{(\partial p / \partial V)_T} \quad . \tag{99c}$$

Inserting Eqs. (99), (98a) and (99c) into Eq. (97) and neglecting $T(\partial p / \partial T)$ against p we obtain

$$\left(\frac{\partial p}{\partial V}\right)_E = \left(\frac{\partial p}{\partial V}\right)_T \left[1 - \frac{1}{3}\frac{p(\partial^2 p / \partial V^2)_T}{(\partial p / \partial V)_T^2}\right] - \frac{2}{9}\frac{p}{V} \quad . \tag{100}$$

The last term is certainly negative (and small). In order that condition III be fulfilled, it is therefore sufficient that the bracket be positive. This is almost certainly the case because solids obey very nearly the Tait equation, choosing $V_o = V_T$, Eq. (79) becomes

$$p = B\left[e^{(V_o - V)/\kappa} - 1\right] \quad , \tag{100a}$$

$$\left(\frac{\partial p}{\partial V}\right)_T = -(B/\kappa)\, e^{(V_o - V)/\kappa} \quad , \tag{100b}$$

$$\left(\frac{\partial^2 p}{\partial V^2}\right)_T = (B/\kappa^2)\, e^{(V_o - V)/\kappa} \quad . \tag{100c}$$

Therefore

$$p\frac{\partial^2 p}{\partial V^2} < \left(\frac{\partial p}{\partial V}\right)^2 \tag{100d}$$

and the square bracket in Eq. (100) is positive.

(f) *Phase Transitions*. This is the only case for which we have found condition III not always satisfied. In the two-phase region, p depends on T only, that is, $(\partial p / \partial V)_T = 0$. The specific heat, C_V, is always positive. Therefore condition III, (Eq. 97), reduces to

$$\left(\frac{\partial p}{\partial T}\right)_V \left(\frac{\partial E}{\partial V}\right)_T > 0 \quad . \tag{101}$$

Using Eq. (98a), the Clapeyron equation (82b), and (84c) this gives

$$\frac{\Delta S}{\Delta V}\left(T\frac{\Delta S}{\Delta T} - p\right) = \frac{\Delta E \Delta S}{\Delta V^2} > 0 \quad , \tag{101a}$$

or simply

$$\Delta E\, \Delta S > 0 \quad . \tag{102}$$

The condition is therefore that the energy and the entropy should change in the same direction. This is fulfilled for practically all phase transitions, but there are a few exceptions, such as the transformations ice I - ice II, or ice III - ice V. [Dorsey,[7] p. 617].

Summarizing, we find that condition III, like the other two conditions, seems to be satisfied for practically all homogeneous (one-phase) systems but to break down for a few phase transformations. The condition seems to be more generally fulfilled than condition I.

17. Discussion of a hypothetical case: a material which satisfies conditions I and II but not III

We have shown in Sec. 16 that probably all materials satisfying condition I will also satisfy condition III. However, since no general proof could be given, it may still be worth while to discuss the consequences of a violation of condition III in a material which satisfies I and II.

Condition III was used in Sec. 8 to show that the energy increases monotonically with the entropy. Condition III is only a sufficient, not a necessary condition for this; in order that the energy have a maximum, $(\partial p / \partial V)_E$ must not only be positive but also must satisfy Eq. (43). If Eq. (43) is fulfilled for some point on the shock curve, the energy must have a maximum at this point, must then decrease, have a minimum, and finally increase again with S at extremely high temperatures. Of course E may have several maxima and minima.

As we have shown in connection with Eq. (47), the *pressure* will increase with the entropy at least as long as the energy does. Only if E_2 decreases sufficiently rapidly with increasing entropy and increasing V_2,

$$p_2 = -p_1 + \frac{2(E_2 - E_1)}{V_1 - V_2} \tag{47}$$

will reach a maximum and then decrease. To find the condition for a maximum of p_2, we proceed similarly as in Eq. (42), only considering p and V as independent variables. Then we have

$$\left(\frac{\partial E}{\partial p}\right)_V dp + \left(\frac{\partial E}{\partial V}\right)_p dV = -\frac{1}{2}(p_2 + p_1)dV + \frac{1}{2}(V_1 - V_2)dp \quad . \tag{103}$$

In order that $dp / dV = 0$, we must have

$$\left(\frac{\partial E}{\partial V}\right)_p = -\frac{1}{2}(p_2 + p_1) \quad . \tag{103a}$$

The left-hand side can be transformed, using the thermodynamic relations given by Eqs. (11) and (12)

$$\left(\frac{\partial E}{\partial V}\right)_p = \left(\frac{\partial E}{\partial V}\right)_S + \left(\frac{\partial E}{\partial S}\right)_V \left(\frac{\partial S}{\partial V}\right)_p = -p_2 - T_2 \frac{(\partial p / \partial V)_S}{(\partial p / \partial S)_V} \tag{103b}$$

so that condition (103a) reduces to

$$\Delta p \left(\frac{\partial p}{\partial E}\right)_V = -2\left(\frac{\partial p}{\partial V}\right)_S \quad . \tag{104}$$

There is no general thermodynamical or statistical reason why Eq. (104) should not be fulfilled for some substance. If it is fulfilled, p_2 will have a maximum. If this is the case, the extrema of the various variables of state must follow each other, in the order of increasing entropy, like this: first a minimum of the volume, then a maximum of the pressure, a minimum of the energy and finally a maximum of the volume.

We shall now investigate the consequences of the maxima and minima of energy and pressure for the stability of shock waves. We have shown in Sec. 11 that a shock wave cannot split into two waves going in opposite directions if, for all values $p_2 < p_4$, we have

$$(p_4 - p_1)(V_1 - V_4) > (p_2 - p_1)(V_1 - V_2) \tag{105}$$

[reverse of Eq. (62)]. We showed in Sec. 11 that condition (105) is certainly fulfilled if the energy increases monotonically with increasing pressure and entropy. We showed further that a maximum of the energy on the shock curve is a necessary but not a sufficient condition for a splitting of shock waves.

However, it can easily be seen that a maximum of the *pressure* on the shock curve is a sufficient (not necessary) condition for instability of shock waves. If a shock wave with "initial" state 1 and "final" state 4 is to split, Eq. (57) must be fulfilled. To show that this is possible, we compare the expressions

$$\delta = u_{C,4} - u_{C,3} \tag{106}$$

and

$$D = (u_{0,1} - u_{0,4}) - (u_{A,1} - u_{A,2}) \quad . \tag{106a}$$

The first of these is very easy to calculate if the pressures p_4 and $p_3 = p_2$ differ only slightly because then wave C is a small shock wave and can be treated as almost adiabatic (cf.§3); we have [see Eqs. (58) and (10)]

$$\delta = u_{C,4} - u_{C,3} = \frac{V_4 - V_3}{V_4} u_{C,4} \approx \frac{1}{V_4}\left(\frac{\partial V}{\partial p}\right)_S (p_4 - p_3)\mathrm{a}_4 = \frac{V_4}{\mathrm{a}_4}(p_2 - p_4) \tag{107}$$

This is always negative if $p_4 > p_2$.

On the other hand, we have from Eqs. (61) and (61a):

$$D = \sqrt{(p_4 - p_1)(V_1 - V_4)} - \sqrt{(p_2 - p_1)(V_1 - V_2)} \quad . \tag{108}$$

If condition (105) is fulfilled, D will be positive and therefore

$$D > \delta \quad . \tag{108a}$$

On the other hand, if p_4 and p_2 are near the maximum of p, we have $p_4 - p_1 \approx p_2 - p_1$, so that

$$D \approx -\frac{1}{2}(V_4 - V_2)\sqrt{\frac{p_4 - p_1}{V_1 - V_4}} \quad . \tag{108b}$$

This is negative (because, just before the maximum of p, V increases with increasing p) and $|D|$ can be made as large as we wish in comparison with $|\delta|$ [Eq. (107)]. Therefore, near the pressure maximum, we have

$$D > \delta \quad . \tag{108c}$$

Therefore there must be a certain intermediate value of p_4 (between the energy maximum and the pressure maximum) for which $D = \delta$, so that Eq.(57) is fulfulled. Shock waves in which the pressure behind the wave lies within a certain range, close to the pressure maximum, can split into two waves going in opposite directions, as described in the beginning of Sec. 11.

Let us investigate the way in which the split occurs, as a function of p_4. We shall assume that the pressure does not vary much over the region in which a split may occur. The Eq. (107) for δ will be sufficiently accurate, while D can be obtained by expanding Eq. (108):

$$D = \frac{(V_2 - V_4)(p_4 - p_1) - (p_2 - p_4)(V1 - V_4)}{2(u_{0,1} - u_{0,4})} \quad . \tag{109}$$

Condition (57) requires δ = D; that is,

$$p_2 - p_4 = \frac{(p_4 - p_1)(V_2 - V_4)}{V_1 - V_4 + 2V_4(u_{0,1} - u_{0,4})/a_4} = c(V_2 - V_4) \quad , \tag{109a}$$

where c is a constant, depending very slightly on the state 4. Graphically, we can interpret Eq. (109a) as follows (see Fig. 5). On the one hand, we draw the shock curve which gives p_2 as a (complicated) function of V_2. On the other hand, we consider the straight line Eq. (109a) whose slope, c, can be calculated from known quantities. The intersections of the two curves will give the possible solutions p_2,V_2 for given p_4,V_4 (see Fig. 5). If p has a maximum and a minimum, then for states p_4 near these extrema, there will be three intersections of the straight line with the p(V) curve, including the intersection of p_4,V_4; for values of p_4 far from the extrema, there will be only one intersection. In the latter case, shock waves corresponding to p_4,V_4 cannot split.

It is easy to see that three intersections may occur without p having a maximum and minimum if only at some point of the shock curve

$$dp / dV < c \quad , \tag{109b}$$

where dp / dV is taken along the shock curve, and c is the quantity defined in Eq. (109a).

If there are three intersections, a split is possible as far as the kinematics of the shock waves is concerned. Now, for thermodynamic reasons, the split will occur in such a way that the entropy at the instant immediately after the split is as large as possible. This requires obviously the selection of the highest of the three intersections: For then the entropy of state 2 is greater than for either of the other two intersections, and moreover state 2 propagates most rapidly into medium 1 (see Sec. 9). In addition, state 3, having the highest pressure compatible with the kinematic conditions, will also have a higher entropy than 4, and will spread into 4. Therefore, if there should be a shock wave corresponding to one of the lower intersections (A and B) in Fig. 5, it would split into one shock wave corresponding to the highest intersection (C), and another small shock wave going in the opposite direction. The shock wave would thus increase in amplitude rather than decrease by its "split". The newly established shock wave of higher entropy change and velocity would then be stable.

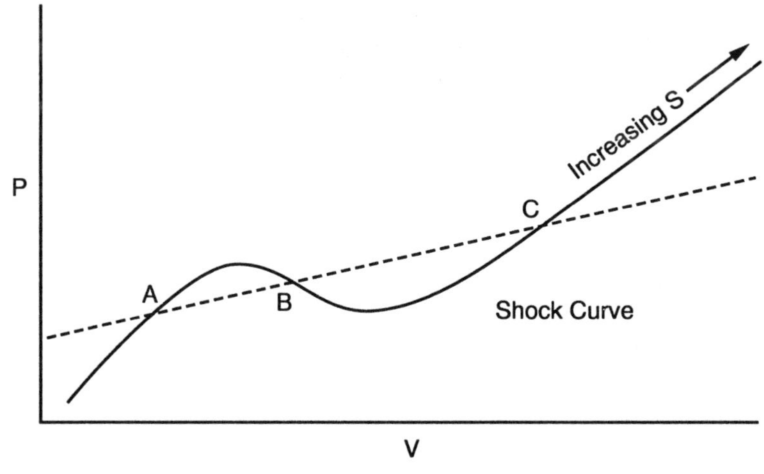

Fig. 5. Splitting of shock waves. The solid curve represents the shock curve in a p,V diagram; the curve has a maximum and a minimum. The dotted line is the straight line $p - p_4 = c(V - V_4)$ defined by Eq. (109a). The intersections A and B represent unstable shock waves, C a stable wave.

C. Conclusion

18. Relation to the theory of Duhem

The theory of shock waves in a medium with a general equation of state has been discussed previously by Duhem.[11] The principal difference between his treatment and the present one is that he did not make any assumptions about the equation of state, while we have made several, namely, the conditions I, II, III, and the facts about the equation of state at high temperatures (Sec. 5). Duhem was therefore not able to come to such general conclusions as we did.

Duhem did recognize the importance of condition I. He possessed the main results of (Sec. 3) of this paper, viz. that small but finite compressional waves correspond to an increase of entropy if $\partial^2 p / \partial V^2 > 0$, while finite rarefaction waves would be stable if $\partial^2 p / \partial V^2 < 0$ (pp. 177-178 of his paper). He also found that for $\partial^2 p / \partial V^2 > 0$, the velocity of small compression waves, with respect to the less dense material, is greater than the velocity of sound while relative to the denser material, the shock wave moves more slowly than sound [Eq. (15) of this paper, p. 178 of Duhem's paper]. However, he could claim the validity of his statements only for a rather restricted range of densities behind the shock wave, without being able to state the conditions of validity in physical terms.

Moreover, Duhem found already that the entropy change for small shock waves is proportional to a higher power of the density change than the first [his Eq. (19)], but he did not find the third power law [Eq. (13) of our paper]. Generally speaking, his paper contains part of the results of our Sec. 3, but none of the later sections. In particular, he did not discuss the stability problems (Secs. 10 and 11) and he could not obtain the general results of Secs. 4, 7 and 9.

19. Summary

We have shown in this paper:

1. General theorems can be derived about shock waves in any substance whose equation of state satisfies the three conditions.

$$\left(\frac{\partial^2 p}{\partial V^2}\right)_S > 0 \quad , \qquad \text{(I)}$$

[11] P. Duhem, Zeits. für Physik. Chemie *69*, 169 (1909). I am indebted to Professor J. Von Neumann for drawing my attention to this paper.

$$V\left(\frac{\partial p}{\partial E}\right)_V > -2 \quad , \tag{II}$$

$$\left(\frac{\partial p}{\partial V}\right)_E < 0 \quad . \tag{III}$$

2. The three conditions are satisfied for all single-phase systems which we have investigated (Secs. 12, 15 and 16), namely, ideal gases with constant or variable specific heat, gases obeying Van der Waals' equation, dissociating gases, liquids and solids at normal temperatures and solids at extremely low temperatures. Condition I is violated from most phase changes (§13). Phase changes can, however, be excluded from consideration because they require too long a time to occur in shock waves (Sec. 1).

3. If condition I is satisfied for a substance, then all compressive waves of small but finite amplitude are connected with an increase of entropy (Sec. 3).

4. The increase of entropy for small volume change ΔV, is proportional ΔV^3 [Eq. (13)].

5. If, in addition to condition I, also condition II is fulfilled, that is, $V(\partial p / \partial E)_V > -2$ everywhere, then all compression waves of whatever amplitude are connected with an increase of entropy and are therefore thermodynamically stable (Sec. 4). Rarefaction waves of finite amplitude are unstable and dissolve into trains of infinitesimal waves.

6. If condition I is *not* satisfied, as at the boundary of phases rarefaction waves of finite amplitude may be stable (Sec. 14) provided the phase transition can occur, and compression waves may, under certain circumstances, consist of two shock waves of different velocities, one traveling behind the other.

7. If conditions I and II are satisfied, and if the state of the material in front of the shock wave is given (V_1, S_1), there is one and only one solution for any value of the entropy, S_2, behind the shock wave, between S_1 and infinity (Sec. 7). If the state of the material *behind* the shock wave is given (V_2, S_2), there is one and only one solution for any value of the entropy S_1, in front of the shock wave, from S_B to S_1 where S_B is either zero or corresponds to a phase boundary (Sec. 17).

8. With increasing entropy of the material behind the shock wave the specific volume V_2, decreases to a minimum, $V_{2\min}$, which is ordinarily reached at temperatures of the order of 10,000°. For still higher temperatures, V_2 increases again to $V_1 / 4$ (Secs. 5 and 8). From some substances V_2 may

have several minima and maxima, possibly including some at lower temperatures.

If the state behind the shock wave is given (V_2, S_2), then the specific volume of the material in front of the wave, V_1, increases monotonically with decreasing entropy, S_1 (Sec. 8).

9. If condition III (as well as I and II) is satisfied, the specific energy, E_2, and the pressure, p_2, of the material behind the shock wave increase monotonically with the entropy S_2, for given V_1, S_1. If the state behind the shock wave, V_2, S_2, is given, the specific energy in front, E_1, decreases monotonically with , even if condition III is not fulfilled. No similar theorem holds for the pressure, p_1, in front. In any case, the energy and pressure behind a shock wave are higher than the same quantities in front of the wave (Sec. 8).

10. If conditions I and II are fulfilled, the velocity of any shock wave with respect to the material in front of it is always greater than the velocity of sound in that material. For a given state in front of the shock wave, V_1, S_1, there is one and only one possible shock wave for every shock wave velocity u_1 greater than the velocity of sound, a_1. Increasing shock wave velocity u_1 corresponds to increasing entropy behind the shock wave, S_2 (Sec. 9).

11. The velocity of the shock wave relative to the material behind it, u_2, is always less than the sound velocity, a_2, in that material. If the state behind the wave is given, there is one and only one possible shock wave for every velocity u_2 between a_2 and a certain minimum velocity, u_{2B}. If the velocity is u_{2B}, the state of the material in front of the wave either lies on a phase boundary or has zero absolute temperature. Decreasing velocity u_2 corresponds to decreasing entropy in front of the shock wave, S_1 (Sec. 9).

12. In a material satisfying conditions I and II, a shock wave will overtake any waves, of infinitesimal or finite amplitude, which precede it, and will be overtaken by any wave following it.

13. If conditions I and II are satisfied, no shock wave can start from the same point at the same time and in the same direction as any other wave, whether of infinitesimal or of finite amplitude (Sec. 10). No shock wave can split into any number of waves going in the same direction.

14. In a material fulfilling also condition III, no one-dimensional shock wave can split in any way whatsoever (Sec. 11).

Weyl
courtesy of AIP Emilio Segrè Visual Archives

HERMANN WEYL

Weyl was a German mathematician and student of Hilbert at the University of Göttingen. He was born in Elmshorn, near Hamburg, Germany, on Nov. 9, 1885. He emigrated to the United States in 1933, where he worked at Institute for Advanced Study at Princeton University until his retirement in 1951. He died in 1955.

His work on shock waves in arbitrary fluids was originally published as a report in 1944 and was a treatment of much the same material as considered by Bethe in his 1942 report to the National Defense Research Council. Weyl developed his ideas without the benefit of Bethe's report. This parallels Hugoniot's development of the theory of discontinuous flow without knowledge of Rankine's publication.

Shock Waves in Arbitrary Fluids

Hermann Weyl, *Communications on Pure and Applied Mathematics*, Vol. II, Nos. 2-3, pp. 103-122 (1949)

Introduction

The following investigation tries to clear up the general hydrodynamical and thermodynamical foundations of the shock phenomenon.[1] The *first part,* Sections 2–5, answers the question: What are the conditions for the equation of state of a fluid under which shocks with their distinctive qualitative features may be produced. These conditions, enumerated in Section 3, are partly of differential, partly of global nature. The *second part,* Sections 6–7, investigates the physical structure of the shock layer whose "infinitesimal" width is of the order of magnitude ε provided heat conductivity and viscosity are small of the same order. Initial state and final state are singular points for the differential equations of the shock layer, and it is shown that they are of such a nature as to make one expect the problem to have a unique solution.

Part I. Thermodynamics and the Shock Phenomenon

1. Shocks

At any place and time a fluid is in a definite *thermodynamical state* Z and in a definite *state of motion,* the latter being characterized by the components u, v, w of the velocity $\vec{u}$ in a Cartesian system of coordinates x, y, z. The possible thermodynamical states form a two-dimensional manifold $\mathcal{Z}$; examples of quantities having definite values in a definite thermodynamical state are density r, pressure p, temperature T; volume $\tau = \rho^{-1}$, potential energy e, heat content or enthalpy $i = e + p\tau$, and entropy S per unit mass. On $\mathcal{Z}$ the fundamental thermodynamical equation

$$de = T\,dS - p\,d\tau \tag{1}$$

[1] This paper was originally submitted to the Applied Mathematics Panel in 1944 as a report. The same problem is treated in, *The theory of shock waves for an arbitrary equation of state,* H.A. Bethe, Division B, NDRC, OSRD, Report No. 545, to which the author had no access.

holds. It is once and for all assumed that we are dealing with a stationary state $(Z, \vec{u})$; i.e., that neither Z nor $\vec{u}$ vary in time, but depend on x, y, z only.

In the limit for vanishing viscosity and heat conductivity, a discontinuity is possible such that $(Z, \vec{u})$ has a constant value $(Z_0, \vec{u}_0)$ for $x < 0$ and another constant value $(Z_1, \vec{u}_1)$ for $x > 0$. The symbols [0], [1] refer to the two regions $x < 0$, $x > 0$ which are divided by the shock front $x = 0$. Because of conservation of mass, we have $\rho_0 u_0 = \rho_1 u_1$. Denoting by M this common flow of mass, we may write

$$u_0 = M\tau_0, \qquad u_1 = M\tau_1.$$

Then the laws of conservation of momentum and energy give the relations

$$\begin{gathered} Mu_0 + p_0 = Mu_1 + p_1, \\ Mv_0 = Mv_1, \qquad Mw_0 = Mw_1, \\ M\left\{i_0 + \tfrac{1}{2}\left(u_0^2 + v_0^2 + w_0^2\right)\right\} = M\left\{i_1 + \tfrac{1}{2}\left(u_1^2 + v_1^2 + w_1^2\right)\right\} \end{gathered}$$

while the law of increasing entropy ("More entropy flows in than out") requires

$$MS_1 \geq MS_0.$$

The phenomenon that results for $M \neq 0$ is called a *shock*. The problem then splits into two parts, one involving only the normal velocity component u, the other referring to the tangential components and stating that they cross the shock front unchanged:

$$u_0 = M\tau_0, \qquad u_1 = M\tau_1, \tag{2.1}$$

$$Mu_0 + p_0 = Mu_1 + p_1, \tag{2.2}$$

$$i_0 + \tfrac{1}{2}u_0^2 = i_1 + \tfrac{1}{2}u_1^2, \tag{2.3}$$

$$v_0 = v_1, \qquad w_0 = w_1. \tag{3}$$

If, however, M vanishes, we obtain a *vortex sheet* or *slip stream* characterized by the relations

$$u_0 = u_1 = 0, \qquad p_0 = p_1.$$

The fluid in [1] glides tangentially past that in [0] and the pressure is the same on both sides of the discontinuity. Our object will largely be the study of shocks in an arbitrary fluid.

After introducing the values (2.1) for the velocities, (2.2) gives

(4) $$M^2 = \frac{p_1 - p_0}{\tau_0 - \tau_1},$$

and thereupon (2.3) yields a relation between the two thermodynamical states Z_0 and Z_1; namely,

$$i_1 - i_0 = \tfrac{1}{2}M^2(\tau_0 - \tau_1)(\tau_0 + \tau_1) = \tfrac{1}{2}(p_1 - p_0)(\tau_0 + \tau_1) \quad \text{or}$$
$$e_1 - e_0 = \tfrac{1}{2}(p_1 + p_0)(\tau_0 - \tau_1).$$

Hence the problem of shocks is reduced to a study of this relation between two states Z_0, Z_1:

(5) $$H \equiv H(Z_1, Z_0) \equiv (e_1 - e_0) - \tfrac{1}{2}(p_1 + p_0)(\tau_0 - \tau_1) = 0$$

(Hugoniot equation).

We are only interested in shocks in one and two dimensions. If in the latter case the normal unit-vector of the shock front in the direction $[0] \to [1]$ is (α, β) instead of $(1, 0)$ the equations (2.1) and (3) for the velocities read as follows:

$$\alpha u_0 + \beta v_0 = M\tau_0, \qquad \alpha u_1 + \beta v_1 = M\tau_1,$$
$$-\beta u_0 + \alpha v_0 = -\beta u_1 + \alpha v_1.$$

A simple analysis (which will presently occupy us in much greater detail) reveals that the value of M resulting from (4) is of the order ρc where c is the acoustic velocity. Thus we are dealing with a situation like this: A certain quantity might be zero, but if it is not, it is even ≥ 1. Clearly quite different circumstances must be responsible for the two phenomena of shock and slip stream, and $M = 0$ is in no way to be considered here the limiting case of a non-vanishing M. The problem of the shock layer will be compared to that of the slip layer in Section 6 by taking viscosity and heat conductivity into account and then letting them tend to zero. In particular, this passage to the limit will explain why a shock is not conservative with respect to entropy although it conserves energy.

2. *Thermodynamical Assumptions*

Next we specify our assumptions concerning the thermodynamical behavior of our fluid. They will be enumerated, I–IV.

I. *Infinitesimal adiabatic increase of pressure produces compression:*

$$\left(\frac{d\tau}{dp}\right)_{ad} < 0.$$

II. *The rate of compression — $d\tau/dp$ diminishes in this process*[2]:

$$\left(\frac{d^2\tau}{dp^2}\right)_{ad} > 0.$$

These local hypotheses will be supplemented by two assumptions of "global" character.

III. *In the continuous process of adiabatic compression one can raise pressure arbitrarily high.*

IV. *The state Z is uniquely specified by pressure and specific volume, and the points (p, τ) representing the possible states Z in a (p, τ)-diagram form a convex region.*

[It would be more natural but less elementary to divide IV into a local and a global part, the local postulate asserting that in the neighborhood of a given state Z_0 the variables p, τ can be used as parameters for the specification of states; in other words that the projection $Z \to (p,\tau)$ of the manifold $\mathcal{Z}$ of states Z upon a (p,τ)-plane is locally one-to-one. The global part would assert that the projection of $\mathcal{Z}$ covers a convex region $\overline{\mathcal{Z}}$ in the (p,τ)-plane and that Z_0 being any given states and (p_0,τ_0) its projection, one never runs against an obstacle when, on starting at Z_0, one lets Z vary so that its projection (p,τ) follows a given path in Z begining at (p_0,τ_0) ("continuation"). All this means that $\mathcal{Z}$ is a covering surface over the convex $\overline{\mathcal{Z}}$ without singularities and relative boundaries. But the convex $\overline{\mathcal{Z}}$ being simply connected, this covering surface necessarily consists of a single sheet, or (p,τ) determines Z uniquely. We thus fall back on the more elementary formulation of IV. The assumption of convexity for the region $\overline{\mathcal{Z}}$ will prove to be quite essential for our investigation. Henceforth we simply denote by Z the projection of Z in the (p,τ)-plane.]

Because of IV, entropy S is a (single-valued) function of p and τ, $S = S(p,\tau)$. The adiabatic process of compression is thus defined by

$$\frac{\partial S}{\partial p}dp + \frac{\partial S}{\partial \tau}d\tau = 0.$$

From the fact that (1) is a total differential we have

[2] It can be made plausible that condition II is essential for the formation of a shock wave.

$$\frac{\partial}{\partial p}\left(T\frac{\partial S}{\partial \tau}-p\right)=\frac{\partial}{\partial \tau}\left(T\frac{\partial S}{\partial p}\right)$$

or

(6)
$$\frac{\partial T}{\partial p}\frac{\partial S}{\partial \tau}-\frac{\partial T}{\partial \tau}\frac{\partial S}{\partial p}=1.$$

Hence $\partial S/\partial p$, $\partial S/\partial \tau$ cannot vanish simultaneously, and thus hypothesis I requires that $\partial S/\partial p$, $\partial S/\partial \tau$ are of the same sign:

$$\frac{\partial S}{\partial p}>0,\quad \frac{\partial S}{\partial \tau}>0 \quad \text{or} \quad \frac{\partial S}{\partial p}<0,\quad \frac{\partial S}{\partial \tau}<0.$$

As the region Z is connected, the one set or the other will hold *everywhere*. We assume the first:

(I′)
$$\frac{\partial S}{\partial p}>0,\quad \frac{\partial S}{\partial \tau}>0.$$

(The other alternative would make little difference.) Infinitesimal adiabatic compression is now described by

$$dp=a^*\,dt,\quad d\tau=b^*\,dt;\quad a^*=\frac{\partial S}{\partial \tau},\quad b^*=-\frac{\partial S}{\partial p}$$

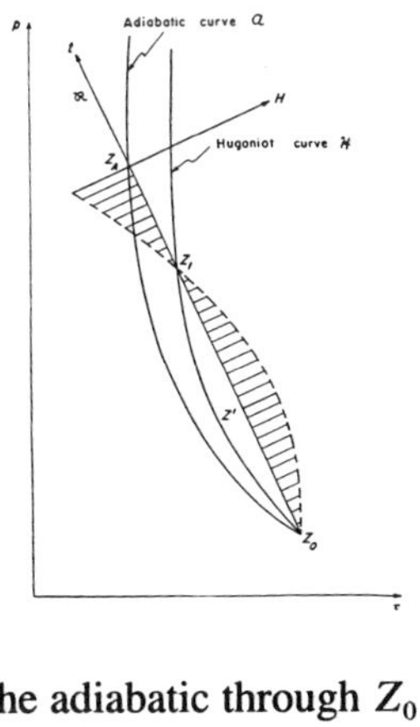

Fig. 1. This diagram shows the adiabatic through Z_0 and the Hugoniot line in a p,τ-diagram; moreover, in the form of a shaded profile, H as function of the distance t along a ray: H reaches its maximum at Z', passes through zero at Z_1, and is negative at Z_A.

(a *positive* factor of proportionality dt corresponding to an *increase* of pressure). Hypothesis II yields

$$\frac{d}{dt}\left(\frac{b*}{a*}\right) > 0 \quad \text{or} \quad b*\frac{da*}{dt} - a*\frac{db*}{dt} < 0,$$

i.e.

$$b*\left(\frac{\partial a*}{\partial p}a* + \frac{\partial a*}{\partial \tau}b*\right) - a*\left(\frac{\partial b*}{\partial p}a* + \frac{\partial b*}{\partial \tau}b*\right)$$
$$= -a*^2\frac{\partial b*}{\partial p} + a*b*\left(\frac{\partial a*}{\partial p} - \frac{\partial b*}{\partial \tau}\right) + b*^2\frac{\partial a*}{\partial \tau} < 0$$

or

$$\text{(II}'\text{)} \qquad \frac{\partial^2 S}{\partial p^2}\left(\frac{\partial S}{\partial \tau}\right)^2 - 2\frac{\partial^2 S}{\partial p \partial \tau}\frac{\partial S}{\partial \tau}\frac{\partial S}{\partial p} + \frac{\partial^2 S}{\partial \tau^2}\left(\frac{\partial S}{\partial p}\right)^2 < 0.$$

In the future we shall use hypotheses I and II in their analytic forms (I′) and (II′).

Our next move consists in developing a number of consequences from the assumptions I–III. Let $(a,b) \neq (0,0)$ be two given numbers which determine the direction of a straight line through $Z_0 = (p_0, \tau_0)$,

$$p = p_0 + at, \qquad \tau = \tau_0 + bt,$$

the half-line or ray being obtained by the restriction $t \geq 0$ on the parameter t. We follow the straight line or the ray as long as it stays within Z, and form

$$\frac{dS}{dt} = \frac{\partial S}{\partial p}a + \frac{\partial S}{\partial \tau}b = S'$$
$$\frac{dS'}{dt} = \frac{\partial S'}{\partial p}a + \frac{\partial S'}{\partial \tau}b = \frac{\partial^2 S}{\partial p^2}a^2 + 2\frac{\partial^2 S}{\partial p \partial \tau}ab + \frac{\partial^2 S}{\partial \tau^2}b^2.$$

Lemma 1. *S' is positive (negative) and hence S increasing (decreasing) along the whole straight line, provided $a \geq 0$, $b \geq 0$ ($a \leq 0$, $b \leq 0$).*

This follows from assumption (I′).

Lemma 2. *If $S' = 0$ for a certain value of t, then $dS'/dt < 0$ for the same value.*

Proof. Substitute the values of a and b derived from $S' = 0$ into dS'/dt and use the inequality (II′).

Lemma 3. *If $S' \leq 0$ for $t = 0$, then $S' < 0$ for $t > 0$.*

Proof. First assume $S' < 0$ for $t = 0$. Should S' change sign as Z travels along the ray, it would have to vanish somewhere. Suppose this occurs for the first time at $t = t_1$. As $S' < 0$ before t reaches this value, S' passes the zero level at t_1 ascending; hence $dS'/dt \geq 0$ for $t = t_1$. But these two relations $S' = 0$, $dS'/dt \geq 0$ for $t = t_1$ contradict Lemma 2.

If S$' = 0$ for $t = 0$ we have $dS'/dt < 0$ for $t = 0$; consequently S' becomes negative immediately after the point Z has started on its way from Z_0, and from then on, as we have seen, S' must remain negative.

We turn to III. Starting at a given point $Z_0 = (p_0, \tau_0)$ and always raising p, we follow the adiabatic through Z_0: we thus obtain a continuous monotonely descending function $\tau(p)$, and according to III we can make p travel over the entire infinite interval $p \geq p_0$. While this happens, the directional coefficient

(7) $$s = \frac{p - p_0}{\tau_0 - \tau}$$

of the straight line joining (p_0, τ_0) with the point (p, τ) on the adiabatic $S = S_0$ increases monotonely from a certain positive value m_0 to $+\infty$. m denotes the adiabatic derivative $-dp/d\tau$ and m_0 its value to the point Z_0. Consequently, there is exactly one point on the upper branch $p > p_0$ of the adiabatic that lies on the ray from (p_0, τ_0) in a given direction s, provided s lies between m_0 and ∞.

The analytic proof for these intuitively evident statements runs as follows. Form the (non-total) differential

$$dr = (\tau_0 - \tau)\, dp + (p - p_0) d\tau,$$

which vanishes along any ray through Z_0. We want to show that along the adiabatic dr is positive for a positive increment dp:

(8) $$\frac{dr}{dp} = R = (\tau_0 - \tau) + (p - p_0)\frac{d\tau}{dp} > 0 \quad \text{for} \quad p > p_0.$$

By hypothesis II

(9) $$\frac{dR}{dp} = (p - p_0)\frac{d^2\tau}{dp^2} > 0 \quad \text{for} \quad p > p_0,$$

and since R vanishes for $p = p_0$ it must indeed be positive for $p > p_0$:

(10) $$R = \int_{p_0}^{p} (p - p_0) \frac{d^2 \tau}{dp^2} dp > 0.$$

Using (7) we may write

(11) $$dr = (\tau_0 - \tau)^2 ds$$

and thus the monotone behavior of s is exhibited by the equation

$$\frac{ds}{dp} = R/(\tau_0 - \tau)^2.$$

It is clear that s tends to m_0 for $p \to p_0$, and the combination of hypothesis III with the trivial $s > (p - p_0)/\tau_0$ leads to $s \to \infty$ for $p \to \infty$.

3. The Fundamental Inequality for the Direction $Z_0 Z_1$

For the differential of

(12) $$H(Z, Z_0) = (e - e_0) - \tfrac{1}{2}(p + p_0)(\tau_0 - \tau)$$

we find by (1)

$$dH = de - \tfrac{1}{2}(\tau_0 - \tau)\, dp + \tfrac{1}{2}(p_0 + p)\, d\tau = T\, dS - \tfrac{1}{2}\{(\tau_0 - \tau)\, dp + (p - p_0)\, d\tau\},$$

or

(13) $$dH = T\ dS - \tfrac{1}{2} dr,$$

and hence *along any ray*

(14) $$dH = T\ dS.$$

Let Z_0, Z_1 be two distinct points satisfying the Hugoniot equation $H(Z_1, Z_0) = 0$. Because of the convexity of the region Z (hypothesis IV) we may join these two points by a straight segment lying in Z. Integrating (14) along it we obtain

(15) $$H(Z_1, Z_0) = \int_{Z_0}^{Z_1} T\, dS = 0.$$

From this simple relation we are going to deduce

Theorem 1. *For any two distinct states* Z_0, Z_1 *linked by Hugoniot's equation* $H = 0$ *the following inequalities hold*

(16) $$W = (p_1 - p_0)(\tau_0 - \tau_1) > 0,$$

(17) $$(p_1 - p_0) + m_0(\tau_1 - \tau_0) > 0, \quad (p_1 - p_0) + m_1(\tau_1 - \tau_0) < 0.$$

Before proving the theorem let us discuss its physical significance. We have seen that elimination of the velocities from the conditions for a shock lead to the *Hugoniot equation* and the inequality $W \geq 0$, cf. (4). We now realize that we may omit the supplementary relation $W \geq 0$ because it follows, in fact in the sharper form $W > 0$, from the Hugoniot equation.

The adiabatic derivative

$$\frac{dp}{d\rho} = -\tau^2 \frac{dp}{d\tau} = \tau^2 m$$

is the square of the "acoustic velocity" c. Assume $\tau_0 > \tau_1$. Then the two relations (17) and relation (4) give

(18) $$m_0 < M^2 = \frac{p_1 - p_0}{\tau_0 - \tau_1} < m_1$$

or $\rho_0 c_0 < |M| < \rho_1 c_1$, thus confirming a statement made in Section 2 on the magnitude of M. In terms of the velocities u_0, u_1 given by (2.1) or

$$u_0^2 = \frac{p_1 - p_0}{\tau_0 - \tau_1}\tau_0^2, \quad u_1^2 = \frac{p_1 - p_0}{\tau_0 - \tau_1}\tau_1^2,$$

(18) may be written as

(19) $$|u_0| > c_0, \quad |u_1| < c_1:$$

Relative to the shock front the flow in [0] *is supersonic, in* [1] *subsonic.*

Proceeding to the proof of our theorem, let us travel from Z_0 along that ray which passes through Z_1 and hence set in Lemma 1,

$$a = p_1 - p_0, \quad b = \tau_1 - \tau_0.$$

Were $a \geq 0,\ b \geq 0$, then S' would be positive and hence S monotone increasing along the segment $Z_0 Z_1$, which contradicts the relation (15). Consequently this combination (and in the same way the other, $a \leq 0,\ b \leq 0$) is ruled out, and therefore $ab < 0$, or (16) must hold.

Next consider

$$S' = a\frac{\partial S}{\partial p} + b\frac{\partial S}{\partial \tau}.$$

Were $S' \leq 0$ for $t = 0$, then S' would always be negative by Lemma 3 and hence S would monotonely decrease while Z travels along the straight segment from Z_0 to Z_1, in contradiction to equation (15). Therefore

$$a\left(\frac{\partial S}{\partial p}\right)_0 + b\left(\frac{\partial S}{\partial \tau}\right)_0 > 0,$$

and this is equivalent to the first of the inequalities (17). The second follows from the first by interchanging Z_0 and Z_1.

But our argument shows much more. While Z travels along the ray from Z_0 passing through Z_1, S' starts with a positive value; because of (15) it must change sign before Z reaches Z_1. But S' remains negative from the moment it vanishes for the first time. Any rise and fall in $H = H(Z, Z_0)$ along the ray is coupled with the same in S by the relation $dH = T\ dS$. Hence H first rises monotonely to a positive maximum and then decreases, passing on the descent through 0 for $Z = Z_1$. This description shows that H is positive from Z_0 to Z_1 and negative thereafter. In particular:

On the ray from Z_0 through Z_1 the point $Z = Z_1$ is the *only one* aside from Z_0 itself which satisfies the Hugoniot equation $H(Z, Z_0) = 0$.

According to (16) either the inequalities $p_1 > p_0,\ \tau_0 > \tau_1$ or $p_0 > p_1,\ \tau_1 > \tau_0$ hold for two distinct states $Z_0,\ Z_1$, satisfying the Hugoniot equation $H(Z_1, Z_0) = 0$. We shall indicate these alternatives by $Z_1 > Z_0,\ Z_1 < Z_0$ respectively. By the *Hugoniot contour* $\mathcal{H}$ (for a given Z_0) we understand the locus of all points $Z_1 = (p_1, \tau_1)$ for which

$$H(Z_1, Z_0) = 0 \quad \text{and} \quad Z_1 > Z_0.$$

It is quite essential for our argument to pick this *upper* branch $Z_1 > Z_0$. Our above result may then be stated thus: On the Hugoniot contour

$s=(p_1-p_0)/(\tau_0-\tau_1)$ lies between m_0 and ∞; s is a uniformizing parameter inasmuch as the value of s specifies uniquely to point Z_1. But it must be borne in mind that so far we have not yet proved that to every preassigned value of $s>m_0$ there actually corresponds a point on the Hugoniot contour; we only know that there cannot be more than one.

4. Entropy and Parameterization of the Hugoniot Line

For a moment let us return to the upper branch a of the adiabatic through Z_0, on which p increases monotonely from p_0 to $+\infty$ and $s=(p_1-p_0)/(\tau_0-\tau_1)$ from m_0 to ∞. Moving along a we have by (13)

(20) $$dH=-\tfrac{1}{2}\,dr=-\tfrac{1}{2}R\,dp$$

and hence (10) implies

Lemma 4. $H=H(Z,Z_0)$ *is negative along* a.

As a matter of fact, we even know that H is falling, $R>0$, and by condition (9) falling with increasing rapidity, $dR/dp>0$. The explicit formula giving $H(Z_A,Z_0)$ for any point Z_A on the adiabatic a is by (20), (8) and (9),

$$\begin{aligned}H(Z_A,Z_0)&=-\frac{1}{2}\int_{a\,p_0}^{p_A}R\,dp=-\frac{1}{2}R(p-p_A)\Big|_{p_0}^{p_A}-\frac{1}{2}\int_{a\,p_0}^{p_A}(p_A-p)\frac{dR}{dp}\,dp\\&=-\frac{1}{2}\int_{a\,p_0}^{p_A}(p-p_0)(p_A-p)\frac{d^2\tau}{dp^2}\,dp<0.\end{aligned}$$

This lemma is instrumental in establishing the following propositions:

Theorem 2. *For any two distinct states* Z_0, Z_1 *satisfying the Hugoniot equation one has* $S_1>S_0$ *or* $S_1<S_0$ *according to whether* $Z_1>Z_0$ *or* $Z_1<Z_0$.

Theorem 3. *The Hugoniot contour is a simple line starting from* Z_0 *on which s and S are monotone increasing* (s traveling from m_0 to $+\infty$, S from S_0 to an unknown destination).

Proof. Assume the arrangement $Z_1>Z_0$ for the two points Z_0, Z_1 linked by the Hugoniot equation. Draw the ray $\mathcal{R}$ from Z_0 passing through Z_1; its directional coefficient $s=(p_1-p_0)/(\tau_0-\tau_1)$ lies between m_0 and $+\infty$, and hence $\mathcal{R}$ meets the upper branch of the adiabatic line a in a point $Z_A=(p_A,\tau_A)$ where, according to Lemma 4,

$$H = \int_{\mathcal{R}\, z_0}^{z_A} T\, dS = -\frac{1}{2}\int_{a\, p_0}^{p_A} R\, dp < 0\,.$$

Hence Z on its road along the ray must have passed the point Z' where it reaches it maximum, and running downhill have passed a point Z_1 for which $H = \int_{Z_0}^{Z_1} T\, dS = 0$ before coming to Z_A. Its value S_1 at the point Z_1 of the Hugoniot contour is therefore higher than its value S_0 at Z_A (Theorem 2).

Again, let s be any value between m_0 and ∞. The ray going out from Z_0 in the direction s meets a in a point Z_A where H is negative. Hence H must vanish at a certain point Z_1 between Z' and Z_A (where S is already on the down grade), and thus a point on the Hugoniot contour is obtained corresponding to this preassigned value of $s = (p_1 - p_0)/(\tau_0 - \tau_1)$. We know it is the only one. The question raised at the end of Section 4 is answered in the affirmative, and we may now speak of the Hugoniot contour as a simple line $Z_1 = \Phi(s)$ with the uniformizing parameter s.

Because the point $Z_1 = \Phi(s)$ is thought of as varying along the Hugoniot line, let us drop the index 1. Along $\mathcal{H}$ we may use the differential relation $dH = 0$ or, according to (13) and (11),

$$T\, dS - \tfrac{1}{2}\, dr = T\, dS - \tfrac{1}{2}(\tau_0 - \tau)^2\, ds = 0\,.$$

Because

$$\frac{dS}{ds} = \frac{(\tau_0 - \tau)^2}{2T}$$

thus turns out to be positive, S increases all the way along the Hugoniot line with s. This gives a new proof for the inequality $S_1 > S_0$ by integration along that line,

$$S_1 - S_0 = \int_{\mathcal{H}\, m_0}^{s_1} \frac{(\tau_0 - \tau)^2}{2T}\, ds \qquad (s_1 > m_0)$$

Thus Theorem 3 is proved, and Theorem 2 by even two different methods.

Part II. The Problem of the Shock Layer

5. Formulation of the Problem

For the sake of brevity we now denote partial derivatives with respect to (p, τ) by subscripts as in

$$\frac{\partial S}{\partial p} = S_p, \qquad \frac{\partial S}{\partial \tau} = S_\tau.$$

The rest of our argument will be based on the thermodynamical assumptions I–IV, Section 3, but in its course it will become clear that one more condition of highly plausible nature is to be added to our list:

Ia. *Heating a quantity of fluid at constant volume raises its pressure and temperature.*

It must be remembered that the analytic formulation (I′), (II′) of the hypotheses I and II depended on the fixation of a sign. For the previous investigation this did not matter greatly, but it is essential for the behavior of the shock layer. We therefore require explicitly by the statement concerning pressure in Ia, that $(dp/TdS)_{\tau=\text{const}}. > 0$ or $S_p > 0$. The other part of Ia then adds the inequality $T_p/S_p > 0$ or $T_p > 0$.

In a stationary field the state $(Z, \vec{u})$ of a fluid of given thermodynamical nature depends on the spatial coordinates x, y, z only, and not on time. The five equations which, in the absence of an external force, express conservation of mass, momentum and energy, take heat conductivity and viscosity into account by the flow of heat $\vec{j} = -\lambda \operatorname{grad} T$ and a stress tensor S of the form

$$S_{ik} = p\delta_{ik} - S^*_{ik}, \qquad S^*_{ik} = \mu' \operatorname{div} \vec{u} \cdot \delta_{ik} + \mu\left(\frac{\partial u_i}{\partial x_k} + \frac{\partial u_k}{\partial x_i}\right)$$

(in the writing down of which we have for the moment adopted the subscript notation for the coordinates x_i and velocity components u_i). $\lambda = \lambda[Z]$, μ, μ' are given functions of the thermodynamical state Z; $\lambda > 0$, $\mu > 0$, $\mu' + \frac{2}{3}\mu > 0$. As they are small we intend to let them pass to zero. In two ways such a passage to the limit $\lambda, \mu, \mu' \to 0$ can give rise to a discontinuity in the plane $x = 0$:

(A) Write ϵx for x and $\epsilon\lambda$, $\epsilon\mu$, $\epsilon\mu'$ for λ, μ, μ' leaving all other quantities untouched, and then let the positive constant factor ϵ tend to zero *(shock layer)*.

(B) Let ϵ tend to zero after writing ϵx, ϵu for x, u and $\epsilon^2\lambda$, $\epsilon^2\mu$, $\epsilon^2\mu'$ for λ, μ, μ' *(slip layer).*

The different effect is clearly exemplified by the continuity equation

(21) $$\frac{\partial}{\partial x}(\rho u)+\frac{\partial}{\partial y}(\rho v)+\frac{\partial}{\partial z}(\rho w)=0.$$

Substitution (A) changes it into

$$\frac{\partial}{\partial x}(\rho u)+\epsilon\frac{\partial}{\partial y}(\rho v)+\epsilon\frac{\partial}{\partial z}(\rho w)=0$$

which for $\epsilon \to 0$ reduces to

$$\frac{\partial}{\partial x}(\rho u)=0,$$

whereas process (B) leaves the three-term equation (21) unchanged. This is typical. None but the derivatives $\partial/\partial x$ survive the limiting process (A), the others $\partial/\partial y$, $\partial/\partial z$, disappear from the equations. The shock layer is thus seen to be a one-dimensional problem, no interaction taking place between the one-dimensional fibers of space y = constant, z = constant. The slip layer on the other hand constitutes a far more complicated three-dimensional problem, identical with that of Prandtl's boundary layer. We are concerned here only with the shock layer.

With the abbreviation $\mu^* = \mu' + 2\mu$, $\mu^* > 0$, the limiting process (A) leads to the following equations:

(22.1) $$\rho u = \text{constant} = M,$$

(22.2) $$Mu + p - \mu^*\frac{du}{dx} = \text{constant},$$

(22.3) $$Mv - \mu\frac{dv}{dx} = \text{constant}, \qquad Mw - \mu\frac{dw}{dx} = \text{constant},$$

(22.4) $$\left(p - \mu^*\frac{du}{dx} + \rho e^*\right)u - \mu\left(v\frac{du}{dx} + w\frac{dw}{dx}\right) - \lambda\frac{dT}{dx} = \text{constant},$$

where e^* denotes the total energy $e + \frac{1}{2}\left(u^2 + v^2 + w^2\right)$. Here the quantities λ, μ, μ' are no longer small. We assume that $(Z,\vec{u})$ approaches definite constant values $(Z_0, \vec{u}_0)$, $(Z_1, \vec{u}_1)$ for $x \to -\infty$ and $x \to +\infty$ respectively, whereas the derivatives d/dx tend to zero.

The energy equation is equivalent to the entropy equation

$$(23)\quad T\frac{d}{dx}\left(\rho u S - \frac{\lambda}{T}\frac{dT}{dx}\right) = \mu^*\left(\frac{du}{dx}\right)^2 + \mu\left\{\left(\frac{dv}{dx}\right)^2 + \left(\frac{dw}{dx}\right)^2\right\} + \frac{\lambda}{T}\left(\frac{dT}{dx}\right)^2,$$

from which one learns that the flow of entropy

$$\rho u S - \frac{\lambda}{T}\frac{dT}{dx} = MS - \frac{\lambda}{T}\frac{dT}{dx}$$

is a monotone increasing function of x and therefore $MS_1 \geq MS_0$. The equality sign could prevail only if u, v, w, T and hence by (22.2) also p were constant. Excluding this trivial case of a constant state[3] $(Z, \vec{u})$ we thus obtain the *strict inequality*

$$(24)\qquad MS_1 > MS_0$$

implying $M \neq 0$. The latter remark shows how misleading it is to link the alternative $M \neq 0$, $M = 0$ to the discrimination of shock and slip layer; rather do these two possibilities correspond to the two ways (*A*) and (*B*) of passing to the limit of vanishing viscosity and heat conductivity. Being sure now that M does not vanish, we fall back upon our old relations connecting $(Z_0, \vec{u}_0)$ and $(Z_1, \vec{u}_1)$, including the Hugoniot relation. Let us assign the labels 0, 1 to the two states Z_0, Z_1, and accordingly orient the x-axis, so that $Z_0 < Z_1$. Then Theorem 2 combined with the law of increasing entropy (24) gives $M > 0$. The meaning of this inequality is that the shock front moves in the direction $[1] \to [0]$, or that Z_0 and not Z_1 is the initial state of the fluid not yet reached by the propagating shock.

---

[3] Indeed, in general, p = constant, T = constant, imply τ = constant. There is, however, one extremely exceptional case in which the conclusion does not stand, namely if $M = 0$ and if for some constant value of p the temperature T as function of τ stays constant in an entire τ-interval. Maybe this is a warning against the danger of a singularity in the shock layer problem which one steers clear of with certainty only by assuming $T_\tau \neq 0$ or, what is the same if the hypotheses I–IV, Ia are preceded by another to the effect that p, T may serve as local parameters on Z everywhere.

6. Character of the Two Singular Points Z_1, Z_0: Node and Saddle[4]

It is easy to dispose of the equations (22.3) for v and w. Single out the one for v,

$$\mu \frac{dv}{dx} = M(v-b), \qquad b = \text{constant},$$

and introduce the variable

$$\xi = M\int_0^x \frac{dx}{\mu(x)} \qquad \text{where } \mu(x) = \mu[Z(x)].$$

We find

$$v(x) - b = Ce^{\xi}, \qquad C = \text{constant}.$$

$\mu(x)$ tends to positive values $\mu_0 = \mu[Z_0]$, $\mu_1 = \mu[Z_1]$ for $x \to -\infty$, $x \to +\infty$ respectively; therefore

$$\xi = \frac{M}{\mu_0}x + o(x) \qquad \text{for } x \to -\infty,$$
$$\xi = \frac{M}{\mu_1}x + o(x) \qquad \text{for } x \to +\infty.$$

Considering that $M > 0$ one realizes that v cannot tend to a finite limit v_1 for $x \to +\infty$ unless $C = 0$. Thus v and w must be constant throughout the shock layer.

(22.1) is satisfied by $u = M\tau$. Denoting by $M^2 a$ the constant at the right of (22.2) and substituting from (22.2) the value of $p - \mu^* \, du/dx$ into (22.4) we are left, as is readily seen, with the following two ordinary differential equations of first order for the state $Z = Z(x)$:

$$\text{(25)} \qquad \begin{aligned} \mu^* M \frac{d\tau}{dx} &= M^2(\tau - a) + p \\ \frac{\lambda}{M}\frac{dT}{dx} &= e - \frac{1}{2}M^2(\tau - a)^2 - c. \end{aligned}$$

---

[4] For ideal gases of constant specific heat with certain special values of the adiabatic exponent γ, cf. R. Becker, Zeitschrift für Physik 8 (1922), 321–362; in particular, pp. 339–347.

The constants M^2, a, c are determined by

$$M^2\tau_0 + p_0 = M^2\tau_1 + p_1 = M^2 a,$$
$$e_0 - \tfrac{1}{2}M^2(\tau_0 - a)^2 = e_1 - \tfrac{1}{2}M^2(\tau_1 - a)^2 = c.$$

As the right members of (25) vanish for $Z = Z_0$ and $Z = Z_1$, the two points Z_0, Z_1 turn out to be singular points for the differential system (25). If we care only for the trajectory in the (p,τ)-diagram along which $Z(x)$ moves while x runs from $-\infty$ to $+\infty$ we are faced with one ordinary differential equation of first order in the (p,τ)-plane,

$$\frac{\lambda\, dT}{e - \frac{1}{2}M^2(\tau - a)^2 - c} = \frac{\mu^*\ d\tau}{(\tau - a) + \left(p/M^2\right)}.$$

But contrary to the customary viewpoint taken in the classical investigations on singular points, which were started by Briot and Bouquet almost one hundred years ago and are reproduced in all treatises on differential equations, or interest is in the *parameterized* trajectory $Z = Z(x)$.

Let us briefly recount the well-known facts about singular points relevant for our purposes.[5] Placing the singular point at the origin 0 we study two simultaneous differential equations for the "vector" $\varphi = (\varphi_1(x),\ \varphi_2(x))$ in a $\varphi_1,\ \varphi_2$ - plane,

$$\frac{d\varphi_1}{dx} = F_1(\varphi_1,\ \varphi_2), \qquad \frac{d\varphi_2}{dx} = F_2(\varphi_1,\ \varphi_2)$$

where $F_1,\ F_2$ are given functions of $\varphi_1,\ \varphi_2$ vanishing at the origin. From the outset it is evident that any translation $x \to x + h$ carries a solution $\varphi = (\varphi_1(x),\ \varphi_2(x))$ into a solution ("equivalent solutions"). Suppose the Taylor expansions of $F_1,\ F_2$ begin with the terms

$$F_1(\varphi_1,\varphi_2) \sim R_{11}\varphi_1 + R_{12}\varphi_2, \qquad F_2(\varphi_1,\varphi_2) \sim R_{21}\varphi_1 + R_{22}\varphi_2.$$

The equations

$$\varphi_1' = R_{11}\varphi_1 + R_{12}\varphi_2, \qquad \varphi_2' = R_{21}\varphi_1 + R_{22}\varphi_2$$

[5] See for direct constructive proofs: H. Weyl, *Concerning a classical problem in the theory of singular points of ordinary differential equations*. Actas de la Academia Nacional de Ciencias de Lima, Volume 7, 1944, pp. 21–60.

define a linear mapping $\varphi \to \varphi' = R\varphi$. We determine its two eigen-values k and corresponding eigen-vectors $\varphi = \alpha$ (or rather eigen-directions) by $R\alpha = k\alpha$, explicitly

$$(k - R_{11})\alpha_1 - R_{12}\alpha_2 = 0,$$
$$-R_{21}\alpha_1 + (k - R_{22})\alpha_2 = 0,$$

and assume that the two roots k_1, k_2 of the secular equation

$$\begin{vmatrix} k - R_{11}, & -R_{12} \\ -R_{21}, & k - R_{22} \end{vmatrix} = 0$$

are real, distinct and $\neq 0$. In a suitable affine coordinate system in the φ_1, φ_2 - plane our mapping is then described by $\varphi_1' = k_1\varphi_1$, $\varphi_2' = k_2\varphi_2$, so that F_1, F_2 are of the form

$$F_1(\varphi_1, \varphi_2) = k_1\varphi_1 + \cdots,$$
$$F_2(\varphi_1, \varphi_2) = k_2\varphi_2 + \cdots.$$

We distinguish three cases according to the signs of the eigen-values k_1, k_2. In the case (i) of two negative eigen-values, $k_2 < k_1 < 0$, every solution φ which comes sufficiently near to the origin plunges into it with x tending to $+\infty$; its asymptotic behavior is described by

$$\varphi_1(x) \sim a \cdot e^{k_1 x}, \qquad \varphi_2(x) \sim 0 \cdot e^{k_1 x} \qquad (x \to \infty)$$

in the sense that

$$e^{-k_1 x}\varphi_1(x) = a + O(e^{-\epsilon x}),$$
$$e^{-k_1 x}\varphi_2(x) = O(e^{-\epsilon x})$$

where ϵ is a positive constant, the same for all solutions, while the constant a is specific (*node*, see Figure 2). In the case (ii) of one negative, one positive eigen-value, $k_1 < 0$, $k_2 > 0$, there exist exactly two solutions (1_+, 1_- in Figure 3) approaching 0 with $x \to +\infty$; their asymptotic behavior is described by

$$\varphi_1(x) \sim \pm e^{k_1 x}, \qquad \varphi_2(x) \sim 0 \cdot e^{k_1 x} \qquad (x \to +\infty)$$

(*saddle*). Of course this statement is to be interpreted so that equivalent solutions are considered as one and the same. The term saddle is better understood by observing that in this case (ii) two other solutions, 2_+ and 2_-, plunge into 0 in the direction of the φ_2 - axis, while x goes to $-\infty$. In case (iii) of two positive eigen-values there is no solution whatsoever which approaches the origin as x tends to $+\infty$, except the trivial one $\varphi_1 = \varphi_2 = 0$ (whereas, of course, every solution that comes sufficiently near 0 takes the plunge when x moves toward $-\infty$).

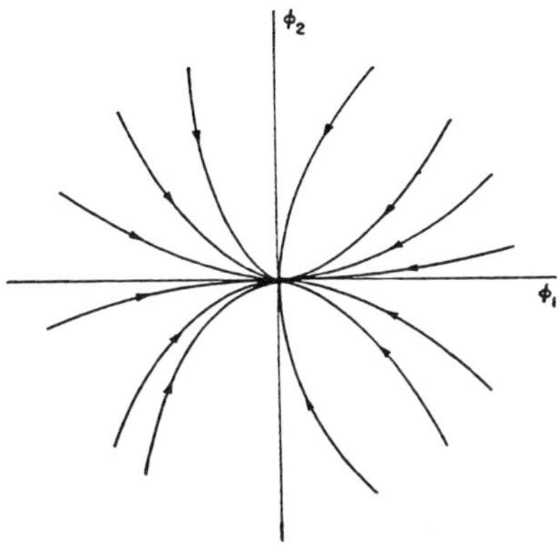

Fig. 2. Node.

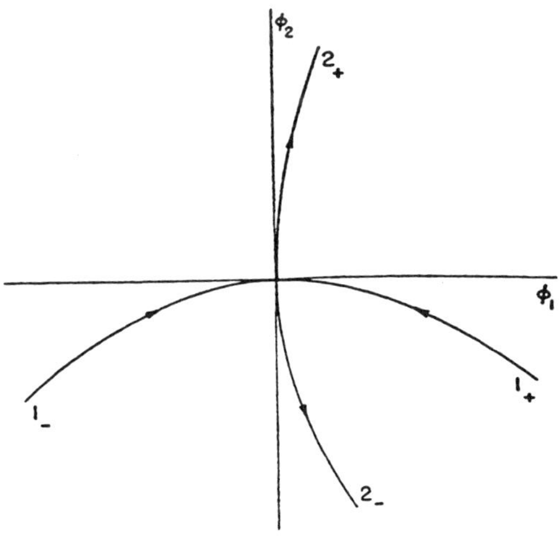

Fig. 3. Saddle.

With these facts in mind we now advance our decisive

Theorem 6. *For the differential equations of the shock layer determining the transition from a state* Z_0 *to a state* $Z_1 > Z_0$, *the initial state* Z_0 *is a node, the end state* Z_1 *a saddle.*

Clearly this behavior in the neighborhood of the two singular points Z_0, Z_1 is such as to favor the existence of a unique solution of the shock layer problem. All other combinations would make one expect either no solution or an infinity of solutions.

Again, the proof depends, above all, on the fundamental inequalities (18) for $M^2 = (p_1 - p_0)/(\tau_0 - \tau_1)$. Since S_τ/S_p is the adiabatic derivative $m = -dp/d\tau$ we may write them as

$$M^2 S_p^{(0)} - S_\tau^{(0)} > 0, \qquad M^2 S_p^{(1)} - S_\tau^{(1)} < 0. \tag{26}$$

They must be combined with

$$M > 0; \qquad S_p > 0, \qquad T_p > 0. \tag{27}$$

First investigate the neighborhood of Z_0 and therefore write

$$M^2(\tau - a) + p = M^2(\tau - \tau_0) + (p - p_0).$$

For infinitesimal $\delta p = p - p_0$, $\delta\tau = \tau - \tau_0$

$$\delta e = T_0\ \delta S - p_0\ \delta\tau, \qquad \delta\tfrac{1}{2}(\tau - a)^2 = (\tau_0 - a)\ \delta\tau, \qquad \text{hence}$$
$$\delta\left\{e - \tfrac{1}{2}M^2(\tau - a)^2\right\} = T_0\ \delta S - \left\{p_0 + M^2(\tau_0 - a)\right\}\ \delta\tau = T_0\ \delta S,$$

therefore in neglecting terms of higher order

$$e - \tfrac{1}{2}M^2(\tau - a)^2 - c \sim T_0(S - S_0).$$

The following approximate linear system for $\delta\tau$, δp results:

$$\frac{d}{dx}(\delta\tau) = g\left(M^2\ \delta\tau + \delta p\right),$$
$$\frac{d}{dx}(\delta T) = g_0'\ \delta S$$

where

$$g = 1/M\mu^* > 0, \qquad g' = TM/\lambda > 0$$

and δS, δT stand for

$$\delta S = S_p^{(0)}\,\delta p + S_\tau^{(0)}\,\delta\tau, \qquad \delta T = T_p^{(0)}\,\delta p + T_\tau^{(0)}\,\delta\tau.$$

In order to avoid unnecessary encumbrances, we drop the index 0. Then the equations determining the eigen-values k and corresponding eigen-directions $(\delta\tau,\ \partial p)$ at Z_0 become

$$k\cdot\delta\tau = g\left(M^2\delta\tau + \delta p\right), \qquad k\cdot\delta T = g'\delta S$$

giving rise to the following secular equation for k,

$$\begin{vmatrix} k - gM^2, & -g \\ kT_\tau - g'S_\tau, & kT_p - g'S_p \end{vmatrix} = 0 \qquad \text{or}$$

(28)
$$T_p\cdot k^2 - \left\{g\left(M^2T_p - T_\tau\right) + g'S_p\right\}\cdot k + gg'\left(M^2S_p - S_\tau\right) = 0.$$

Compute the discriminant

$$\begin{aligned}\Delta &= \left\{g\left(M^2T_p - T_\tau\right) + g'S_p\right\}^2 - 4gg'T_p\left(M^2S_p - S_\tau\right)\\ &= \left\{g\left(M^2T_p - T_\tau\right) - g'S_p\right\}^2 - 4gg'\left(T_pS_\tau - T_\tau S_p\right)\end{aligned}$$

and thus verify by (6) that Δ is positive,

$$\Delta = \left\{g\left(M^2T_p - T_\tau\right) - g'S_p\right\}^2 + 4gg' \geq 4gg' = \frac{4T}{\lambda\mu^2} > 0.$$

Consequently the eigen-values are real and distinct, at the singular point Z_0 as well as at Z_1.

Their product equals

$$gg'\cdot\left(M^2S_p - S_\tau\right)/T_p,$$

and because of (27) and the fundamental inequalities (26) that product is positive at Z_0, negative at Z_1. Hence the two eigen-values are of opposite sign at Z_1, of equal sign at Z_0. Whether the latter sign is positive or negative can be decided by the sum of the two k's for which (28) gives the value

$$\left\{g\left(M^2 T_p - T_\tau\right) + g_0' S_p\right\} / T_p.$$

All constituents of this expression are positive, including

$$M^2 T_p^{(0)} - T_\tau^{(0)} > \left(S_\tau^{(0)} / S_p^{(0)}\right) T_p^{(0)} - T_\tau^{(0)} = 1 / S_p^{(0)} > 0.$$

Thus we arrive at the result that the two roots k at Z_0 are *positive*. Our theorem has now been proved: the fundamental inequalities (17) not only show that by a shock the flow changes from supersonic to subsonic but also that one of the two points Z_0, Z_1 is a node, the other a saddle.

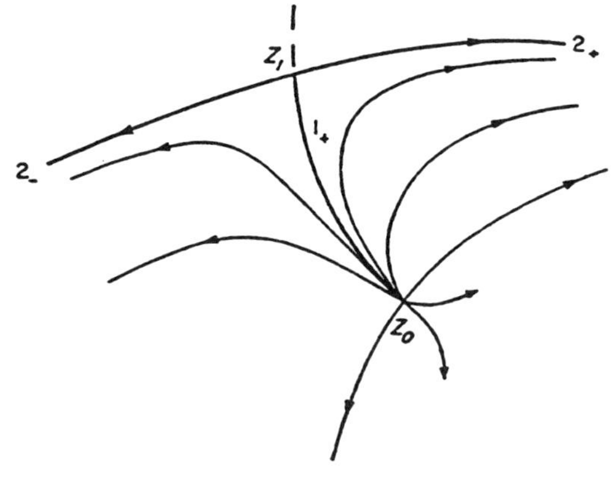

Fig. 4. The topological picture of the trajectories in the (p, τ)-plane.

In order actually to prove the existence of a unique solution of the shock layer problem, one has to bridge the gap between the neighborhoods of the two singular points Z_0, Z_1 by some sort of topological argument. In that respect one fact is of decisive importance:

Theorem 7. The differential equations of the shock layer have now singular point besides Z_0, Z_1.

Indeed if there existed three distinct singular points Z_0, Z_1, Z_2 they would satisfy the equations

$$M^2\tau_0 + p_0 = M^2\tau_1 + p_1 = M^2\tau_2 + p_2 = M^2 a,$$
$$e_0 - \tfrac{1}{2}M^2(\tau_0 - a)^2 = e_1 - \tfrac{1}{2}M^2(\tau_1 - a)^2 = e_2 - \tfrac{1}{2}M^2(\tau_2 - a)^2.$$

The first set states that Z_0, Z_1, Z_2 lie on a straight line of negative inclination $-M^2$, and we may assume the arrangement $Z_0 < Z_1 < Z_2$. Then the second set requires that both $Z = Z_1$ and $Z = Z_2$ satisfy the Hugoniot equation $H(Z, Z_0) = 0$, in contradiction to a statement previously proved by which there cannot be more than one point Z on any ray Z_0 satisfying that equation. The observation is urged upon us that whereas the Hugoniot relation is not transitive, the relation between Z_0, Z_1 defined by the *simultaneous* equations

$$M^2\tau_0 + p_0 = M^2\tau_1 + p_1, \qquad H(Z_1, Z_0) = 0$$

is (M^2 being a given constant).

The topological picture of the trajectories which one expects is indicated by Figure 4. If this is correct, the region covered by all trajectories running into Z_0 for $x \to -\infty$ would be bounded by the saddle line through Z_1 consisting of the two branches 2_+ and 2_- (Figure 3). The outlook is encouraging.

Index